Tutor in a Book's

GEOMETRY

By Jo Greig

TUTOR IN A BOOK®

Berkeley, California

Compositor and Interior Design: Jo Greig
Mathematics Editor: James R. Shilleto, Ph.D.
Cover Designer: Michael Tanamachi
Editor: Christine E. McGowan
Artwork: Ka-Wai Lui
Problems Editor: D. William McPhee, Jr.

International Standard Book Number: 978-0-9786390-5-1

Library of Congress Control Number: 2012945132

First published 2007. Second Edition 2012.

Copyright © 2012
Published by
The Geometry Store
Berkeley, California 94702

Contents

Author's Note — I have tutored students in mathematics for more than 30 years. A good tutor can help a struggling student pass a class that the student might otherwise fail. A good tutor can help a strong student do much better in a class than he or she might otherwise do. I have written this book with the hope of giving all students the advantage of a good tutor. I have attempted to include every explanation, every drawing, every hint, every memory tool, every problem that students always seem to struggle with and every bit of enthusiasm that I try hard to impart to my private tutoring students. Good luck with Tutor in a Book's Geometry!   Jo Greig, August 1, 2012

A Note to Parents on How to Use this Book

If the semester is just beginning: The student should read the next day's assignment in this book (check in the textbook for the correct topic) prior to each school day. The student should be sure to do the exercises in this book to reinforce his or her understanding of the topic.

If the school year is well under way and the student is really struggling: He or she needs to start at the beginning of this book and read through to the point where his or her class is at the present time. This is important because geometry is sequential and you cannot build understanding on a shaky foundation. This book is much shorter than a textbook and much more informal and accessible. Be sure the student reads each section, carefully studies the example problems and then does the exercises. In addition to reinforcing the review and proper learning of the material, doing the exercises successfully is very empowering to the student. Step-by-step solutions to each exercise with appropriate explanations and illustrations are included in the Answer Section at the back of the book.

If the school year is well under way and the student understands some material: He or she can use this book like a cookbook, carefully going over the chapters that cover the troublesome topics. However, the student should study any necessary chapters in their entirety, paying careful attention to the examples, and complete all exercises in those chapters.

Suggestions for All Students

Make a flashcard for every symbol, term, definition, theorem, postulate and corollary as each is introduced in your textbook. Memorize the information. Test yourself on your flashcards everyday. Only use those postulates, theorems, corollaries and definitions that are introduced in your textbook.

Always read your class's next day's topic in this book the night *before* class. Carefully study the example problems or proofs. Do the related exercises in this book and be sure to check your answers. The time you spend will end up saving you study time in the long run. In class, try to sit in the front of the room. Listen to your teacher's lecture and take careful notes including the problems done in class.

Carefully read your textbook and do the class work and homework when it is assigned. Don't fall behind. Geometry cannot be crammed in the night before an exam. Be diligent and stay positive. I've had students who received low grades on the first exams but who worked diligently and never gave up and ended up with A's for both semesters.

General Notes to Parents — Here are some hints on how to help your child succeed in geometry:

If you were good in geometry and are trying to help your son or daughter but are just put off by the 700+ page textbook, this is the book for you. The language is simple and the illustrations and review charts tie related information together to quickly refresh your memory. The standard problems (in both senses of the word) complete with fully explained and illustrated solutions are here.

If you didn't like geometry, it's probably better not to mention this to your child. After all, you would never tell your child how much you hated to read! In fact, doing so gives your child permission to fail. Instead, encourage your child to follow the many tips in this book, to be respectful to his or her teacher and to do all required assignments when they are assigned. Always ask to see each day's completed homework. You will be able to tell if each problem was attempted. Meet your child's teacher and let the teacher know that you support him or her. (Most will sincerely appreciate it).

A major reason that students do poorly in geometry is they don't realize that geometry is like a combination of a foreign language and a mathematics course. Students need to learn the basics (the symbols, how to name an angle, etc.), and then the vocabulary, *all* of the theorems and so forth, in order to be successful. To compound the problem, textbooks give a piece of information one time, then expect the student to recall the information for the rest of the school year. Some teachers allow students to make and refer to notes or theorem sheets, but this is rarely helpful because the student has no idea where to look or what to look for. Experience has shown, that all of this information needs to be learned and learned thoroughly. In this book, key information is stated, stressed and restated. You can help your child by insisting on flashcards and then quizzing him or her on the information.

Regarding proofs, over the years, proofs have gained an unfair reputation for difficulty. However, proofs follow fairly predictable patterns and once a student learns the patterns and check steps, proofs become much easier. This book includes dozens of thoroughly explained proofs including many visuals and tips to help the student recognize and learn the patterns. An interesting fact is that beginning proofs are very subtle and therefore more difficult; later proofs are actually easier.

Unless your school provides an integrated algebra/geometry course, successful completion of first-year algebra is the prerequisite for geometry. This is necessary because there is a good deal of algebra in a geometry course, especially in the second semester. If you want your child to accelerate, ask the school counselor if the school will allow geometry and second-year algebra to be taken together.

Sometimes, a student feels that he or she just does not click with a particular teacher and/or the textbook. In my experience, switching teachers is rarely helpful. However, if your school will allow a change, go ahead, but then insist that your child take a new approach with the course (being very attentive to this "better" teacher, doing all homework promptly and so forth). As for textbooks, they do differ in quality and sometimes a concept that is perfectly clear to a team of authors is beyond the reach of a teenager. I have tried to address this issue in this book by using less formal teaching methods — ones that I've developed over the years and which are teen-age tested and approved.

Tutor in a Book®'s
Geometry©

Jo Greig

1. POINTS, LINES AND PLANES

Points are the abstract building blocks of geometry. In fact, all of the objects that we learn about in geometry are made up of points. We do not formally **define** points, but we know a lot about them:

> 1. Each point is a unique location.
> 2. Points have no size at all.
> 3. Points are fixed, they cannot move.
> 4. We draw a dot • to show the location of a point.
> 5. We name a point with a single capital letter.
>
> •*A*   •*C*
>
> •*B*

Space is defined in mathematics, as the collection of all points. We can move through space, but the points are fixed and cannot move.

Lines are not formally **defined**, but we know a lot about them:

> 1. Lines are fixed and straight; they can have slope but cannot bend.
> 2. Think of a line as being formed from the points through which it passes.
> 3. Each line is distinct.
> 4. Lines have a single dimension, but it helps to think of a line as being one point wide.
> 5. All lines go on forever, that is, infinitely, in two directions.
> 6. To draw a line, we add some width so that we can see the line.
> 7. To show that a line is infinite, we put an arrowhead on each end.

There are two ways to **name** a line:

1. A line may be named by a single lowercase letter with a double arrowhead on top of the letter:

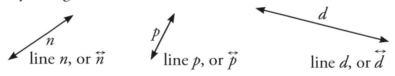

line *n*, or $\overleftrightarrow{n}$ line *p*, or $\overleftrightarrow{p}$ line *d*, or $\overleftrightarrow{d}$

2. A line may also be named by using the names (the letters) of *any two* points through which the line passes, together with a double arrowhead on top. The two points may be named in *either order*.

The line on the left may be named by any of the following:

$$\overleftrightarrow{AB},\ \overleftrightarrow{BA},\ \overleftrightarrow{BC},\ \overleftrightarrow{CB},\ \overleftrightarrow{AC},\ \text{or}\ \overleftrightarrow{CA}.$$

Lines and Points Together:

We say a line *includes* or *contains* any point(s) through which it passes.

We say a point *lies on* or *in*, or is *on* or *in* or is *captured by* or *contained in* the line.

Quick Review — Whenever you see the word line in geometry, it means a straight, infinite object. Lines cannot be divided in half; lines have no satisfactory midpoint. Each line is infinite and in this sense cannot be measured. And remember, like points, lines are *not* (formally) defined.

Answer true or false: A line is defined as a straight, infinite object.

Answer: False! (A line is not defined).

Collinear — Points that lie on a single line are *collinear*.

Any **2** points are collinear. That is, exactly one line exists (though it might not be shown) that passes through the two points. However, when considering three or more points, you need to stop and ask yourself, could a (straight) line pass through this set of points?

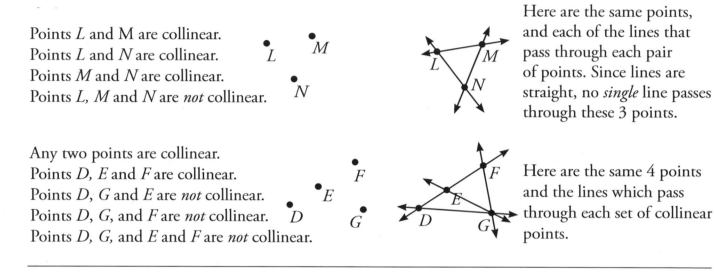

Points *L* and M are collinear.
Points *L* and *N* are collinear.
Points *M* and *N* are collinear.
Points *L*, *M* and *N* are *not* collinear.

Here are the same points, and each of the lines that pass through each pair of points. Since lines are straight, no *single* line passes through these 3 points.

Any two points are collinear.
Points *D*, *E* and *F* are collinear.
Points *D*, *G* and *E* are *not* collinear.
Points *D*, *G*, and *F* are *not* collinear.
Points *D*, *G*, and *E* and *F* are *not* collinear.

Here are the same 4 points and the lines which pass through each set of collinear points.

Between — The word *between* means 2 things: 1. Collinear with and,
                                                   2. In between.

When a problem says that a point is between two points you are being given both conditions.

When you are trying to decide if a point is between two points you must check for both conditions.

Here are some examples:

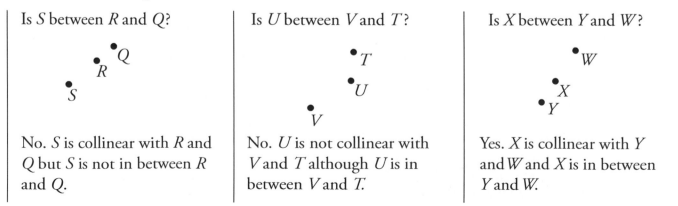

Is *S* between *R* and *Q*?	Is *U* between *V* and *T*?	Is *X* between *Y* and *W*?
No. *S* is collinear with *R* and *Q* but *S* is not in between *R* and *Q*.	No. *U* is not collinear with *V* and *T* although *U* is in between *V* and *T*.	Yes. *X* is collinear with *Y* and *W* and *X* is in between *Y* and *W*.

Segments — A (line) segment is a part of a line. A segment has a beginning and an end; this means that a segment is *finite*, it can be measured. Unlike lines, segments are defined:

> A segment is: 1. Its two endpoints, and
> 2. All of the points between its two endpoints.

The symbol for a segment is the names of the two endpoints (the capital letters in either order) with a bar over the top. Here's an example:

The segment at right is named $\overline{AB}$ or $\overline{BA}$.

What does $\overline{AB}$ include?

1. $\overline{AB}$ includes its two endpoints, A and B, and
2. All of the points between (collinear with and in between) A and B.

Note that $\overline{BA}$ includes the same collection of points. That's why $\overline{AB}$ is the same as $\overline{BA}$.

Here are more segments and the symbols that could be used to name them:

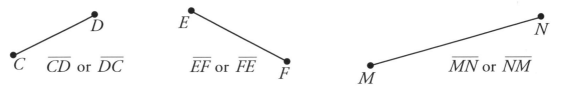

D
E
N
C $\overline{CD}$ or $\overline{DC}$ $\overline{EF}$ or $\overline{FE}$ F M $\overline{MN}$ or $\overline{NM}$

Length and Distance

The *length* of a segment is the *distance* between its endpoints.

1. The symbol for distance is the names of the two endpoints (the capital letters) *in either order*. "Plain" RS or SR means the length of segment $\overline{RS}$.

$\xleftarrow{} RS \xrightarrow{}$
R S

$\overline{RS}$ is RS units long!

2. Length is always positive, so its measure, that is, the distance between two points, is always positive. For example, you cannot be **minus 5' 6"** tall!

$\overset{4}{\rule{3cm}{0.4pt}}$
V W
$WV = VW = 4$

Number Lines   1. The letters are the names of the points on the line.
2. The numbers are the *coordinates* of the points.

B C D E
-1 0 1 2

A coordinate tells us the distance and direction that a point is from zero. The distance between any two points is the number of units from the first point to the second. Distance isn't about direction or which side of zero either point is on. It's about how far you would have to go to get from one point to the other.

Distance is always positive.   Is distance always a *whole* number? No.

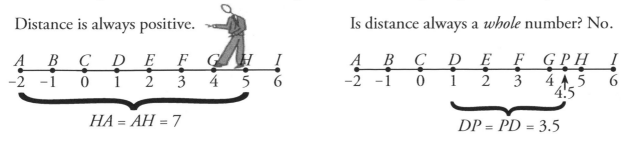

A B C D E F G H I
-2 -1 0 1 2 3 4 5 6

$HA = AH = 7$

A B C D E F G P H I
-2 -1 0 1 2 3 4 5 6
4.5

$DP = PD = 3.5$

How Do You Find Distance on a Number Line?

Three ways:

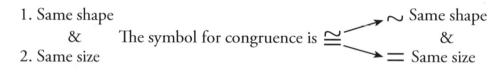

1. Subtract the coordinate on the left from the coordinate on the right: $KL = 5.7 - {}^-1.8 = 5.7 + 1.8 = 7.5$, or,

2. Take the absolute value of the difference of the two coordinates: $KL = |5.7 - {}^-1.8| = 7.5$, or,

3. Just count carefully. If the points are on different sides of zero, count carefully from zero, putting the pieces together: $KL = 1.8 + 5.7 = 7.5$

Equality and Congruence

A segment's size is measured by its length. If two segments have the same size, their lengths are equal.

$$W \overset{5}{\bullet\!\!-\!\!\bullet} X \qquad Q \overset{5}{\bullet\!\!-\!\!\bullet} P \qquad WX = QP = 5$$

If we say two objects are *congruent*, we are saying 2 things:

1. Same shape
       &   The symbol for congruence is $\cong$ → $\sim$ Same shape
                                                          &
2. Same size   → $=$ Same size

Yet all segments have the same shape. This means, for segments, same shape and same size mean the same thing. In fact, for segments, congruence and equal length can be used interchangeably:

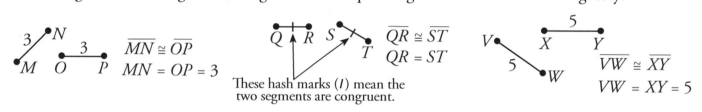

These hash marks (*I*) mean the two segments are congruent.

Midpoints of Segments

Segments are finite; they have a beginning and an end. Therefore, a segment has a midpoint. In fact, a segment is the only geometric object that can have a midpoint.

DEFINITION OF A MIDPOINT: *The point on a segment which divides the segment into two congruent parts.*

Getting Ready for Proofs

A tip **Know your definitions.** For example, if a proof gives: N is the midpoint of $\overline{FG}$, you are supposed to think and respond with $\overline{FN} \cong \overline{NG}$. Other facts are true, but they are not the definition of a midpoint. Conversely, if this proof had given $\overline{FN} \cong \overline{NG}$, you think and respond with, N is the midpoint of $\overline{FG}$.

$$F \overset{\;\;\;\;I\;\;\;\;\;\;\;\;\;I}{\bullet\!-\!\!\bullet\!-\!\!\bullet} G$$
$$\overline{FN} \cong \overline{NG}$$

How Do You Find a Particular Point and Its Coordinate?

A	B	C	D	E	F	G	H	I
-2	-1	0	1	2	3	4	5	6

Example: Find the midpoint of $\overline{IA}$ and the coordinate of the midpoint.

In order to locate the point which divides $\overline{IA}$ into two equal parts, we must first find IA, (the total length of the segment), $IA = |6 - {}^-2| = 8$ (or just count).

So, each part must be ½ of the total length, in this example ½ of 8, or 4 units long.

This means the midpoint must be 4 units from either endpoint. Subtract 4 from 6 (or count 4 units back from 6), which equals 2.

Since the number 2 is paired with point E, the midpoint of $\overline{IA}$ is E and its coordinate is 2. $\checkmark$*

*Note: In this book $\checkmark$ means, this is the correct answer.

Now You Try It — Using the number line on the right find:

A	P	B	C	D	E	F	G	W	H	L	I
-2		-1	0	1	2	3	4		5		6
	-1.2							4.5		5.7	

1. AE 2. WP 3. EA 4. GA

5. The midpoint and the coordinate of the midpoint of $\overline{HB}$.

Solutions for all **Now You Try It** problems are shown in the back of the book.

THE SEGMENT ADDITION POSTULATE:

If B is between (collinear with and in between) A and C, then

$$AB + BC = AC$$

A^+_{tip} **Know your Postulates and Theorems**

Even if your teacher allows you to use notes or theorem sheets on tests, you still need to memorize and understand each postulate and theorem as it is introduced.

Use the Segment Addition Postulate in 2 ways:

1. In a proof when you need to show that a segment is equal to the sum of its parts, or that the sum of its parts is equal to the (entire) segment. Remember, AB means the length of $\overline{AB}$.

2. In algebra problems like this one: If $MO = 24$, find x.

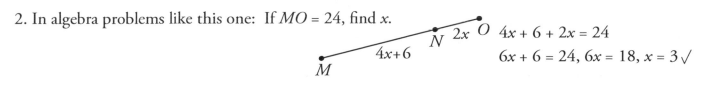

$4x + 6 + 2x = 24$

$6x + 6 = 24, 6x = 18, x = 3 \checkmark$

Rays — A *ray* is like a half-line. A ray has one endpoint and goes on forever in a single direction.

Here are the rules for naming a ray:
1. A ray is named with 2 points:

    1.) The *endpoint* is always named *first*.

    2.) Then, another point through which the ray passes is named.

2. A *single* arrowhead pointing to the *right*, is drawn over the 2 letters.

Using a single arrowhead for a ray makes sense because the ray goes on forever, but only in one direction. Here are some more examples of rays and their names:

$\vec{JK}$ or $\vec{JL}$. Either is correct and both name the same ray, that is, the same collection of points. $\vec{KL}$ is a different ray.	$\vec{HG}$. The endpoint is *H*. No matter which way the ray itself is going, the arrowhead on the symbol goes to the right.	$\vec{DC}$. The endpoint (*D* in this case) is always named first and the arrowhead on the symbol goes to the right.

DEFINITION OF A RAY: *$\vec{AB}$ includes (segment) $\overline{AB}$ and every other point, we'll call each one X for a moment, such that B is between X and A.* Remember, objects in geometry are collections of points. Think about how the definition of a ray includes every collinear point on one side of the endpoint.

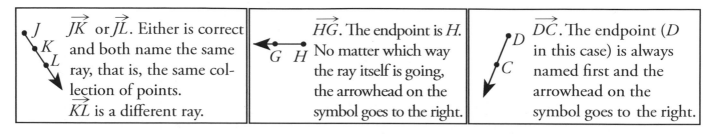

Opposite Rays — 2 rays are said to be *opposite rays* if:

    1. They share a common endpoint, and

    2. That endpoint is between (collinear with and in between) another point on *each* ray. Here are some examples:

$\vec{VU}$ and $\vec{VW}$ are opposite rays.

$\vec{LK}$ and $\vec{LM}$ are opposite rays.

Now You Try It — Give the definition of each symbol shown below:

1. $\vec{AB}$ 2. AB 3. $\overleftrightarrow{AB}$ 4. $\overline{AB}$

Planes — A *plane* is another geometric object that is not formally defined. A plane has two dimensions, length and width, but no depth. A plane is flat and has no edges because it goes on forever. To get an idea of what a plane would look like, think of the surface of a perfectly still ocean, or the floor of a large gym. To represent planes, we draw figures like these, naming them with a single capital letter:

Plane *R*   Plane *M*   Plane *A*

A second way to name a plane is to name any three (or more) *non-collinear* points which are included in the plane. For example, in the figure on the right, plane *EFG* includes the small rectangle *EFGH* that we see in the drawing, but it goes on forever. Other names for plane *EFG* are *EFGH, EFH, FGH, HFG* and so on. But **not** *EFHG*, (no zigzagging). There are five other planes shown in the figure. Try to name them.

It is also important to imagine planes that are not shown, for example, plane *HEJK* (see Figure 1 below). Try to sketch in plane *GHIJ* in Figure 2.

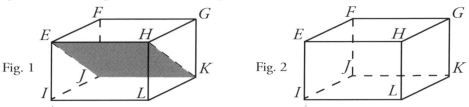

Fig. 1   Fig. 2

Coplanar — If one plane contains 2 (or more) objects we say the objects are *coplanar*. For example, lines $\overleftrightarrow{EF}$ and $\overleftrightarrow{HG}$ above are coplanar. Lines $\overleftrightarrow{EF}$ and $\overleftrightarrow{LK}$ are also coplanar (although the plane is not shown).

Finding and Naming the **Intersection** of Geometric Objects

The intersection of two or more objects means the *points that are shared* by the objects.

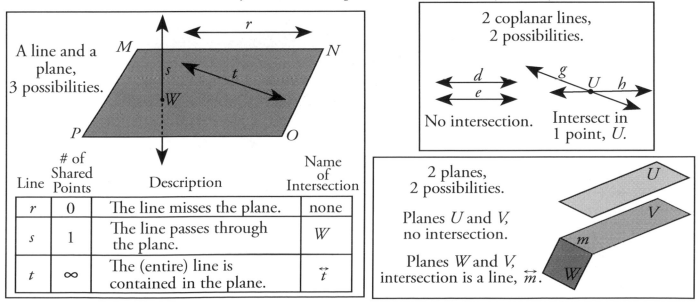

A line and a plane, 3 possibilities.

Line	# of Shared Points	Description	Name of Intersection
r	0	The line misses the plane.	none
s	1	The line passes through the plane.	*W*
t	∞	The (entire) line is contained in the plane.	$\overleftrightarrow{t}$

2 coplanar lines, 2 possibilities.

No intersection.   Intersect in 1 point, *U*.

2 planes, 2 possibilities.

Planes *U* and *V*, no intersection.

Planes *W* and *V*, intersection is a line, $\overleftrightarrow{m}$.

Questions having to do with points, lines, planes and intersections:

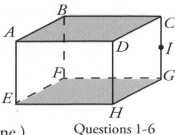

Questions 1-6

1. Name a line that's not shown.
   Here are a few, $\overleftrightarrow{AC}$, $\overleftrightarrow{BD}$, $\overleftrightarrow{EG}$, $\overleftrightarrow{FH}$, $\overleftrightarrow{AI}$, $\overleftrightarrow{CE}$ and $\overleftrightarrow{AH}$.

2. What is the intersection of planes *ABC* and *DCG*?
   (Remember that 3 noncollinear points name and determine a unique plane.)
   Whenever two planes intersect, they intersect in a *line*, in this case, $\overleftrightarrow{DC}$.

3. Name the intersection of planes *ABCD*, *DCGH* and *BCGF*.
   An intersection means the point(s) shared by *all* of the named objects. The answer is point *C*.
   Look up at the corner of your room to see another example of this type of intersection.

4. Name 3 planes that are not shown.
   Here are four: *ACGE*, *BCHE*, *CDEF*, *ADGF*. Recall that *ACG* names the same plane as *ACGE*.
   In fact, *any three* non-collinear points identify and name (exactly) one plane, because only one
   plane can pass through any set of 3 non-collinear points.

5. What is the intersection of planes *ABGH* and *DCGH*?
   Whenever two planes intersect, they intersect in a *line*, in this case, $\overleftrightarrow{GH}$.

6. What is the intersection of $\overleftrightarrow{CG}$ and *I*? What is the intersection of *ABGH* and *I*?
   The intersection of a point and another object is either the point or none. In the case of
   $\overleftrightarrow{CG}$ the answer is point *I*. In the case of *ABGH*, the answer (and the intersection) is none.
   Some textbooks and teachers will say 0 or ∅ or { } or the "null set" to describe no intersection.

Segment Bisectors

A (line) segment has a midpoint. The midpoint of a segment divides the segment into
two congruent parts. Lines, rays and planes cannot have midpoints and cannot be
bisected. Remember, each is an infinite object. However, a line, a ray or a plane can be
a bisec*tor*, that is, they can do the bisec*ting*.

DEFINITION OF A SEGMENT BISECTOR: *A segment bisector is a line, ray, plane or another segment that*
passes through the midpoint of a segment.

Given $\overline{AB}$ with midpoint *M*:

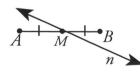

Line $\overleftrightarrow{n}$ bisecting
segment $\overline{AB}$.

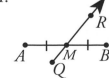

Ray $\overrightarrow{QR}$ bisecting
segment $\overline{AB}$.

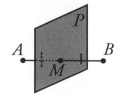

Plane *P* bisecting
segment $\overline{AB}$.

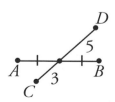

Segment $\overline{CD}$ bisec*ting*
segment $\overline{AB}$, but not
being bisec*ted* by $\overline{AB}$.

Postulates and Theorems about Points, Lines and Planes

Your textbook might have a number of postulates and theorems about points, lines and planes and it is important that you study, understand and memorize them. As you read the postulates and theorems below, remember that each point is distinct (different) from any other point; each line is distinct from any other line; each plane is distinct from any other plane.

The following are some of the postulates and theorems that your textbook might include:

POSTULATE: *Two points determine a (unique) line.*

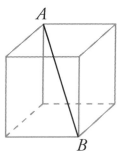

Study the drawing. Points *A* and *B* which are on opposite corners of the box, are fixed. Lines are straight and cannot move. Only one line passes through both point *A* and point *B*. Given any two points, only one (unique) line is determined.

"Two points determine a *unique* line" means two things: that the line exists and that only one such line exists. Other ways to state these two ideas would be to say one and only one, or exactly one.

POSTULATE: *Through any three points there is at least one plane and through any three noncollinear points there is a unique plane.*

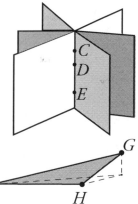

An infinite number of planes pass through any given line. If three points happen to be on the same line (that is, if they are collinear), then an infinite number of planes pass through the line and therefore, include the three points as well. The figure at right shows three (of the infinite number) of planes that pass through the line which contains points *C*, *D* and *E*.

If three points do not lie on the same line (that is, if they are not collinear), then only one plane passes through all three points. The figure at right shows the unique plane that passes through noncollinear points *G*, *H* and *I*.

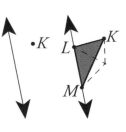

THEOREM: *Through a line and a point not on the line there is a unique plane.*

Choose any two points on the line and then visualize the plane through those two points and the third point which is not on the line. See the 2 figures on the right.

THEOREM: *If two lines intersect, then one and only one plane contains them.*

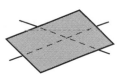

A good way to visualize this theorem is to imagine 2 Pick Up Sticks, cross them and then imagine a piece of paper that is laid on top of the crossed sticks. The paper represents the unique plane that contains both lines.

Angles

What makes an angle?

1. The sides of an angle are 2 rays.
2. The 2 rays share a common endpoint.
3. The common endpoint is the *vertex* of the angle.
4. Since the vertex is a point, it is named by a single capital letter.
5. Note that an angle is a plane object and also includes the infinite "interior region" between the rays.

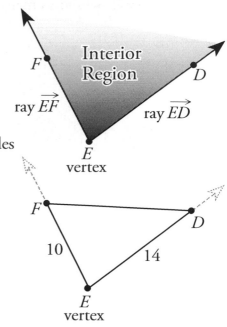

Sometimes a figure will show segments as an angle's sides, however, we understand that the angle's actual sides are rays which go on forever. Nevertheless, the lengths of the segments can be important pieces of information for the problem we are solving.

Naming Angles — There are two symbols for an angle, $\angle$ and $\measuredangle$. This book will use $\angle$.

In naming angles, make sure the name you use, refers only to the angle you intend to name.

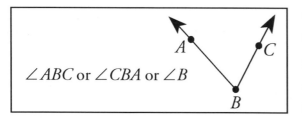

$\angle ABC$ or $\angle CBA$ or $\angle B$

The angle on the left can be called $\angle ABC$ or $\angle CBA$. When we use 3 letters to name an angle, **the vertex is the middle letter.** We could also call the angle simply $\angle B$ since there would be no confusion as to which angle we had in mind.

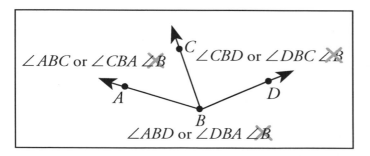

$\angle ABC$ or $\angle CBA$ $\not\angle B$ $\angle CBD$ or $\angle DBC$ $\not\angle B$

$\angle ABD$ or $\angle DBA$ $\not\angle B$

Look at the angles on the left. We could call the left most angle either $\angle ABC$ or $\angle CBA$ and everyone would know which angle we had in mind. However, what is $\angle B$? Point B is the vertex of more than one angle, so to call any angle $\angle B$ would be confusing. Similarly, the right most angle could be called $\angle CBD$ or $\angle DBC$, but not $\angle B$. The large outer angle could be called $\angle ABD$ or $\angle DBA$, but not $\angle B$.

Sometimes your textbook will use numbers to name angles, especially in complicated figures where many angles are shown.

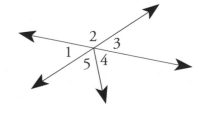

Now You Try It — Give as many correct names as you can for the angles below:

1.

2.

3.

4.

Measuring Angles

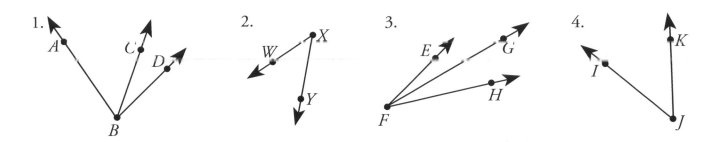

The *lengths* of the sides of an angle have nothing to do with the size of the angle itself. For example, the measures of all of the angles shown on the right are equal.

What the size of an angle measures is rotation. The vertex of an angle is the center of the angle. The vertex is the point around which the rotation takes place. An angle's measure is the amount of rotation around the vertex beginning at one ray (sometimes called the initial ray*****) and ending at the other ray (sometimes called the terminal or terminating ray*****).

*If your textbook uses these terms, remember that it doesn't matter which is the initial ray and which is the terminal ray. One side of the angle is initial and the other side is terminal (or terminating).

Angles are measured in degrees. The symbol for degree is °. Each of the angles at right measures 56°. A complete rotation (as in a circle) measures 360°. It is traditional in beginning geometry, when the rotation is more than 180°, to measure in the other direction. So, all angles studied will be 180° or less.

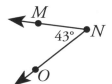

This is $\angle MNO$.
$m\angle MNO = 43$.

"m" means *degree measure*.
$m\angle ABC = 56$.
$m\angle ABC$ is a **number**.
"plain"$\angle ABC$ is the angle itself.

This is $\angle ABC$.
$m\angle ABC = 56$.

Measuring Angles With a Protractor

How To Use a Protractor
Task: Measure ∠*DEF* on the right.
Before you start, be sure that your
protractor is face up. The numbers
should be oriented as shown in the
figure below.

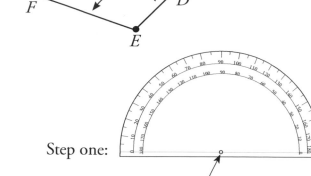

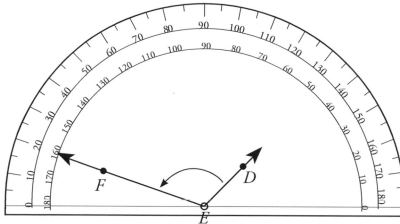

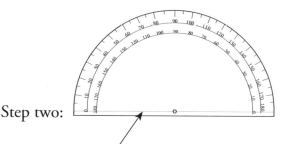

Step one:

First notice the small circle in the center of
the bottom of the protractor. Center the
circle over the vertex, point *E*, of the angle.

Step two:

Now notice the line through the circle
at the bottom of the protractor. Turn
the protractor so that this line lines up
with the right side of the angle, $\overrightarrow{ED}$,
being careful to keep the circle centered
over the vertex. Now read the measure
on the inside row of numbers where the
second side of the angle $\overrightarrow{EF}$, crosses the
protractor (or would cross it if you were
to extend the ray).

∠*DEF* measures approximately 113°. When we measure angles with
a protractor our answer is just an approximation. It's a good practice
to always follow the inside row of numbers (the row that starts out on
the right with double digits) on your protractor. When you always do
things in a particular way, you're less likely to make mistakes.

Now You Try It — Pictures of angles do not always show the points through which the rays pass. In fact, there are many different ways in which your teacher and your textbook might draw and label angles. Several of these are shown in the measuring exercises below.

Using a protractor:

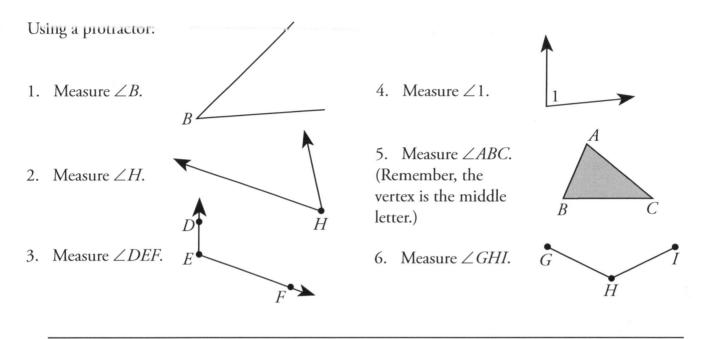

1. Measure ∠B.

2. Measure ∠H.

3. Measure ∠DEF.

4. Measure ∠1.

5. Measure ∠ABC. (Remember, the vertex is the middle letter.)

6. Measure ∠GHI.

Putting Angles into Categories

Angles are put into categories based on their measures. The names of the categories below are used in problems and it is important that you memorize them.

> **Acute** angle: measures between 0° and 90°. (Think of "a cute little" angle.)
> **Right** angle: measures exactly 90°.
> **Obtuse** angle: measures between 90° and 180°. (Obtuse means thick.)
> **Straight** angle: measures exactly 180°.

An acute angle can be as close to 90° as we wish, provided that it is less than 90°. For example, an angle measuring 89.9999° is acute. An angle measuring 90.0001° is obtuse.

Now You Try It — Identify each of the following as an acute, right, obtuse or straight angle.

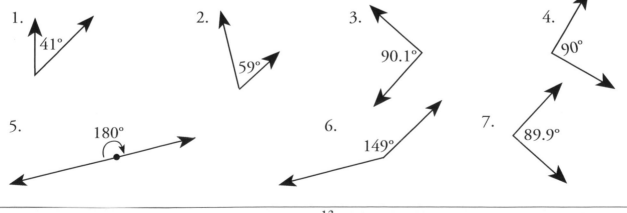

1. 41°

2. 59°

3. 90.1°

4. 90°

5. 180°

6. 149°

7. 89.9°

More About Angles

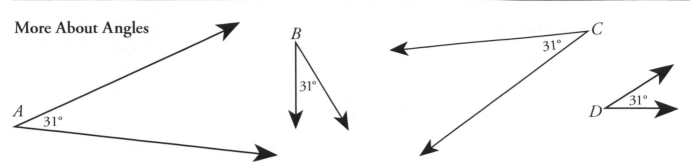

Because their degree measure is the same, the four angles above are equal. Angles of equal degree measure have the same shape (remember, although it's not shown in the figures, the sides of each angle are rays, and all rays go on forever). Therefore, every angle of say, 31°, is the same size. So for angles, *congruence is the same as equal measure.* Therefore, we can write:

$$\angle A \cong \angle B \cong \angle C \cong \angle D \quad \text{or} \quad m\angle A = m\angle B = m\angle C = m\angle D$$

When angles in figures are marked with the same number of marks, it means the angles are congruent.

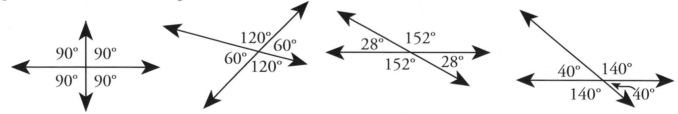

Special *Pairs* of Angles

Vertical Angles

One of the few things we can assume in geometry, is that lines that appear straight are straight. Whenever two lines cross, two pairs of equal, opposite angles are formed. *Pairs* of opposite angles are called *vertical angles*. Remember, there is no such thing as *a* vertical angle, they always come in pairs. Here are a few examples:

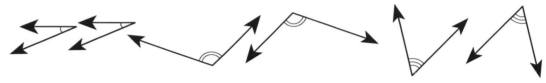

In each case, opposite angles are equal, and together, they make a *pair* of vertical angles.

DEFINITION OF VERTICAL ANGLES: *Two opposite angles formed by the intersection of two lines.*

VERTICAL ANGLE THEOREM: *Vertical angles are congruent.* $\angle EBD \cong \angle ABC$ **VAT***

The Vertical Angle Theorem (VAT) is frequently used in proofs and problems. Whenever you see two crossed lines in a figure, immediately mark at least one of the pairs of vertical angles equal. This is often a piece of information that you need in order to solve the problem. $\mathbf{A}^{+}_{\text{tip}}$

*Note: Although this book uses the abbreviation VAT, only use it in your class, if your teacher uses it.

More Special *Pairs* of Angles

Adjacent Angles

Adjacent angles is a term that refers to a *pair* of angles in the same plane (for example, the plane of this piece of paper) that meet the following three conditions:

1. The two angles have the same vertex.
2. One of the angle's terminal side is the other angle's initial side. In other words, they share a common side, and
3. The angles do not overlap.

Figures 1 and 2 are examples of pairs of adjacent angles:

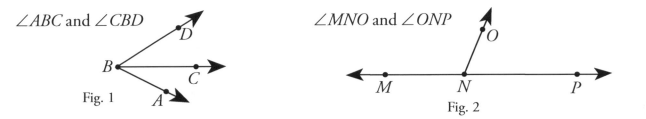

∠ABC and ∠CBD

Fig. 1

∠MNO and ∠ONP

Fig. 2

Here are 3 examples of angles that are NOT adjacent:

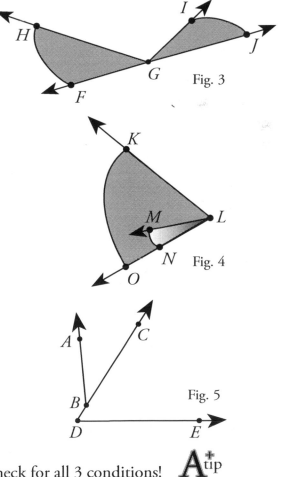

In Figure 3, ∠FGH and ∠JGI are not adjacent angles. The angles share point *G* as their vertex and do not overlap, however, they do not share a common side. They are breaking condition 2.

Fig. 3

In Figure 4, ∠KLO (the larger surrounding angle) and ∠MLN are *not* adjacent angles. They do share a common vertex *L* and they do share a common side $\overrightarrow{LO}$ (which is the same as $\overrightarrow{LN}$), however, the angles overlap. They are breaking condition 3.

Fig. 4

In Figure 5, ∠ABC and ∠CDE are *not* adjacent angles. Although the angles do not overlap, they do not share a common vertex and do not share a common side since a segment of $\overrightarrow{DC}$ is not included in $\overrightarrow{BC}$. The angles are breaking conditions 1 and 2.

Fig. 5

When asked if 2 angles are adjacent, be sure to check for all 3 conditions! A⁺tip

More Special *Pairs* of Angles

Complementary Angles

Complementary 90°

2 angles whose measures total **90** are *complementary*. The angles may or may not be adjacent angles or even near each other. Look at ∠A and ∠B below. Since their measures total 90, they are complementary angles. We say the 2 angles are *complementary* or that one angle *complements* the other or that two angles are *complements*. In each case we mean the same thing: their measures add up to 90.

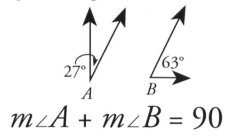

In the figure on the right, angles *MNO* and *ONP* are complementary. In this case they are also adjacent angles and therefore, we know that ∠*MNP* is a right angle.

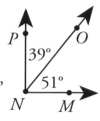

$$m\angle A + m\angle B = 90$$

Supplementary Angles

Supplementary 180°

Two angles whose measures total 180 are *supplementary*. The angles may or may not be adjacent angles or even near each other. Look at ∠*G* and ∠*H* below. Since their measures total 180, they are supplementary angles. We say 2 angles are *supplementary* or that one angle *supplements* the other or that two angles are *supplements*. In each case we mean the same thing: their measures add up to 180.

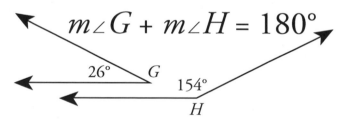

$$m\angle G + m\angle H = 180°$$

In the figure on the right, angles *FDE* and *EDC* are supplementary. In this case, they are also adjacent angles and, therefore, we know that ∠*FDC* is a straight angle.

★MEMORY HINTS★
Complementary and Supplementary sound alike. How do you keep them straight??

Here's one way:  = 180

Here's another: **S**upplementary angles add up to a **S**traight angle.

And another: Complementary starts with a C, which comes before S in the alphabet. 90 comes before 180 on the number line.

More Special *Pairs* of Angles

> Note: Check in the index of your textbook. If "linear pair" and theorems involving linear pairs are not listed, skip this definition.

DEFINITION OF A LINEAR PAIR: *If the non-shared sides of adjacent angles are opposite rays, the two angles form a linear pair.*

Here are two examples:

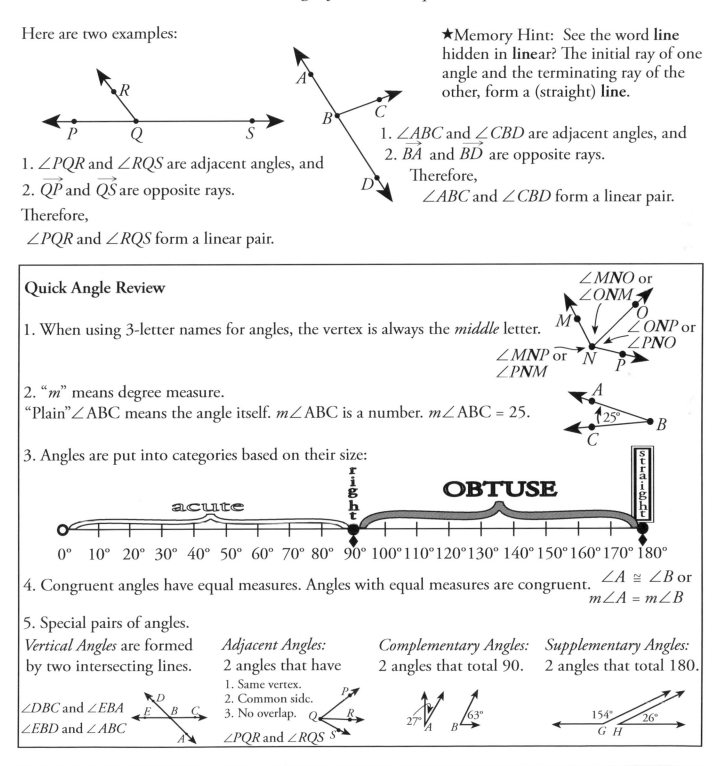

1. ∠PQR and ∠RQS are adjacent angles, and
2. $\overrightarrow{QP}$ and $\overrightarrow{QS}$ are opposite rays.

Therefore,

∠PQR and ∠RQS form a linear pair.

★Memory Hint: See the word **line** hidden in **line**ar? The initial ray of one angle and the terminating ray of the other, form a (straight) **line**.

1. ∠ABC and ∠CBD are adjacent angles, and
2. $\overrightarrow{BA}$ and $\overrightarrow{BD}$ are opposite rays.

Therefore,

∠ABC and ∠CBD form a linear pair.

Quick Angle Review

1. When using 3-letter names for angles, the vertex is always the *middle* letter.

∠MNO or ∠ONM
∠ONP or ∠PNO
∠MNP or ∠PNM

2. "*m*" means degree measure.
"Plain" ∠ABC means the angle itself. *m*∠ABC is a number. *m*∠ABC = 25.

3. Angles are put into categories based on their size:

acute   right   OBTUSE   straight

0° 10° 20° 30° 40° 50° 60° 70° 80° 90° 100° 110° 120° 130° 140° 150° 160° 170° 180°

4. Congruent angles have equal measures. Angles with equal measures are congruent. ∠A ≅ ∠B or *m*∠A = *m*∠B

5. Special pairs of angles.

Vertical Angles are formed by two intersecting lines.

∠DBC and ∠EBA
∠EBD and ∠ABC

Adjacent Angles:
2 angles that have
1. Same vertex.
2. Common side.
3. No overlap.
∠PQR and ∠RQS

Complementary Angles:
2 angles that total 90.

27° 63°

Supplementary Angles:
2 angles that total 180.

154° 26°

Tips for Solving Problems About Angles

1. You know that lines are straight, so look for them and remember, a straight angle equals 180°.

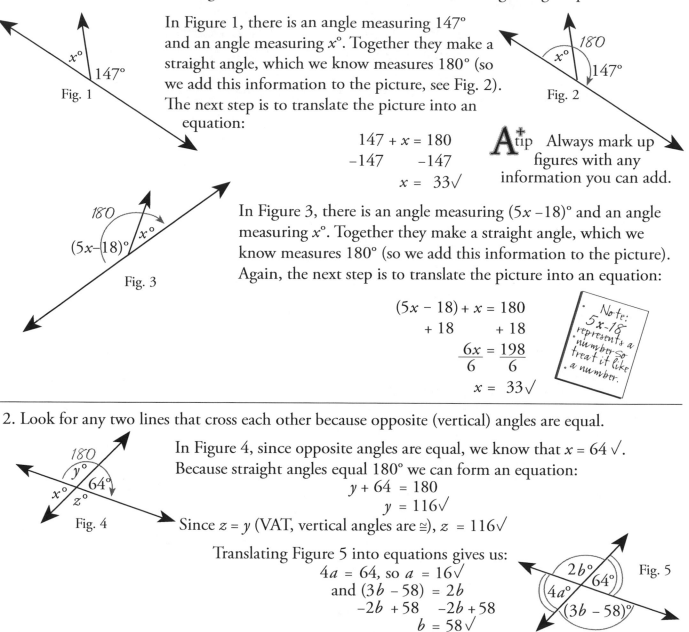

In Figure 1, there is an angle measuring 147° and an angle measuring $x°$. Together they make a straight angle, which we know measures 180° (so we add this information to the picture, see Fig. 2). The next step is to translate the picture into an equation:

$$147 + x = 180$$
$$\underline{-147 \qquad -147}$$
$$x = \ 33 \checkmark$$

A^{*}_{tip} Always mark up figures with any information you can add.

In Figure 3, there is an angle measuring $(5x - 18)°$ and an angle measuring $x°$. Together they make a straight angle, which we know measures 180° (so we add this information to the picture). Again, the next step is to translate the picture into an equation:

$$(5x - 18) + x = 180$$
$$\underline{+ 18 \qquad\qquad + 18}$$
$$\frac{6x}{6} = \frac{198}{6}$$
$$x = \ 33 \checkmark$$

Note: 5x-18 represents a number so treat it like a number.

---

2. Look for any two lines that cross each other because opposite (vertical) angles are equal.

In Figure 4, since opposite angles are equal, we know that $x = 64 \checkmark$. Because straight angles equal 180° we can form an equation:
$$y + 64 = 180$$
$$y = 116 \checkmark$$
Since $z = y$ (VAT, vertical angles are ≅), $z = 116 \checkmark$

Translating Figure 5 into equations gives us:
$$4a = 64, \text{ so } a = 16 \checkmark$$
$$\text{and } (3b - 58) = 2b$$
$$\underline{-2b + 58 \qquad -2b + 58}$$
$$b = 58 \checkmark$$

---

3. Solving word problems: Give algebraic "names" to quantities, then translate words into an equation.
Example: Find the measure of an angle if the supplement of the angle is 7 times the complement of the angle.

1st name the quantities:	2nd use the names to translate the words in the problem into the correct equation:	3rd solve the equation:
Let the angle = x Then the supplement of the angle = $180 - x$ And the complement of the angle = $90 - x$	$180 - x = 7(90 - x)$	$180 - x = 7(90 - x)$ $180 - x = 630 - 7x$ $6x = 450, \ x = 75 \checkmark$

Now You Try It

1. a. List the 3 conditions for angles to be adjacent.   b. Sketch a pair of adjacent angles.

       (1)

       (2)

       (3)

2. Are $\angle 1$ and $\angle 2$ in the figure adjacent angles? If not, why not? Do you know anything else about these two angles?

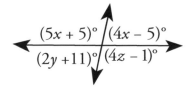

3. Find x, y and z in problems a and b.

a.
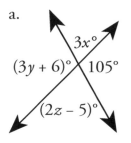

b.
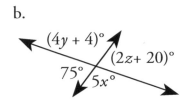

4. Find x and y and z. Think about the relationship between two angles that make up a straight angle.

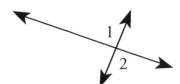

5. a. An angle's complement is five times the measure of the angle. Find the angle's measure.

  b. If the supplement of an angle is 20 more than three times its complement find the angle's measure.

6. Name and define 4 special pairs of angles. Sketch an example of each, naming the angles that meet the definition.

THE ANGLE ADDITION POSTULATE:

If point K lies in the interior region of ∠JOL then m∠JOK + m∠KOL = m∠JOL.

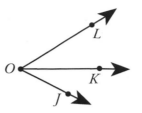

Use the Angle Addition Postulate in two ways:

1. In a proof when you need to show that two adjacent angles equal the larger (surrounding) angle, or that the larger (surrounding) angle can be broken up and is equal to the sum of the two smaller angles.

2. The Angle Addition Postulate provides the theoretical basis that allows us to solve an algebra problems like this one:

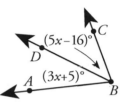

Using the figure on the right, given $m\angle ABC = 77°$, find $m\angle DBC$:

Since point D is in the interior region of $\angle ABC$, the Angle Addition Postulate tells us:

$$m\angle ABD + m\angle DBC = m\angle ABC$$
$$(3x + 5) + (5x - 16) = 77$$
$$8x - 11 = 77°, \quad 8x = 88, \quad x = 11$$

Now substitute in the value of x:

$$m\angle DBC = (5x - 16) = (5(11) - 16) = 39 \checkmark$$

DEFINITION OF AN ANGLE BISECTOR:

An angle bisector is the ray that divides an angle into 2 congruent adjacent angles.

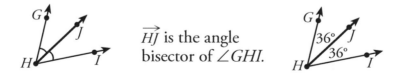

$\overrightarrow{HJ}$ is the angle bisector of $\angle GHI$.

Note: An angle bisector is always a *ray*.
In a proof, if given that $\overrightarrow{HJ}$ is an angle bisector, respond with $\angle GHJ \cong \angle JHI$. If asked to prove $\overrightarrow{HJ}$ is an angle bisector, you would need to show $\angle GHJ \cong \angle JHI$.

In an algebra problem the definition might be used like this:

Given $\overrightarrow{RT}$ is the angle bisector of $\angle QRS$, we know that $2x = 58$ so $x = 29$. $\checkmark$

Getting Started with Proofs

What Is a Proof?

Given: $AE = CE$, $EB = ED$
Prove: $CD = AB$.

A proof is a problem in which you are given some information and then asked to reach a certain valid conclusion (the prove) in a logical way and according to certain rules.

The Two Parts of a Two-Column Proof

1. Statements

The statements are listed on the left side of the page. The statements are where you state your case using information about this particular problem. The last statement is always the prove.

2. Reasons

The reasons are listed on the right side of the page. For each statement, you must say why, the reason (the "because"). And in geometry the reasons are limited to:

1. What is given
2. Definitions
3. Postulates (and the Properties)
4. Theorems (and sometimes mini-theorems called corollaries).

Some textbooks will call these by other names such as conjectures or facts or arguments or rules, but you'll recognize them; they are the reasons that you are allowed to give in a proof.

Remember, you are only allowed to use those reasons that are listed in your textbook and/or given by your teacher!

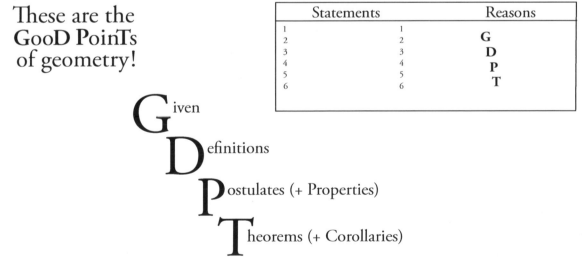

When you're searching for reasons, the **GooD PoinTs** are where you search.

Knowing the Concepts

In order to do proofs, you need to know and understand the language of geometry. This means, the **DPT**s (the definitions, postulates, theorems, etc.). Even if your teacher allows you to use a theorem or note sheet on exams (as some teachers do), what good will it be if you have no idea what any of the information is, what it means or how it should be used?

The most effective way to learn the DPTs is, as each one is introduced in your textbook, work to understand and memorize it. Making and studying accurate flash cards is a very useful learning method.

A⁺ tips for Proofs

— If the teacher or textbook has done parts of the proof, you may not change those parts.

— If you are being asked to do the entire proof, remember, there is usually more than one correct way to do a proof.

— Don't be overly concerned with how many steps you should do, that can depend on how you choose to do the proof.

— In a proof, if a defined term (e.g. "midpoint") is used in the given or prove, the definition of the term will usually be one of the reasons you will use in the proof.

— All definitions are "biconditionals". This means that in a proof, you can use them in 2 different ways:

> A right angle measures 90°, or
> An angle that measures 90° is a right angle.

The beginning part of each sentence is what you have previously shown. The ending part is what you are showing in the current step.

— Most theorems and postulates apply to particular objects, for example segments *or* angles. Based on what the proof you are working on is about, only consider the DPTs that concern that sort of object.

— It helps to remember that homework proofs emphasize the DPT's that were introduced in the current section.

— When preparing for a chapter exam, keep in mind that proofs on the test will emphasize the new concepts introduced in the chapter.

— The last statement in a proof is *always* the "prove".

The Properties

Properties are very much like postulates, they are true facts that we accept without proof, and they are part of the **DPT**s. We use properties in proofs primarily to combine measures of objects and to manipulate values in such a way so as to arrive at our prove. Pay special attention to the Addition, Substitution, and Reflexive Properties because they are the properties most frequently used in proofs.

The Properties and How to Use Them in Proofs

Use the Equality Properties with segment lengths and angle measures since both are real numbers.

ADDITION PROPERTY OF EQUALITY.

If $x = y$ and $a = b$, then $x + a = y + b$.

This property allows you to add two equations together, left side to left side, right side to right side. For example:

If: $AB = DE$ and,

$BC = EF$

then, $AB + BC = DE + EF$

Remember, AB is a length, that is, a number.

SUBTRACTION PROPERTY OF EQUALITY:

If $x = y$ and $a = a$ (Refl.) then $x - a = y - a$, for example:

If $m\angle ABC + m\angle CBD = m\angle DBE + m\angle CBD$

and $m\angle CBD = m\angle CBD$ (Reflexive),

then $m\angle ABC = m\angle DBE$.

Remember, $m\angle ABC$ is a number.

SUBSTITUTION PROPERTY OF EQUALITY:

If $a = b$ then either a or b can be substituted for the other in any equation. For example,

If $AB + BC = DE + EF$

and $AB + BC = AC$, $DE + EF = DF$

then, $AC = DF$

Substitution is calling the same quantity by a different name. It's as if, using the information in the 2nd line, you went back to the 1st line and crossed out the quantities, replacing them with their other names:

$$\overset{AC}{\cancel{AB + BC}} = \overset{DF}{\cancel{DE + EF}}$$

DIVISION PROPERTY OF EQUALITY.

If $x = y$ and $c \neq 0$, then $\frac{x}{c} = \frac{y}{c}$

MULTIPLICATION PROPERTY OF EQUALITY:

If $x = y$, then $cx = cy$.

The Multiplication Property allows you to multiply *both* sides of an *equation* by the same number.

DISTRIBUTIVE PROPERTY:

$a(b + c) = ab + ac$

SYMMETRIC PROPERTY OF EQUALITY:

If $x = y$, then $y = x$.

For example:

If $AB = CD$, then, $CD = AB$.

Symmetric is also a congruence property, so:

If $\angle A \cong \angle B$, then $\angle B \cong \angle A$ and

if $\overline{AB} \cong \overline{CD}$ then $\overline{CD} \cong \overline{AB}$

The last statement of a proof must match the prove exactly. Use the Symmetric Property to "flip" an equation or congruence if necessary.

REFLEXIVE PROPERTY:

$m\angle D = m\angle D$, $AB = AB$, $x = x$.

Before you add or subtract the measure of an angle or a segment to or from both sides of an equation, stop and state that the measure of the object is equal to itself by "Reflexive Property". Notice that this creates the second equation that you need in order to use the Addition or Subtraction Property.

Reflexive is also a congruence property, so:

$\angle A \cong \angle A$, and $\overline{BC} \cong \overline{BC}$

TRANSITIVE PROPERTY:

Equality is transitive: If $AB = CD$ and $CD = EF$ then $AB = EF$.

Congruence is transitive: If $\angle A \cong \angle B$ and $\angle B \cong \angle C$ then $\angle A \cong \angle C$.

Same eye color is transitive: If my eyes are the same color as your eyes and your eyes are the same color as your mother's eyes then my eyes are the same color as your mother's eyes.

Some characteristics "transfer" and some don't. In proofs, if a quantity or object is skipped over you are probably looking at the Transitive Property. (*CD* was skipped over in the equality example, $\angle B$ was skipped over in the congruence example and you were skipped over in the eye color example.)

Know the properties and know how to use them!

Putting it All Together - Beginning Proofs

Segments	vs.	Angles

DEFINITION

Midpoint of a Segment: The point that divides a segment into *2 congruent parts*.

Angles do not have midpoints.

DEFINITION

Segment Bisector: A line, ray, plane or segment that *intersects a segment at its midpoint*.

DEFINITION

Angle Bisector: A ray which divides an angle into *2 congruent adjacent angles*.

*Midpoint Theorem:

If M is the midpoint of $\overline{AB}$, then $AM = \frac{1}{2}AB$ and $MB = \frac{1}{2}AB$

*Angle Bisector Theorem:

If $\overrightarrow{BD}$ is the bisector of $\angle ABC$, then $\angle ABD \cong \frac{1}{2}\angle ABC$ and $\angle DBC \cong \frac{1}{2}\angle ABC$

*Only learn and use these theorems if your textbook includes them. If it does, pay attention to the details: The definitions of a midpoint and an angle bisector state that the 2 parts are congruent. The theorems state that the 2 parts are ½ of the whole.

Definitions → parts ≅	Memory Tip
Theorems → parts ½ whole	

POSTULATE

Segment Addition Postulate: If B is between A and C, then $AB + BC = AC$.

Recall that we can use the Segment Addition Postulate to state that the 2 parts equal the whole or that the whole equals the 2 parts. This is a method often used to introduce the prove into the proof.

POSTULATE

Angle Addition Postulate: If point K lies in the interior region of $\angle JOL$ then $m\angle JOK + m\angle KOL = m\angle JOL$.

Recall that we can use the AngleAddition Postulate to state that the 2 parts equal the whole or that the whole equals the 2 parts. This is a method often used to introduce the prove into the proof.

ADDITION PROPERTY OF EQUALITY:
For segments:
    If: $AB = DE$ and,
       $BC = EF$
then, $AB + BC = DE + EF$

For angles:
If $m\angle GHL = m\angle MNP$, &
  $m\angle LHK = m\angle PNQ$, then
$m\angle GHL + m\angle LHK = m\angle MNP + m\angle PNQ$

SUBSTITUTION PROPERTY OF EQUALITY:
For segments: If $AB + BC = DE + EF$
        and $AB + BC = AC$,
        $DE + EF = DF$
        then, $AC = DF$

For angles:
If $m\angle GHL + m\angle LHK = m\angle MNP + m\angle PNQ$
and $m\angle GHL + m\angle LHK = m\angle GHK$
and $m\angle MNP + m\angle PNQ = m\angle MNQ$
then, $m\angle GHK = m\angle MNQ$

Thinking Through a Proof

Given: $AE = CE$, $EB = ED$
Prove: $CD = AB$.

1. Draw a sketch. Add the information from the "Given" to your sketch

2. The prove is *always true*. Make notes as you try to figure out why it is true.

3. Jot down your reasoning. The path of your thoughts *is* the proof.

4. Add any new information that you were able to figure out to your sketch.

Seg Add Post?

$I + II = I + II$

If *why* the prove is true still isn't clear to you, try assigning lengths to the segments. For example, let AE and $CE = 2$, and EB and $ED = 5$. Studying the sketch, $CD = 7$ and $AB = 7$, so yes, $AB = CD$. This is *not a proof*, but it shows that if the parts are equal, the wholes are equal and suggests the idea of adding the parts together.

Getting Started

We will repeat the given and prove for clarity:

Given: $AE = CE$, $EB = ED$.
Prove: $AB = CD$.

The **shaded** letters are to help us track the different elements:

Statements	Reasons
1. $AE = CE$, $EB = ED$.	1. Given.
2. $AE + EB = CE + ED$.	2. Addition Property of Equality.
3. $AE + EB = AB$, $CE + ED = CD$.	3. Segment Addition Postulate.
4. $AB = CD$.	4. Substitution. (From Stmts. 2 & 3).

Adding the two equations from Statement 1 together.

Using the Segment Addition Post. to state that the sum of the parts equals the whole. This brings AB and CD into the proof.

The **last** statement in any proof, is always the prove.

Calling the same things (from Stmt. 2) by different names (the ones determined in Stmt. 3).

More explanation: In Statement 2, we showed that $AE + EB = CE + ED$

In Statement 3, we showed that $AE + EB = AB$
and $CE + ED = CD$

Statement 3 gave us "new names" for the quantities that Statement 2 showed were equal. Combining this information, allowed us to reach the prove, $AB = CD$ in step 4. Think of it as using the information from Statement 3 to go back and change the names in

Statement 2: $\underbrace{AE + EB}_{} = \underbrace{CE + ED}_{}$ which is the prove,
       AB CD

Still confused by substitution? Here's another way to help you recognize when to use it:	If the **left** sides of 2 equations are equal, the **right** sides are equal too! $AE + EB = AB$, $CE + ED = CD$ AB CD Reason? Substitution.

Now You Try It

Given: $AC = EC$ and $AB = ED$
Prove: $BC = DC$

Study the figure and ask yourself, "Why is the prove true?" Mark up the sketch with the given information. This proof is asking us to show that if the wholes are equal and one pair of parts are equal, then the other pair of parts must also be equal. This suggests taking something away, or subtraction. And in order to subtract something, we'll need to show that the whole is equal to its parts and the Segment Addition Postulate does that. As you think about the proof, jot down the properties or facts you will probably use. Check your answer in the back of the book.

Seg Add Post?
Subtr Prop?

Statements	Reasons
1.	1.
2.	2.
3.	3.
4.	4.
5. $BC = DC$.	5.

Proofs about Angles

Many initial proofs about angles make use of the Addition Property of Equality together with the Angle Addition Postulate. Here's an example:

Given: $m\angle BAE = m\angle ABD$ and $m\angle EAC = m\angle DBC$
Prove: $m\angle BAC = m\angle ABC$

Study the figure. We are given that 2 pairs of parts are equal and asked to prove that the wholes are equal. In fact, this proof is logically identical to the proof about segments on page 25.

Proof:

Adding the 2 equations from Statement 1 together.

Using the Angle Add. Post. to state that the sum of the parts equals the whole. Note that this brings $m\angle BAC$ and $m\angle ABC$ into the proof.

Statements	Reasons
1. $m\angle BAE = m\angle ABD$ and $m\angle EAC = m\angle DBC$.	1. Given.
2. $m\angle BAE + m\angle EAC = m\angle ABD + m\angle DBC$.	2. Addition Property of Equality.
3. $m\angle BAE + m\angle EAC = m\angle BAC$, $m\angle ABD + m\angle DBC = m\angle ABC$.	3. Angle Addition Postulate. Be careful. Stop and think. This proof is about *angles* (not segments).
4. $m\angle BAC = m\angle ABC$.	4. Substitution (Stmts. 2 & 3).

The last statement is always the prove.

Calling the same thing (from Stmt. 2) by a different name (the one determined in Stmt. 3).

26

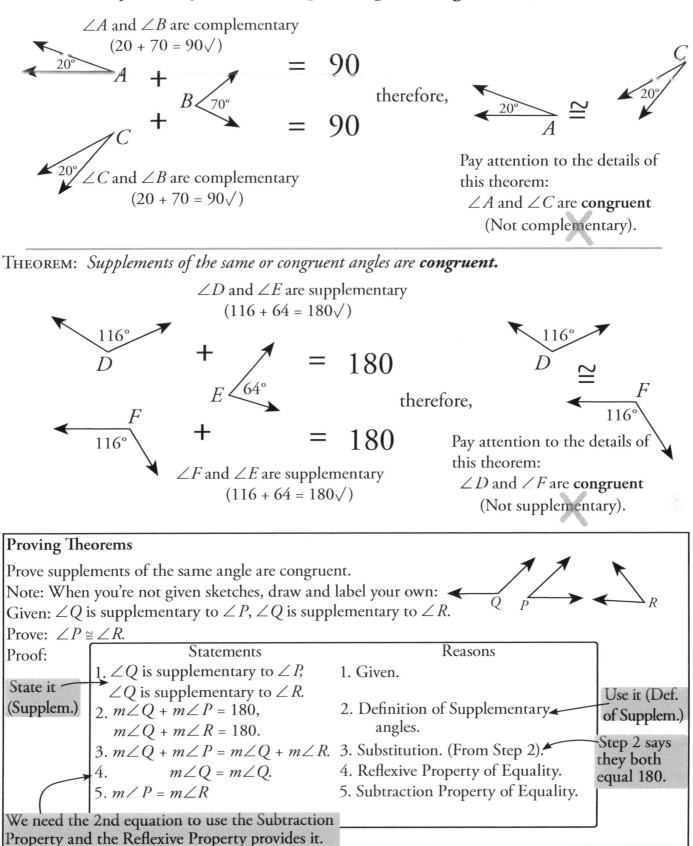

THEOREM: *Complements of the same or congruent angles are* **congruent**.

∠A and ∠B are complementary
(20 + 70 = 90√)

20° A + B 70° = 90

therefore,

+

C

20° = 90

∠C and ∠B are complementary
(20 + 70 = 90√)

20° A ≅ C 20°

Pay attention to the details of
this theorem:
∠A and ∠C are **congruent**
(Not complementary).

THEOREM: *Supplements of the same or congruent angles are* **congruent.**

∠D and ∠E are supplementary
(116 + 64 = 180√)

116° D + E 64° = 180

therefore,

F 116° + = 180

∠F and ∠E are supplementary
(116 + 64 = 180√)

116° D ≅ F 116°

Pay attention to the details of
this theorem:
∠D and ∠F are **congruent**
(Not supplementary).

Proving Theorems

Prove supplements of the same angle are congruent.
Note: When you're not given sketches, draw and label your own:
Given: ∠Q is supplementary to ∠P, ∠Q is supplementary to ∠R.
Prove: ∠P ≅ ∠R.
Proof:

Statements	Reasons
1. ∠Q is supplementary to ∠P, ∠Q is supplementary to ∠R.	1. Given.

2. m∠Q + m∠P = 180,
  m∠Q + m∠R = 180.
	2. Definition of Supplementary
	  angles.

3. m∠Q + m∠P = m∠Q + m∠R.	3. Substitution. (From Step 2).
4. m∠Q = m∠Q.	4. Reflexive Property of Equality.
5. m∠P = m∠R	5. Subtraction Property of Equality.

State it
(Supplem.)

Use it (Def.
of Supplem.)

Step 2 says
they both
equal 180.

We need the 2nd equation to use the Subtraction
Property and the Reflexive Property provides it.

Straight Angles in Problems and Proofs

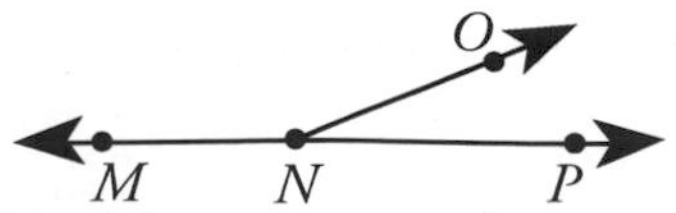

Given the figure above, you are allowed to assume that $m\angle MNP = 180°$.

This assumption allows you to solve certain types of algebra problems, for example:

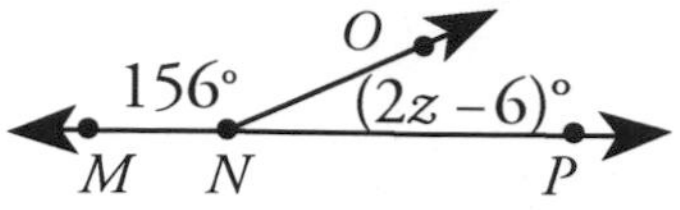

$$156 + 2z - 6 = 180$$
$$2z = 30$$
$$z = 15 \checkmark$$

But, in a proof, if you must show that the two angles are supplementary, it takes 2 steps:

First Way — If your textbook gives this second part of the Angle Addition Postulate:

If $\angle MNP$ is a straight angle and point O is not on $\overleftrightarrow{MP}$, then

$$m\angle MNO + m\angle PNO = 180.$$

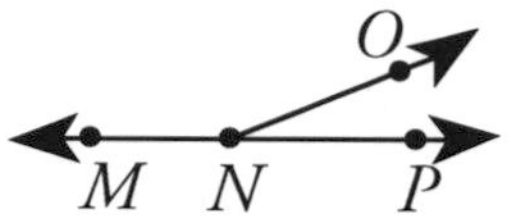

Given a figure like the one above, these are the 2 steps you should use to show 2 angles are supplementary:

Statement	Reason
1. $m\angle MNO + m\angle PNO = 180$.	1. Angle Addition Postulate.
2. $\angle MNO$ & $\angle PNO$ are supplementary.	2. Definition of supplementary angles.

Second Way — If your textbook defines linear pairs and gives this theorem:

Angles that form a linear pair are supplementary.

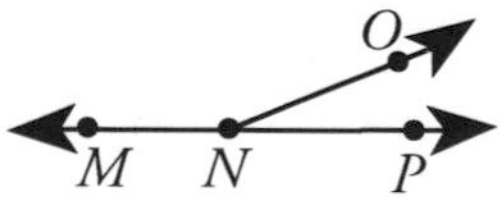

Given a figure such as the one shown, these are the 2 steps you should use to show 2 angles are supplementary:

Statement	Reason
1. $\angle MNO$ & $\angle PNO$ are a linear pair.	1. Definition of a linear pair.
2. $\angle MNO$ & $\angle PNO$ are supplementary.	2. Angles forming a linear pair are supplementary.

Perpendicular Lines and Proofs

DEFINITION: *Perpendicular lines are lines that intersect to form right angles.*

THEOREM: *If two lines form congruent adjacent angles, the lines are perpendicular.*

Given: $\angle ABC \cong \angle CBD$

Prove: $\overleftrightarrow{CE} \perp \overleftrightarrow{AD}$

Proof:

To prove that some object has a particular property, show that the object meets the definition of that property. In fact, if you are stuck on a proof, you might try working backwards, starting with the definition. Below is an example, using arrows to indicate the flow of thought:

The last statement is always the prove. → $\overleftrightarrow{CE} \perp \overleftrightarrow{AD}$ ⟶ Def. of perpendicular lines. ($\perp$ lines form rt. $\angle$'s.)

$\angle ABC$ is a right angle. ⟶ Definition of right angles. (Rt. $\angle$'s = 90°.)

$m\angle ABC = 90$ ⟵⟶ ?

At this point we need to study the problem in order to decide how we can show that $m\angle ABC = 90$, but at least we know what it is that we need to show.

Here's the completed proof going "forwards":

Statements	Reasons
1. $\angle ABC \cong \angle CBD$, or $m\angle ABC = m\angle CBD$.	1. Given.
2. $m\angle ABC + m\angle CBD = 180$.	2. Angle Addition Postulate*.
3. $2(m\angle ABC) = 180$.	3. Substitution (Statements 1&2).
4. $m\angle ABC = 90$. ⟵ state it, then use it	4. Division Property of Equality.
5. $\angle ABC$ is a right angle. ⟶	5. Definition of right angles. (Rt. $\angle = 90°$)
6. $\overleftrightarrow{CE} \perp \overleftrightarrow{AD}$	6. Definition of perpendicular lines. ($\perp$ lines form rt. $\angle$'s)

The last statement is always the prove.   state it, then use it

* If your text defines linear pairs, do the following steps in place of Step 2:

Statements	Reasons
$\angle ABC$ and $\angle CBD$ are a linear pair.	Definition of linear pairs.
$\angle ABC$ and $\angle CBD$ are supplementary.	Linear pairs are supplementary.
$m\angle ABC + m\angle CBD = 180$.	Definition of supplementary.

A⁺tips

On an exam, put down something. Never leave a proof blank. At least put in the given and the prove. Never be afraid to try (teachers appreciate effort). If you think you can recall a pattern, try it.
Most Important Tip: Beginning proofs take getting used to and many students find them hard. This is because you are learning to think in a new way. Proofs will become easier. Be patient!

Now You Try It

1. Given: *AB* = *CD*.
   Prove: *AC* = *DB*.

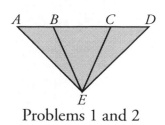

Problems 1 and 2

2. Given: $m\angle AEB = m\angle DEC$.
   Prove: $m\angle AEC = m\angle DEB$.

3. Prove the following:

THEOREM: *If two lines are perpendicular, they form congruent adjacent angles.*
Create an appropriate figure and state the given and the prove.
The theorem at the top of the previous page and this one have a close connection. They are called "converses". Can you see what the connection is?

Be sure to check in the back of the book to see how you did!

Here are some examples of *conditional or "if-then"* statements:

> If you live in Malibu, then you live in California.
> If this is alternative rock, then it is music.
> If you are a soloist in the choir, then you have a good voice.

In symbols we write: If **p** then **q**.

p is called the premise (or hypothesis) and **q** is called the conclusion.

Premise (or hypothesis)		Conclusion
if you live in Malibu, this is alternative rock, you are a soloist in the choir,	**then**	you live in California. it is music. you have a good voice.

The statements above are examples of true if-then statements. Look at Figure 1. Since Malibu is in California, you cannot live in Malibu without also living in California. A point cannot be in the small circle without also being in the surrounding figure.

We could also say, "You must live in California if you live in Malibu" or "You live in Malibu, only if you live in California." Even though the phrases have been switched around, living in Malibu is still the hypothesis and living in California is still the conclusion. Be careful, *"only if"* actually means *"then"*.

Fig. 1

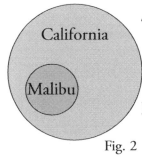

The circles in Figure 2 are an example of a *Venn Diagram*, a logical model which helps us to picture how conditional statements work. You can see that any points in the small circle are automatically in the larger surrounding circle.

Figure 3 demonstrates a pattern. Note that the **p**remise, that is, the "if" part of the statement, is the **p**etite (smaller) inner circle. The conclusion, the "then" part, is the larger circle.

Fig. 2

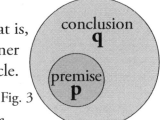

Fig. 3

Truth Value of If-Then Statements and the Law of Detachment

True: An example of a true if-then statement is, if $x = 2$, then $2x = 4$. In a *true* if-then statement, given that the premise, that is, the "if" part of the statement, is true, then the conclusion, that is, the "then" part of the statement, is also true. This is the **LAW OF DETACHMENT**.

False: An if-then statement is *false* if you can show that given a true premise, that is, given that the "if" part of the statement is true, there is *even one* example of the conclusion, that is the "then" part of the statement, being false. Here's an example of a false if-then statement:

> Statement: If a person is a star, then he or she lives in Hollywood.

> Counterexample: Adele lives in Surrey, England.

Since Adele is a star, (that is, the premise is true) and she does *not* live in Hollywood (that is, the conclusion is not true) Adele is a *counterexample* that proves that the if-then statement is false.

Another example of a false if-then statement:

$$\text{Statement: if } x^2 = 25, \text{ then } x = 5$$
$$\text{Counterexample: } x = -5$$

Since $(-5)^2 = 25$, $x = -5$ is a valid counterexample and disproves the if-then statement. Notice that in this case, -5 is the *only* counterexample, but one counterexample is all that we need to disprove an hypothesis.

There are other logical statements that are related to and based upon an if-then statement:

Name	Form	Order and Negations	Memory Helpers
If-then statement	If **p**, then **q**.	First comes the **p**, then comes the **q**.	$p \rightarrow q$
Contrapositive	If not **q**, then not **p**.	The positions of **p** and **q** have been swapped *and* both have been negated (made negative).	$\cancel{q} \rightarrow \cancel{p}$
Converse	If **q** then **p**.	The positions of **p** and **q** have been swapped.	$q \rightarrow p$
Inverse	If not **p**, then not **q**.	The positions remain the same; both **p** and **q** have been negated (made negative).	$\cancel{p} \rightarrow \cancel{q}$

Example 1.

If-then statement:	If you are an Olympic gymnast, then you are an excellent athlete.
Contrapositive:	If you are not an excellent athlete, then you are not an Olympic gymnast.
Converse:	If you are an excellent athlete, then you are an Olympic gymnast.
Inverse:	If you are not an Olympic gymnast, then you are not an excellent athlete.

In this example, the if-then statement and its contrapositive are both true. The converse is false since many excellent athletes are not Olympic gymnasts and the inverse is false for the same reason. Notice that the words "if" and "then" stay put. They are not part of the premise or the conclusion.

Example 2.

If-then statement:	If $x > 1$, then $x > 5$
Contrapositive:	If $x \ngtr 5$, then $x \ngtr 1$
Converse:	If $x > 5$, then $x > 1$
Inverse:	If $x \ngtr 1$, then $x \ngtr 5$

In this example, the if-then statement and its contrapositive are both false, the converse and the inverse are both true.

Example 3.

If-then statement:	If $x = 7$, then $2x = 14$
Contrapositive:	If $2x \neq 14$, then $x \neq 7$
Converse:	If $2x = 14$, then $x = 7$
Inverse:	If $x \neq 7$, then $2x \neq 14$

In this example, the if-then statement, its contrapositive, converse and inverse are all true. In other words *all four* statements are true.

Although there seem to be many different combinations of true and false possibilities, in fact, two pairs of the four conditional statements are logically connected.

if-then ⟷ contrapositive

An if-then statement and its contrapositive are logically linked:
both are true or **both** are false.

If you live in Malibu, then you live in California. **T**

If you don't live in California, then you don't live in Malibu. **T**

converse ⟷ inverse

The con*verse* of a statement and the in*verse* of a statement are logically linked:
both are true or **both** are false.

If you live in California, then you live in Malibu. **F**

If you don't live in Malibu, then you don't live in California. **F**

*Memory Hint: The "*verses*" are logically linked.

The Venn diagrams of each of the four statements, together with its "Truth Value" are below:

STATEMENT	VENN DIAGRAM	TRUTH Value
IF-THEN If you live in Malibu, then you live in California.		TRUE
CONTRAPOSITIVE If you don't live in California, then you don't live in Malibu. Study the figure. Since the conclusion is negated, the object (you!) is outside the larger circle and therefore could not possibly be in the smaller inner circle.		TRUE
CONVERSE If you live in California, then you live in Malibu. Since we can find a counterexample (Hollywood for example), the statement is false.	Maybe you live in Malibu. / Or maybe you live in Hollywood.	FALSE
INVERSE If you don't live in Malibu, then you don't live in California. Since we can find a counterexample, Hollywood, the statement is false.	You might live in Hollywood.	FALSE

THE LAW OF SYLLOGISM

Syllogisms are combinations of two conditionals in which the conclusion of the first statement is the hypothesis of the second. Here's an example,

If you live in Malibu, then you live in California.

If you live in California, then you live in the United States.

} The conclusion of the first statement must be the hypothesis of the second statement.

LAW OF SYLLOGISM: *If **both** conditionals are true, then you may conclude correctly, that given Hypothesis **1**, Conclusion **2** is true.*

This means that the first hypothesis leads directly to the second conclusion. In symbols we can write:

$$p \rightarrow q \text{ and } q \rightarrow r \text{ which can be combined to form: } p \rightarrow q \rightarrow r \text{ or more simply } p \rightarrow r.$$

Using the example above, since *both statements are true*, you may conclude that if you live in Malibu, then you live in the United States.

Here's the Venn diagram of our example.

Now You Try It

1. List the four logical conditionals and using symbol shorthand give an example of each. State the two pairs that are logically connected.

2. Given the statement: If you are in a top band, then you are famous.
a. Form the contrapositive, converse and inverse and state the truth value of each of the 4 statements.

b. If your cousin Sally is not famous, what can you conclude about her?
c. If your cousin Sam is famous, what can you conclude about him?

3. Given the statement: If $x^2 > 16$, then $x > 4$. Form the contrapositive, the converse and the inverse and determine the truth value of each of the four statements.

4. Explain the terms syllogism and the Law of Syllogism. Give an original example of each.

Indirect Proofs

Certain proofs are much easier if done *indirectly*.

Regular Proofs (also called *Direct Proofs*) — We are given some information (the "given") and asked to prove a certain conclusion (the "prove"). We start with the given and add other information based on theorems, postulates, and definitions until we reach the conclusion.

Indirect Proofs — We are given some information (the "given") and asked to reach a certain conclusion (the "prove"). We start by **temporarily** assuming that the prove (*never* the given) is negative (i.e.

```
GooD PoinTs
Given
Definitions
Postulates(+ Properties)
Theorems (+ Corollaries)
```

false). We then add other information based on the **GooD PoinTs** until we contradict either the given or some other known fact (the **DPTs**). Since this is impossible, (because the **GooD PoinTs** are true), the contradiction shows that our temporary assumption was false. Therefore, "the prove" must be true. (Don't worry, indirect proofs are easier than they sound.)

Here's the pattern for Indirect Proofs:
1. Assume temporarily that the prove is false, this means negate (make negative) the *prove*.
2. Study the problem and figure out the effect(s) of your temporary assumption.
3. Based on the assumption, make statements and reasons until you contradict either the given or some other known fact (the **GooD PoinTs** again!).
4. State that since something that is true has been contradicted, your temporary assumption must be false and therefore, the prove must be true.

Note: An Indirect Proof is usually done as a **Paragraph Proof**, which means, a series of sentences. You make a statement, and in parenthesis, you give the reason (the **GooD PoinT**) that supports your statement. A Paragraph Proof is just as rigorous as a two-column proof, it just appears in paragraph form.

To get started with Indirect Proofs, we need to learn how to make the **prove** negative. Here are some examples:

$$\text{If the prove is:} \quad AB = MN$$
$$\text{the } \textit{negative} \text{ of the prove is:} \quad AB \neq MN$$

$$\text{If the prove is:} \quad m\angle C \neq m\angle D$$
$$\text{the } \textit{negative} \text{ of the prove is:} \quad m\angle C = m\angle D$$

$$\text{If the prove is:} \quad \triangle ABD \text{ is an isosceles triangle.}$$
$$\text{the } \textit{negative} \text{ of the prove is:} \quad \triangle ABD \text{ is not an isosceles triangle.}$$

$$\text{If the prove is:} \quad a < b$$
$$\text{the } \textit{negative} \text{ of the prove is:} \quad a \geq b$$
(You must allow for *a* to be greater than *b* or equal to *b*.)

Here's an example*:

Given: $m\angle A \neq m\angle B$, and $\angle A$ is complementary to $\angle C$ and $\angle B$ is complementary to $\angle D$.
Prove: $m\angle C \neq m\angle D$ indirectly using a paragraph proof.

1. "Assume temporarily". 2. Negate the prove. 3. The argument.

Assume temporarily that $m\angle C = m\angle D$. Since $\angle A$ is complementary to $\angle C$ and $\angle B$ is complementary to $\angle D$ (Given), $m\angle A + m\angle C = 90$ and $m\angle B + m\angle D = 90$ (definition of complementary angles). Therefore, $m\angle A + m\angle C = m\angle B + m\angle D$ (Substitution). Since we have assumed temporarily that $m\angle C = m\angle D$ we can subtract them (Reflexive) from the equation. Doing so would mean $m\angle A = m\angle B$ (Subtraction). But this contradicts the Given which is impossible, so our temporary assumption was false, and $m\angle C \neq m\angle D$.

6. Restate the prove. 4. Contradict a Good Point.

5. "temporary assumption was false"

Be Careful! You assume temporarily that the Prove is false. Never the given!

Now You Try It — Hint, follow the steps.
Given: $m\angle I \neq m\angle J$,
Prove: $\angle I$ and $\angle J$ are not both right angles. Do the proof indirectly and use a paragraph proof.

Indirect proofs are a lot like the contrapositive form of an if-then statement. Here's an example:

If-then statement: If the speed limit is 65 miles per hour and you arrived at a concert which is 65 miles away in 45 minutes, then you were speeding.

Contrapositive: If you did not speed, then you could not be at a concert which is 65 miles away if you have been traveling for 45 minutes.

Using an indirect proof, prove: Given that the speed limit is 65 miles per hour and that you arrived at a concert which is 65 miles away in 45 minutes, prove that you have been speeding.
Proof:

First, *assume temporarily* that you were not speeding. But if you were not speeding and have traveled for 45 minutes at the top legal speed of 65 miles per hour you should only be 48.75 miles away (65mph × 45/60 hrs). But you are 65 miles away (the Given). Therefore, since we have contradicted the Given, you must have been speeding!

*Note: For additional examples of Indirect Proofs see page 242 in the Answer Section.

4. PARALLEL LINES

Lines, Transversals and More Special Pairs of Angles

A *transversal* is a line that crosses (transverses) 2 coplanar lines at two different points.

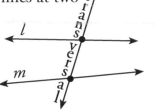

Eight angles are formed by the transversal and the lines it crosses. There are special terms for *pairs* of angles in certain *relative positions*. It is important to understand and memorize these terms.

Special *Pairs* of Angles:

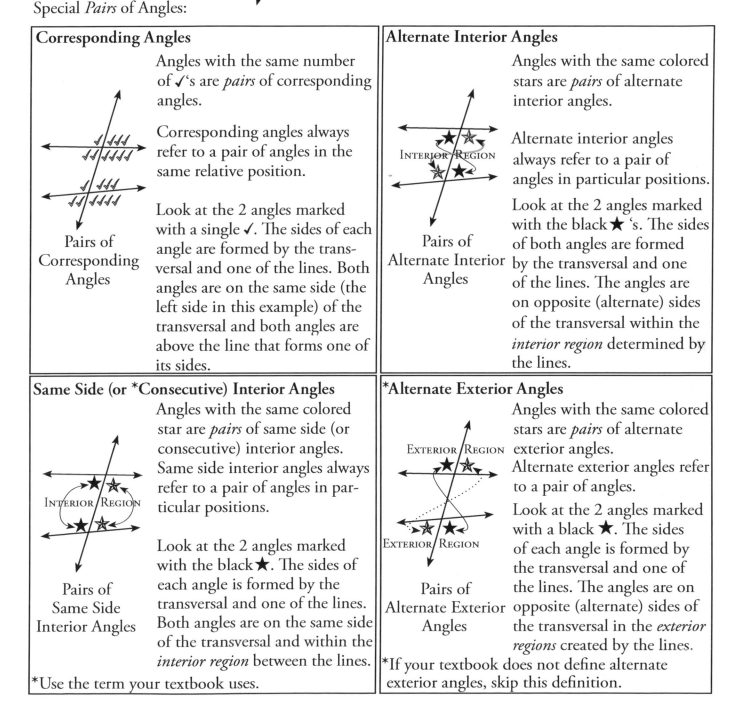

Corresponding Angles

Pairs of Corresponding Angles

Angles with the same number of ✓'s are *pairs* of corresponding angles.

Corresponding angles always refer to a pair of angles in the same relative position.

Look at the 2 angles marked with a single ✓. The sides of each angle are formed by the transversal and one of the lines. Both angles are on the same side (the left side in this example) of the transversal and both angles are above the line that forms one of its sides.

Alternate Interior Angles

Pairs of Alternate Interior Angles

Angles with the same colored stars are *pairs* of alternate interior angles.

Alternate interior angles always refer to a pair of angles in particular positions.

Look at the 2 angles marked with the black ★ 's. The sides of both angles are formed by the transversal and one of the lines. The angles are on opposite (alternate) sides of the transversal within the *interior region* determined by the lines.

Same Side (or *Consecutive) Interior Angles

Pairs of Same Side Interior Angles

Angles with the same colored star are *pairs* of same side (or consecutive) interior angles. Same side interior angles always refer to a pair of angles in particular positions.

Look at the 2 angles marked with the black ★. The sides of each angle is formed by the transversal and one of the lines. Both angles are on the same side of the transversal and within the *interior region* between the lines.

*Use the term your textbook uses.

*Alternate Exterior Angles

Pairs of Alternate Exterior Angles

Angles with the same colored stars are *pairs* of alternate exterior angles.

Alternate exterior angles refer to a pair of angles.

Look at the 2 angles marked with a black ★. The sides of each angle is formed by the transversal and one of the lines. The angles are on opposite (alternate) sides of the transversal in the *exterior regions* created by the lines.

*If your textbook does not define alternate exterior angles, skip this definition.

Consider lines *i* and *j* in the illustration below. Lines *k* and *l* are transversals from the point of view of lines *i* and *j*. However, if we are considering lines *k* and *l*, then lines *i* and *j* are the transversals.

Example: Classify the following pairs of angles as corresponding angles, alternate interior angles, same side interior angles or none of these. If the angles do form a special pair, name the lines and the transversal that form the 2 angles.

1. ∠1 and ∠9 — Corresponding angles, lines *i* and *j*, transversal *k*.

2. ∠1 and ∠5 — Corresponding angles, lines *k* and *l*, transversal *j*.

3. ∠5 and ∠13 — Corresponding angles, lines *i* and *j*, transversal *l*.

4. ∠12 and ∠13 — Alternate interior angles, lines *k* and *l*, transversal *i*.

5. ∠1 and ∠12 — Alternate exterior angles, lines *i* and *j*, transversal *k*.

6. ∠1 and ∠8 — Alternate exterior angles, lines *k* and *l*, transversal *j*.

7. ∠12 and ∠13 — Alternate interior angles, lines *k* and *l*, transversal *i*.

8. ∠4 and ∠10 — Same side interior (or consecutive) angles, lines *i* and *j*, transversal *k*.

9. ∠4 and ∠13 — None (no transversal is shared by both angles).

10. ∠4 and ∠6 — None (this pair of angles are not in any of the correct relative positions).

Now You Try It

Using the figure above, classify the following pairs of angles as corresponding angles, alternate interior angles, same side interior angles or none of these. If the angles do form one of the listed special pairs, name the lines and the transversal that form the 2 angles.

1. ∠3 and ∠7   4. ∠2 and ∠5   7. ∠7 and ∠13

2. ∠4 and ∠5   5. ∠3 and ∠10   8. ∠4 and ∠13

3. ∠15 and ∠6   6. ∠11 and ∠10   9. ∠11 and ∠8

A Closer Look — The angles that make up the special pairs defined on the previous page must both be formed by the *same* transversal:

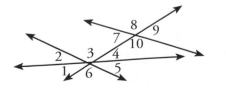

∠7 and ∠2 are *not* corresponding angles because their sides are formed by 4 (entirely different) lines.

∠10 and ∠5 are *not* same side interior angles because their sides are formed by 4 (entirely different) lines.

Parallel Lines — Two lines are parallel if they are: 1. Coplanar and 2. Do not intersect.

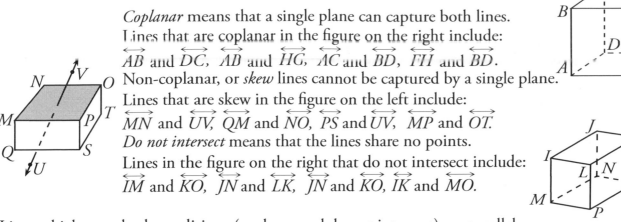

Coplanar means that a single plane can capture both lines. Lines that are coplanar in the figure on the right include: $\overleftrightarrow{AB}$ and $\overleftrightarrow{DC}$, $\overleftrightarrow{AB}$ and $\overleftrightarrow{HG}$, $\overleftrightarrow{AC}$ and $\overleftrightarrow{BD}$, $\overleftrightarrow{FH}$ and $\overleftrightarrow{BD}$. Non-coplanar, or *skew* lines cannot be captured by a single plane. Lines that are skew in the figure on the left include: $\overleftrightarrow{MN}$ and $\overleftrightarrow{UV}$, $\overleftrightarrow{QM}$ and $\overleftrightarrow{NO}$, $\overleftrightarrow{PS}$ and $\overleftrightarrow{UV}$, $\overleftrightarrow{MP}$ and $\overleftrightarrow{OT}$. *Do not intersect* means that the lines share no points. Lines in the figure on the right that do not intersect include: $\overleftrightarrow{IM}$ and $\overleftrightarrow{KO}$, $\overleftrightarrow{JN}$ and $\overleftrightarrow{LK}$, $\overleftrightarrow{JN}$ and $\overleftrightarrow{KO}$, $\overleftrightarrow{IK}$ and $\overleftrightarrow{MO}$.

Lines which meet both conditions (coplanar and do not intersect) are parallel. Pairs of parallel lines in the figure on the right include:

$$\overleftrightarrow{AB} \text{ and } \overleftrightarrow{DC}, \quad \overleftrightarrow{AB} \text{ and } \overleftrightarrow{HG}, \quad \overleftrightarrow{FH} \text{ and } \overleftrightarrow{BD}, \quad \overleftrightarrow{AB} \text{ and } \overleftrightarrow{EF}.$$

The symbol for parallel is $\|$. For example:

$$\overleftrightarrow{AB} \parallel \overleftrightarrow{DC}, \quad \overleftrightarrow{AB} \parallel \overleftrightarrow{HG}, \quad \overleftrightarrow{FH} \parallel \overleftrightarrow{BD}, \quad \overleftrightarrow{AB} \parallel \overleftrightarrow{EF}.$$

Segments are considered to be parallel if the lines that include them are parallel. For example:

$$\overline{AB} \parallel \overline{DC}, \quad \overline{AB} \parallel \overline{HG}, \quad \overline{FH} \parallel \overline{BD}, \quad \overline{AB} \parallel \overline{EF}.$$

The same number of arrowheads on each of two (or more) lines means the lines are parallel.

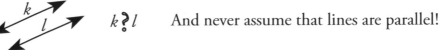

$c \parallel d$ $m \parallel n \parallel o$ $u \parallel v$ $\overleftrightarrow{AB} \parallel \overleftrightarrow{DC}$ $\overleftrightarrow{EA} \parallel \overleftrightarrow{HD}$

Arrowheads at the *end* of lines do *not* mean the lines are parallel.

$k ? l$ And never assume that lines are parallel!

Parallelism is transitive. Here's an example: If $\overleftrightarrow{PS} \parallel \overleftrightarrow{QT}$ and $\overleftrightarrow{QT} \parallel \overleftrightarrow{RU}$ then $\overleftrightarrow{PS} \parallel \overleftrightarrow{RU}$.

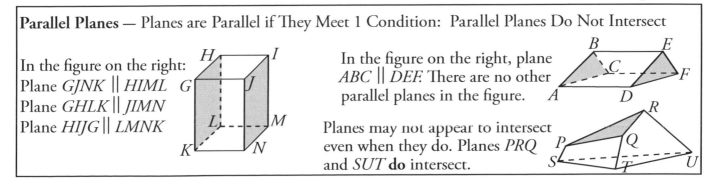

Parallel Planes — Planes are Parallel if They Meet 1 Condition: Parallel Planes Do Not Intersect

In the figure on the right:
Plane *GJNK* $\parallel$ *HIML*
Plane *GHLK* $\parallel$ *JIMN*
Plane *HIJG* $\parallel$ *LMNK*

In the figure on the right, plane *ABC* $\parallel$ *DEF*. There are no other parallel planes in the figure.

Planes may not appear to intersect even when they do. Planes *PRQ* and *SUT* **do** intersect.

Eight Angles Two Measures

Interesting things happen when two *parallel* lines are cut by a transversal. Examine the figure at right. As we learned in the prior section, eight angles are formed. However, because the lines are parallel, the measures of the angles are connected.

If the two lines are parallel, any two of the eight angles are either equal or supplementary. Here are a few examples:

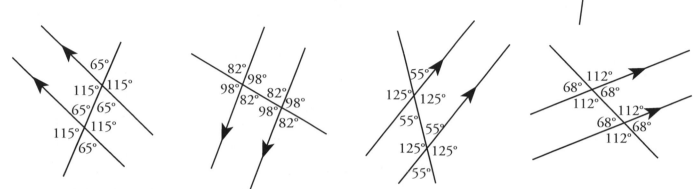

In each of the above figures, there are eight angles but only two angle sizes. In fact, in the special case of the transversal being perpendicular to the lines, all 8 angles will measure 90° and be both equal *and* supplementary, but this is only in the special case. Putting the new information together with the definitions about certain pairs of angles leads to important postulates and theorems.

The figures below show each pair of corresponding angles from the first figure:

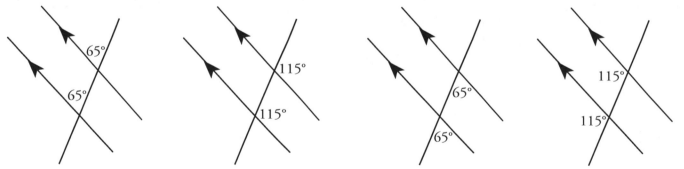

POSTULATE:

If two parallel lines are cut by a transversal, then corresponding angles are equal.

It is important to understand that corresponding angles are equal only if the two lines are parallel. This means that not all pairs of corresponding angles are equal, only those formed by *parallel* lines.

*Note: **PCC** stands for the Postulate above: **P**arallel lines mean **c**orresponding angles are **c**ongruent. It is the first of a number of memory helpers that you will see in this book. They are designed to help you remember certain Postulates and theorems, but don't use these memory helpers on your homework or tests, unless your teacher uses them too.

Investigating further, the figures below show each pair of alternate interior angles from the first figure.

THEOREM:

If two parallel lines are cut by a transversal, then alternate interior angles are equal.

PAIAC is a memory aid for this theorem: **P**arallel lines mean **a**lternate **i**nterior **a**ngles are **c**ongruent.

Now look at the third special pairs of angles, same side (or consecutive) interior angles. Study the two drawings. The connection between the angles is different than the connection of the previous special pairs. Each pair of angles is supplementary. Note: It's important to stop and think when using this theorem, think **supplementary**, (not equal).

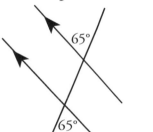 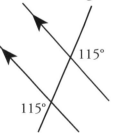 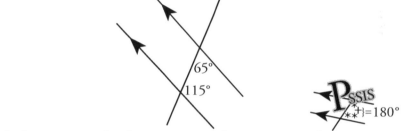

THEOREM:

If two parallel lines are cut by a transversal, then same side (or consecutive) interior angles are **supplementary**.

PSSIS is a memory aid for this theorem: **P**arallel lines mean **s**ame **s**ide **i**nterior angles are **s**upplementary.

The figures below show each pair of alternate exterior angles from the first figure.

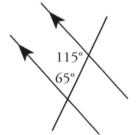

*THEOREM:

If two parallel lines are cut by a transversal, then alternate exterior angles are equal.

PAEAC is a memory aid for this theorem: **P**arallel lines mean **a**lternate **e**xterior angles are **c**ongruent.

*Not all textbooks define exterior angles or state this theorem. Although the theorem is true, do not use any theorem in proofs if your textbook does not state it. (But this one's still a good fact to know!)

This book uses P^{cc} P^{AIAC} P^{SSIS} and P^{AEAC} as abbreviations and memory tools. However, on homework and tests, use the name that your teacher taught you for each postulate and theorem.

Hints for Recognizing Which Type of Special Pair of Angles Are in a Problem:

With corresponding angles, the lines make an F (or ꟻ or Ⅎ or ꟻ).

With alternate interior angles, the lines make a Z (or Ƨ).

With same side interior angles, the lines make a C (or Ɔ).

PROJECT

Draw two parallel lines. Recall that parallel lines have the same slope. Then draw a third line (the transversal). Using your protractor, carefully measure the eight angles formed. Now, draw a transversal with a different slant. Measure the angles formed. Do your findings agree with our new postulate and theorems?

Angle measures for first transversal:

Angle measures for second transversal:

Algebra Review — Equations with Variables on Both Sides

Solve: $5x + 15 = 2x + 75$

What You Do	Why You Do It
$5x + 15 = 2x + 75$	Write The Equation.
$-2x \qquad -2x$	Subtract $2x$ from each side. (You are adding the opposite of $+2x$).
$3x + 15 = 75$	
$-15 \ -15$	Subtract 15 from each side. (You are adding the opposite of $+15$).
$3x = 60$	
$\dfrac{3x}{3} = \dfrac{60}{3}$	Divide through by the coefficient of x.
$x = 20\checkmark$	Done! (Because we have a single positive x all by itself on one side of the equation.)

Notes: Coefficient means the NUMBER in front of the variable.

Now You Try It — Solve for x:

1. $7x - 5 = 4x + 31$

2. $3x + 24 = 4x - 8$

3. $4x + 7 = 5x - 10.5$ (The answer is not an integer.)

4. $20x + 4 = 13x + 39$

Algebra Review — Equations with Variables on Both Sides (continued)

Solve: $4x + 10 = 180 - (2x - 10)$

What You Do	Why You Do It
$4x + 10 = 180 - (2x - 10)$	Write the equation.
$4x + 10 = 180 - 1(2x) - 1(-10)$	Distribute the formerly "invisible" negative one. Remember: $(-)(+) = -$ and $(-)(-) = +$
$4x + 10 = 180 - 2x + 10$	
$+2x \qquad\qquad +2x$	Add $2x$ to both sides. (Adding the opposite of $-2x$).
$-10 \qquad\qquad -10$	Subtract 10. (You are adding the opposite of $+10$).
$6x = 180$	
$\dfrac{6x}{6} = \dfrac{180}{6}$	Divide by the coefficient of x.
$x = 30\checkmark$	Done! (Because we have a single positive x all by itself on one side of the equation.)

Now You Try It — Solve for x:

1. $10x + 5 = 180 - (4x - 7)$
2. $180 - (2x + 3) = 6x - 3$

Putting Algebra to Work with Parallel Line Problems

Example 1. Solve for x:

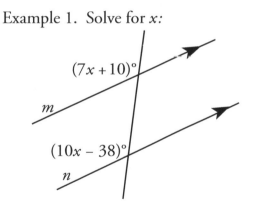

What to Do

Study the drawing. The arrowheads tell us the lines are parallel. Since the two given expressions are for corresponding angles (note that the angles form an Ⅎ), we can apply PCC and set the two expressions equal to each other:

$$7x + 10 = 10x - 38$$
$$-7x + 38 \qquad -7x + 38 \quad \text{Add opposites.}$$
$$48 = 3x$$
$$\frac{48}{3} = \frac{3x}{3} \qquad \text{Divide by the coefficient of } x.$$
$$16 = x\checkmark \qquad \text{Done.}$$

Example 2. Solve for z:

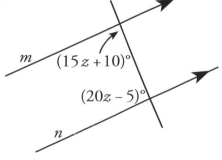

What to Do

The lines are parallel and the given expressions are for same side interior angles (note that they form a ⊃). PSSIS, tells us the angles are supplementary:

$$15z + 10 + 20z - 5 = 180 \qquad \text{Turn English into algebra.}$$
$$35z + 5 = 180 \qquad \text{Like terms collected.}$$
$$-5 \qquad -5 \qquad \text{Adding opposites.}$$
$$35z = 175 \qquad \text{Now divide by coef. of } z.$$
$$z = 5\checkmark \qquad \text{Done.}$$

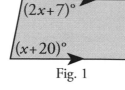
More Parallel Line Problems

Example:

Solve for x.

$(2x+7)°$

$(x+20)°$

Fig. 1

At first glance, this figure doesn't look like the figures we've been seeing, however, if you extend the lines, it does. Now, referring to the 4 methods above, PSSIS tells us that the 2 angles are supplementary:

$(2x+7)°$

$(x+20)°$

Fig. 2

$$2x + 7 + x + 20 = 180$$
$$3x + 27 = 180$$
$$x = 51\checkmark$$

Example:

Solve for x, y and z.

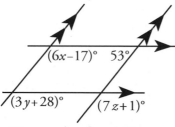

$(6x-17)°$ $53°$

$(3y+28)°$ $(7z+1)°$

The key to this sort of problem is to use one of the methods at the top of the page to create an equation with *one* variable. Study the diagram. PSSIS tells us that the 53° angle must be supplementary to the angle measuring $6x - 17$. This means we can form the following equation:

$$6x - 17 + 53 = 180$$
$$6x = 144$$
$$x = 24\checkmark$$

Now solve for y. Since we know that x = 24, we can substitute 24 for x in the original expression. By PCC and substitution we have:

$$6(24) - 17 = 3y + 28$$
$$127 = 3y + 28$$
$$99 = 3y$$
$$33 = y\checkmark$$

And finally, solve for z, also by PCC and substitution:

$$3(33) + 28 = 7z + 1$$
$$127 = 7z + 1$$
$$126 = 7z$$
$$18 = z\checkmark$$

Notice that there were other ways to solve this problem. Also, the problem asked for the values of x, y and z. Sometimes you might be asked for the measure of each angle. Always be careful to answer the question you were asked to answer.

Answer the *right* question.

Now You Try It

1. Solve for a, b and c.

2. Solve for u, v and w.

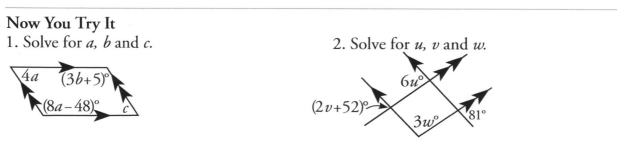

$4a$ $(3b+5)°$

$(8a-48)°$ c

$6u°$

$(2v+52)°$

$3w°$ $81°$

Proofs and Parallel Lines

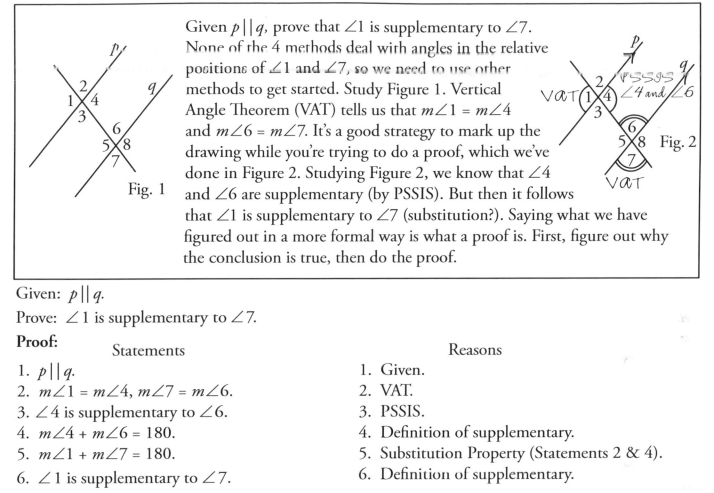

Given $p \parallel q$, prove that $\angle 1$ is supplementary to $\angle 7$. None of the 4 methods deal with angles in the relative positions of $\angle 1$ and $\angle 7$, so we need to use other methods to get started. Study Figure 1. Vertical Angle Theorem (VAT) tells us that $m\angle 1 = m\angle 4$ and $m\angle 6 = m\angle 7$. It's a good strategy to mark up the drawing while you're trying to do a proof, which we've done in Figure 2. Studying Figure 2, we know that $\angle 4$ and $\angle 6$ are supplementary (by PSSIS). But then it follows that $\angle 1$ is supplementary to $\angle 7$ (substitution?). Saying what we have figured out in a more formal way is what a proof is. First, figure out why the conclusion is true, then do the proof.

Fig. 1

Fig. 2

Given: $p \parallel q$.

Prove: $\angle 1$ is supplementary to $\angle 7$.

Proof:

Statements	Reasons
1. $p \parallel q$.	1. Given.
2. $m\angle 1 = m\angle 4$, $m\angle 7 = m\angle 6$.	2. VAT.
3. $\angle 4$ is supplementary to $\angle 6$.	3. PSSIS.
4. $m\angle 4 + m\angle 6 = 180$.	4. Definition of supplementary.
5. $m\angle 1 + m\angle 7 = 180$.	5. Substitution Property (Statements 2 & 4).
6. $\angle 1$ is supplementary to $\angle 7$.	6. Definition of supplementary.

Question: Why couldn't we have substituted right after step 3?

Answer: Substitution is a property of equality and congruence. You can't use substitution with other types of statements.

Now You Try It

Given: $p \parallel q$.

1. Prove: $\angle 2$ is supplementary to $\angle 5$.

2. Prove: $m\angle 1 = m\angle 8$ *without using* PAEAC.

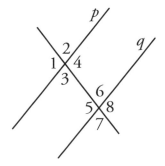

Problems 1 & 2.

Proving Lines are Parallel — In the previous section, we learned that if 2 parallel lines were cut by a transversal, the pairs of angles which were formed were either congruent and/or supplementary. The logic sketches for the 4 methods in the previous section look like this:

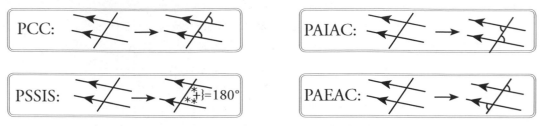

In each case we *start out* knowing that the two lines are parallel (the given). Then we can conclude information about a special pair of angles (the conclusion).

In this section we go backwards. We start out with two coplanar lines cut by a transversal. We are given that a particular pair of angles formed are equal or supplementary. Based upon this fact, we conclude that the lines are parallel.

POSTULATE: *If two coplanar lines are cut by a transversal, and corresponding angles are equal, then the two lines are parallel.*

It is enough to know that a single pair of corresponding angles are equal to conclude that the lines are parallel. **CCP** stands for the Postulate above: Congruent corresponding angles means lines are parallel.

Here is the logic sketch for "CCP":   CCP:

Here is a comparison of the logic sketches for PCC and its **converse**, CCP:

PCC:

CCP:

Here are 3 other new methods that are the **converses** of the theorems shown at the top of this page:

THEOREM: *If two coplanar lines are cut by a transversal, and alternate interior angles are equal, then the lines are parallel.*

THEOREM: *If two coplanar lines are cut by a transversal, and same side (or consecutive) interior angles are **supplementary,** then the lines are parallel.* $\angle 1 + \angle 2 = 180°$

THEOREM*: *If two coplanar lines are cut by a transversal, and alternate exterior angles are equal, then the lines are parallel.*

*Note: Although this theorem is of course true, do not use it in a proof unless it is in your textbook.

P CC | O | CC P
P AIAC | R | CAIA P
P SSIS | ? | SSIS P
P AEAC | | CAEA P

A+ tip When you are doing a proof or a problem about lines and transversals, it is important to choose the correct method. In each case, stop and ask yourself, Am I starting out knowing (being given) that the lines are parallel? If so, you want a method that *begins* with a p (for parallel). Or, are you being asked to show (conclude) that the lines are parallel? If so, you want a method that *ends* with a p (also for parallel). Once you decide which is the right *group* to choose from, add additional markings to the figure; this should help you select the correct method within that group.

Here's an example of how to use one of the methods that ends with a p in a proof:

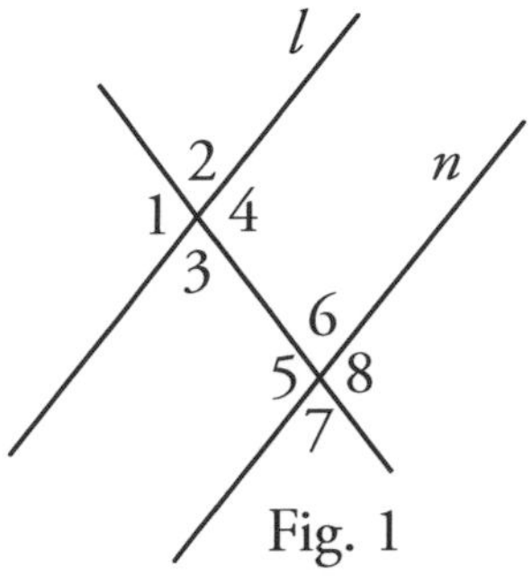

Fig. 1

Given: $\angle 2$ is supplementary to $\angle 8$.
Prove: $l \| n$.
Since we are being asked to conclude that 2 lines are parallel, we definitely want to use one of the new methods (the ones that end with a p). But none of the methods deal with angles in the relative positions of $\angle 2$ and $\angle 8$, so we need to use other methods to get started. Study Figure 1. The Vertical Angle Theorem (VAT) tells us that $m\angle 2 = m\angle 3$ and $m\angle 8 = m\angle 5$. Since $\angle 2$ is supplementary to $\angle 8$, $\angle 3$ must be supplementary to $\angle 5$. This, in turn, would prove that the two lines are parallel by SSISP. As you think about a proof, mark up the figure (see Figure 2). Your "notes" will help you when you do the proof itself.

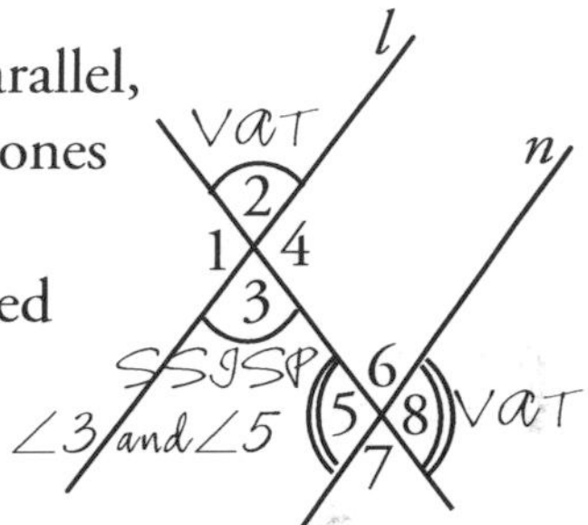

Fig. 2

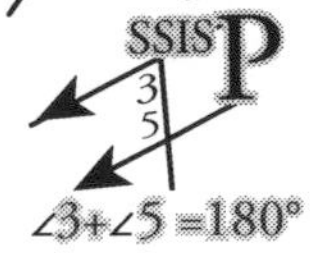

Proof:

Statements	Reasons
1. $\angle 2$ is supplementary to $\angle 8$.	1. Given.
2. $m\angle 2 + m\angle 8 = 180$.	2. Definition of supplementary.
3. $m\angle 2 = m\angle 3$ and $m\angle 8 = m\angle 5$.	3. Vertical Angle Theorem.
4. $m\angle 3 + m\angle 5 = 180$.	4. Substitution Property (Statements 2 & 3).
5. $m\angle 3$ is supplementary to $m\angle 5$.	5. Definition of supplementary.
6. $l \| n$.	6. SSISP.*

*Note: We've used "SSISP" but be sure to name all methods in the way that your teacher requires.

Here's an example of a problem that uses the methods in this section and is the geometry version of a type of problem which appears on many exams and standardized tests:

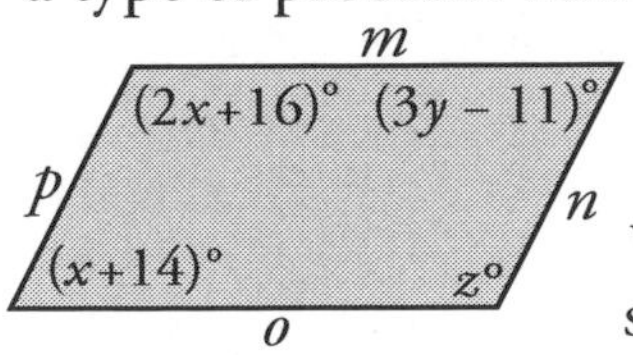

Using the figure, find the values of x, y and z that make $p \| n$ and $m \| o$. Solution: The key to solving a problem like this is to *treat it as if it were so*, which in this case means, assume $p \| n$ and $m \| o$. Then, if lines are parallel, same side interior angles are supplementary (PSSIS). This allows us to form each of the following equations:

$(2x+16)+(x+14) = 180,$
$3x + 30 = 180,$
$3x = 150$ so $x = 50$✓ (Substitute in 50 for x)

$(2x + 16)+(3y - 11) = 180$
$2(50)+16 + 3y - 11 = 3y + 105 = 180$
$3y = 75$ so $y = 25$✓ (Substitute in 25 for y)

$(3y - 11) + z = 180.$
$3(25) - 11 + z = 180$
$z = 116$✓

5. TRIANGLES

THEOREM: *The interior angles of a triangle total 180°.*

This theorem is easy to prove to yourself. Take **any** paper triangle, first cut off the three angles, then place the points together. The result will be a straight (180°) angle. Although this is not a formal proof, it does demonstrate this most basic property of all triangles.

Many problems are based on this simple and well-known theorem. Here are a few examples:

1. Find *x*.

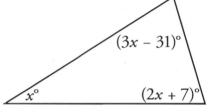

Since the three angles total 180, translate the picture into this equation:

$$x + (2x + 7) + (3x - 31) = 180$$
$$6x + 7 - 31 = 180$$
$$6x - 24 = 180$$
$$\underline{\quad + 24 \quad + 24}$$
$$6x = 204$$
$$\frac{6x}{6} = \frac{204}{6}$$
$$x = 34 \checkmark$$

2. Find *x*.

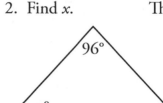

The single arc marks on each of the lower angles tell us that the two angles are equal. Since there are 180° in all triangles, the correct equation will be:

$$x + x + 96 = 180$$
$$2x + 96 = 180$$
$$\underline{\quad - 96 \quad - 96}$$
$$2x = 84$$
$$\frac{2x}{2} = \frac{84}{2}$$
$$x = 42 \checkmark$$

3. Find the measures of angles *a* through *f*.

This is a puzzle problem. A good plan for solving this type of **A⁺tip** problem is to find the angles' measures in the same order (alphabetic in this case) that the angles are named. Usually, this is a good hint about the logical order in which the problem should be solved.

A⁺tip Many problems that look complicated aren't. Each angle in the figure above can be found by using one of the following 3 facts: 1. Vertical angles are congruent (VAT).

2. A straight angle measures 180°.

3. The 3 angles of a triangle add up to 180°.

Complete the table & check your answers in the back of the book.

Letter	Reason	Equation	Answer
a	Straight ∠ = 180	(180 – 103)	77
b	VAT	—	103
c	Angles of a △ = 180	(180–103–58)	19
d			
e			
f			

Seeing all the Triangles – Let's re-examine the triangle in Example 3 from the previous page.

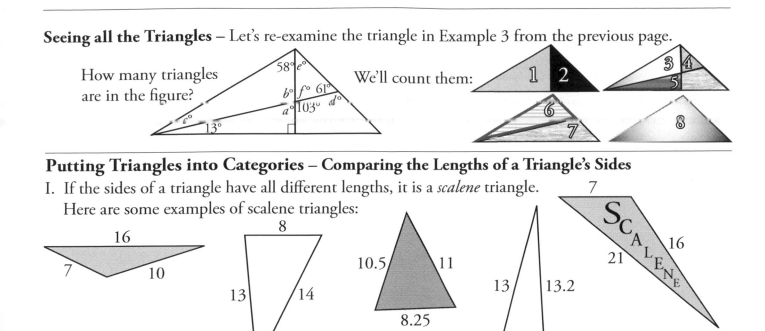

How many triangles are in the figure?

We'll count them:

Putting Triangles into Categories – Comparing the Lengths of a Triangle's Sides

I. If the sides of a triangle have all different lengths, it is a *scalene* triangle. Here are some examples of scalene triangles:

II. If (at least) two of the sides of a triangle have the same length, it is an *isosceles* triangle. Here are some examples of isosceles triangles:

The triangle on the left has three equal sides. This meets the definition of isosceles: *at least two* equal sides.

III. If the lengths of all three sides of a triangle are equal, the triangle is an *equilateral* triangle. Since equilateral means equal sides, this definition should be easy to remember. Below are examples of equilateral triangles. Notice that although they are different sizes, they all look very similar.

A Logic Diagram

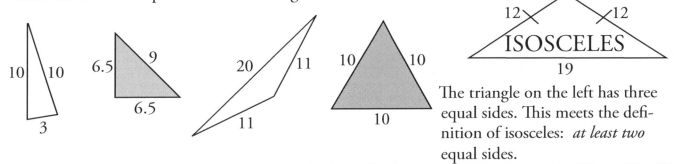

The entire figure represents all triangles. The black triangle represents all isosceles triangles which include all equilateral triangles. All equilateral triangles are isosceles. Is the converse true?

Grouping Triangles by the Sizes of Their Angles

I. If the measures of all of the angles in a triangle are less than 90°, that is, if all the angles are acute, the triangle is *acute*. Think of it as "a cute little triangle". Below are examples of acute triangles. Remember that an angle may be very close to 90° and still be acute.

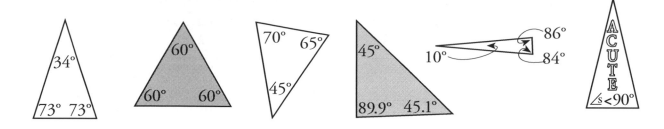

II. If a triangle has a right angle, it is a *right* triangle. The side opposite the right angle is called the hypotenuse of the triangle. The hypotenuse is always the longest side of a right triangle. Here are some examples of right triangles:

This symbol ⌐ means the two legs are perpendicular (⊥).

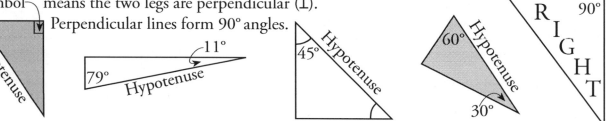

Perpendicular lines form 90° angles.

Rules for Right Triangles:

1. A right triangle is the only triangle that can have a hypotenuse.
2. Since the three interior angles of a triangle total 180°, if one angle measures 90°, there is only 90° left for the other two angles to share. Both angles' measures are positive and together add to 90°. This is why *a triangle can have at most one right angle* and, *the acute angles of a right triangle are complementary.*
3. Never assume a triangle is a right triangle just because it "looks like one".
4. Not all right triangles have two 45° angles. In fact, most don't.

III. If a triangle has an obtuse angle, the triangle is an *obtuse* triangle. Here are some examples of obtuse triangles:

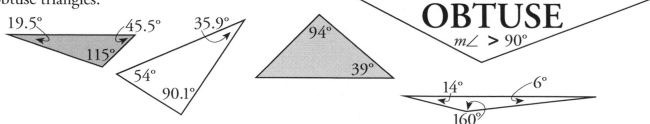

Note that a triangle *can have at most one obtuse angle*. This makes sense because the three angles of a triangle total 180° and if the obtuse angle uses up more than 90°, there are less than 90° left for the other two angles to share.

Exterior Angles of Triangles

Triangles and other polygons have interior angles but they also have exterior angles. Here are the steps to find them:

1. Extend one side of the triangle.

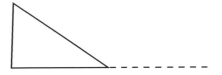

2. The angle formed by the dashed line and the side of the triangle crossing it, is an exterior angle.

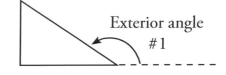

Exterior angle #1

3. Moving in a counter clockwise direction, extend the next side of the triangle. (Turning the piece of paper around as you work makes this easier.) The angle formed by the dashed line and the side of the triangle crossing it, is another exterior angle.

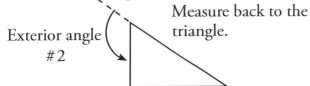

Exterior angle #2

Measure back to the triangle.

4. Again, moving counter clockwise, extend the third side of the triangle and, once again, measure back to the triangle. The angle formed by the dashed line and the side of the triangle crossing it, is the third exterior angle.

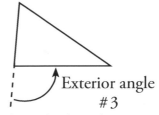

Exterior angle #3

Allowing one at each vertex, each triangle has *three* exterior angles, which is the same as the number of its interior angles. By the way, if we had gone in the other direction (clockwise) we would have ended up with 3 exterior angles having the same measures as those of the exterior angles found above.

Remote Interior Angles

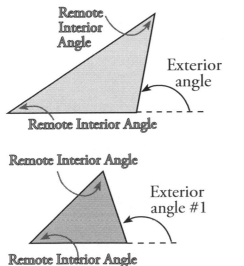

Remote Interior Angle

Exterior angle

Remote Interior Angle

Remote Interior Angle

Exterior angle #1

Remote Interior Angle

Remote means far away. From the exterior angle's point of view, the two interior angles on the other side of the triangle are far away. That is why they are called *remote interior angles*. It's important to remember that what is considered a remote interior angle depends entirely on which exterior angle is being considered.

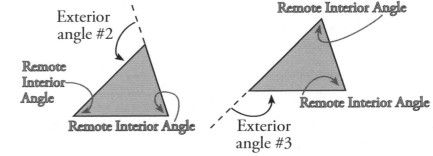

Exterior angle #2

Remote Interior Angle

Remote Interior Angle

Exterior angle #3

Remote Interior Angle

Remote Interior Angle

Exterior Angle Theorem

We already know that:

The interior angles of a triangle total 180° 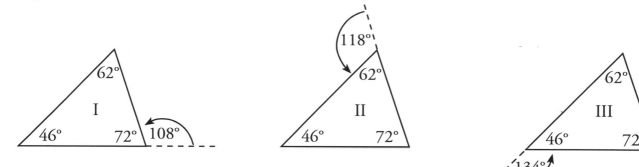 and a straight angle measures 180.°

An exterior angle of a triangle and its adjacent angle make a straight angle. That is, they total 180°.

√√ + √ = 180°

Something interesting happens when we put these ideas together. Study the 3 figures below. Do you see a pattern?

In each case we have:

$$\begin{matrix}\text{Two Remote Angles}\end{matrix} + \begin{matrix}\text{Third Angle}\end{matrix} = 180 \qquad\qquad \begin{matrix}\text{Exterior Angle}\end{matrix} + \begin{matrix}\text{Third Angle}\end{matrix} = 180$$

Since the right sides of the equations are equal, the left sides must also be equal. This allows us to set the left sides equal to each other. Notice how our reasoning follows the pattern of a formal proof.

$$\begin{matrix}\text{Two Remote Angles}\end{matrix} + \begin{matrix}\text{Third Angle}\end{matrix} = \begin{matrix}\text{Exterior Angle}\end{matrix} + \begin{matrix}\text{Third Angle}\end{matrix} \qquad \textbf{SUBSTITUTION PROPERTY}$$

By the reflexive property, the third angle is equal to itself, so we can subtract it from each side of the above equation and therefore:

$$\begin{matrix}\text{Two Remote Angles}\end{matrix} = \begin{matrix}\text{Exterior Angle}\end{matrix} \qquad \textbf{SUBTRACTION PROPERTY}$$

EXTERIOR ANGLE THEOREM: *The measure of an exterior angle of a triangle is equal to the sum of the measures of the two remote interior angles.*

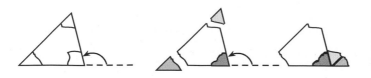

Notice how perfectly the two remote interior angles fit in the place of the exterior angle.

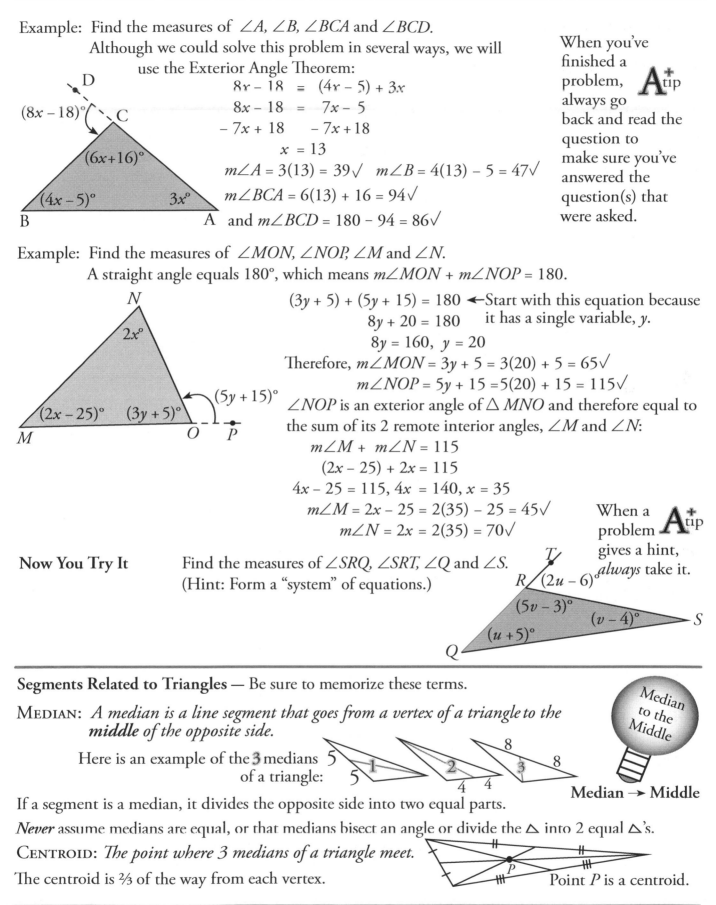

Example: Find the measures of $\angle A$, $\angle B$, $\angle BCA$ and $\angle BCD$.

Although we could solve this problem in several ways, we will use the Exterior Angle Theorem:

$$8x - 18 = (4x - 5) + 3x$$
$$8x - 18 = 7x - 5$$
$$\underline{-7x + 18 \qquad -7x + 18}$$
$$x = 13$$

$m\angle A = 3(13) = 39\checkmark \quad m\angle B = 4(13) - 5 = 47\checkmark$

$m\angle BCA = 6(13) + 16 = 94\checkmark$

and $m\angle BCD = 180 - 94 = 86\checkmark$

> When you've finished a problem, $\mathbf{A}^{+}_{tip}$ always go back and read the question to make sure you've answered the question(s) that were asked.

Example: Find the measures of $\angle MON$, $\angle NOP$, $\angle M$ and $\angle N$.

A straight angle equals 180°, which means $m\angle MON + m\angle NOP = 180$.

$(3y + 5) + (5y + 15) = 180$ ←Start with this equation because it has a single variable, y.
$8y + 20 = 180$
$8y = 160, \ y = 20$

Therefore, $m\angle MON = 3y + 5 = 3(20) + 5 = 65\checkmark$
$m\angle NOP = 5y + 15 = 5(20) + 15 = 115\checkmark$

$\angle NOP$ is an exterior angle of $\triangle MNO$ and therefore equal to the sum of its 2 remote interior angles, $\angle M$ and $\angle N$:

$m\angle M + m\angle N = 115$
$(2x - 25) + 2x = 115$
$4x - 25 = 115, \ 4x = 140, \ x = 35$
$m\angle M = 2x - 25 = 2(35) - 25 = 45\checkmark$
$m\angle N = 2x = 2(35) = 70\checkmark$

> When a problem $\mathbf{A}^{+}_{tip}$ gives a hint, *always* take it.

Now You Try It Find the measures of $\angle SRQ$, $\angle SRT$, $\angle Q$ and $\angle S$. (Hint: Form a "system" of equations.)

Segments Related to Triangles — Be sure to memorize these terms.

MEDIAN: *A median is a line segment that goes from a vertex of a triangle to the **middle** of the opposite side.*

Here is an example of the **3** medians of a triangle:

> Median to the Middle
>
> **Median → Middle**

If a segment is a median, it divides the opposite side into two equal parts.

Never assume medians are equal, or that medians bisect an angle or divide the $\triangle$ into 2 equal $\triangle$'s.

CENTROID: *The point where 3 medians of a triangle meet.*

The centroid is ⅔ of the way from each vertex.

Point P is a centroid.

Segments Related to Triangles (Continued)

ALTITUDE: *An altitude of a triangle is the **perpendicular** line segment that goes from a vertex to the line that includes the opposite side.*

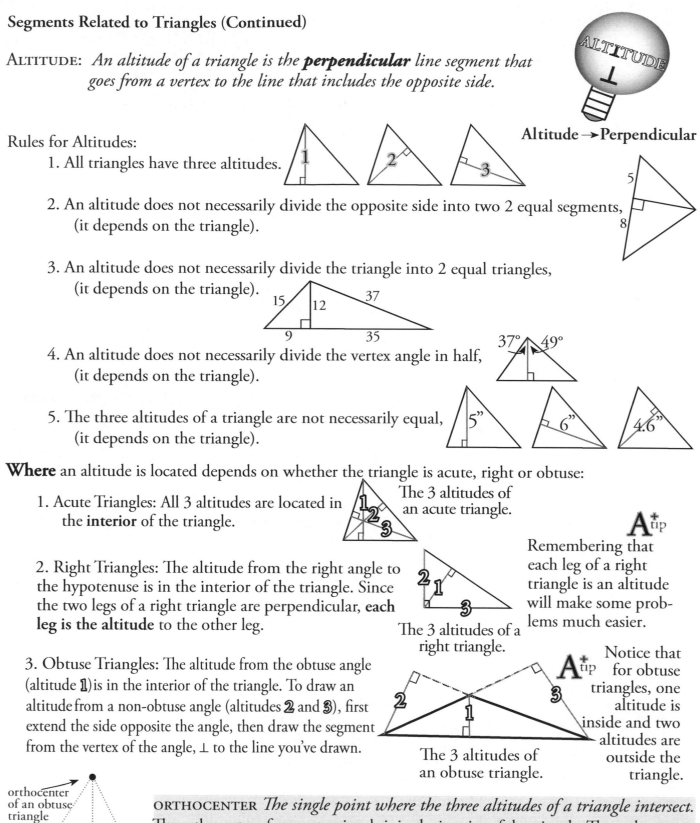

Rules for Altitudes:
1. All triangles have three altitudes.

2. An altitude does not necessarily divide the opposite side into two 2 equal segments, (it depends on the triangle).

3. An altitude does not necessarily divide the triangle into 2 equal triangles, (it depends on the triangle).

4. An altitude does not necessarily divide the vertex angle in half, (it depends on the triangle).

5. The three altitudes of a triangle are not necessarily equal, (it depends on the triangle).

Where an altitude is located depends on whether the triangle is acute, right or obtuse:

1. Acute Triangles: All 3 altitudes are located in the **interior** of the triangle.

The 3 altitudes of an acute triangle.

2. Right Triangles: The altitude from the right angle to the hypotenuse is in the interior of the triangle. Since the two legs of a right triangle are perpendicular, **each leg is the altitude** to the other leg.

The 3 altitudes of a right triangle.

A_{tip}^+ Remembering that each leg of a right triangle is an altitude will make some problems much easier.

3. Obtuse Triangles: The altitude from the obtuse angle (altitude **1**) is in the interior of the triangle. To draw an altitude from a non-obtuse angle (altitudes **2** and **3**), first extend the side opposite the angle, then draw the segment from the vertex of the angle, ⊥ to the line you've drawn.

The 3 altitudes of an obtuse triangle.

A_{tip}^+ Notice that for obtuse triangles, one altitude is inside and two altitudes are outside the triangle.

orthocenter of an obtuse triangle

ORTHOCENTER *The single point where the three altitudes of a triangle intersect.* The orthocenter of an acute triangle is in the interior of the triangle. The orthocenter of a right triangle coincides with the vertex of the right angle. The orthocenter of an obtuse triangle is located outside the triangle.

The Word Congruent

What does it mean for two figures to be congruent? The symbol for congruence is ≅. Pay attention to the two parts of the symbol:

~ The squiggly sign on top means that the figures have the same shape.

= The equal sign on the bottom means that the figures have the same size.

Putting these together tells us that congruent figures have the *same shape and the same size.*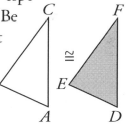

If two figures have the same shape and the same size, it means that one is a perfect copy of the other. That is, they are *exactly alike.*

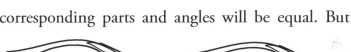

Orientation of Figures

A⁺ₜᵢₚ When two figures are congruent, their corresponding parts and angles will be equal. But it's important to *orient* the figures in the same way because, when figures are oriented properly, you are more likely to compare parts and angles that 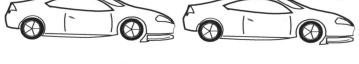 are corresponding and whose measurements are meant to be set equal. For example, given the two figures on the left, it is best to take a few seconds to re-draw the second figure so that both triangles are oriented similarly. This is especially true for figures that are easy to sketch, such as triangles. Be sure to carefully place identifying letters in the correct positions. See the re-drawn figures on the right.

Order Counts

Triangle *ABC* is congruent to triangle *DEF*. In symbols, we write:

$$\triangle ABC \cong \triangle DEF$$

This single *congruence* is giving us **6** pieces of information:

$\angle A \cong \angle D$ $\angle B \cong \angle E$ $\angle C \cong \angle F$ $\overline{AB} \cong \overline{DE}$ $\overline{BC} \cong \overline{EF}$ $\overline{CA} \cong \overline{FD}$

Now study the order of the original congruence: $\triangle ABC \cong \triangle DEF$ Order counts!

By the way, it would be equally correct to say that $\triangle BCA \cong \triangle EFD$, $\triangle CAB \cong \triangle FDE$, $\triangle ACB \cong \triangle DFE$, $\triangle CBA \cong \triangle FED$, $\triangle BAC \cong \triangle EDF$, and so forth. The important idea is that when writing a congruence, name corresponding parts in corresponding positions.

Congruent Triangles

Many problems in geometry involve trying to figure out if two triangles are congruent. If you can show that each of the six pairs of corresponding parts of two triangles are congruent, you've accomplished the task. However, usually you are not given that much information for a problem and, fortunately, you don't need to check all six pairs of parts to prove that the triangles are congruent.

There are **4** different methods we use to prove that any two triangles are congruent:

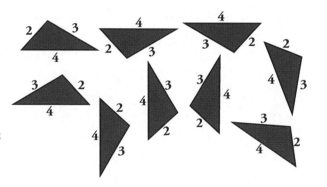

I. SIDE SIDE SIDE CONGRUENCE POSTULATE
If three sides of one triangle are congruent to three sides of another triangle, the triangles are congruent.

Side Side Side Congruency Postulate is true because if the sides of two triangles are the same lengths, it turns out that their corresponding angles are also equal. Since all six parts are congruent, this means the two triangles are congruent. Why is this so? Because, if you are given three lengths and told to make a triangle, there is, at most, one triangle you can make. Here's an example: Given three segments with lengths 2, 3 and 4 respectively, construct as many triangles as possible:

You can put the sides in any order, flip the triangles over, rotate them, it doesn't matter, you always get exactly the same triangle, only the orientation is different. All of the triangles on the right are congruent by SSS≅.

"Using" SSS≅ means that if you can show that two triangles have three pairs of congruent sides, you can state that the triangles are congruent based on SSS ≅ Postulate. Here are some examples of pairs of triangles that are congruent based on SSS ≅:

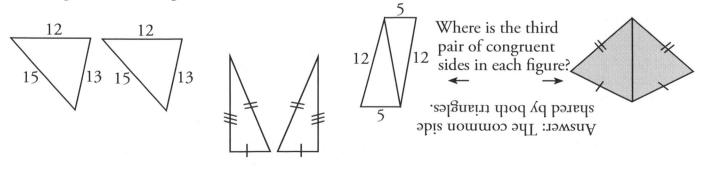

Where is the third pair of congruent sides in each figure?

Answer: The common side shared by both triangles.

Each letter in SSS$\cong$ stands for one *pair* of congruent corresponding sides. Think of it like this: $\overset{\cong}{S}\,\overset{\cong}{S}\,\overset{\cong}{S}$.

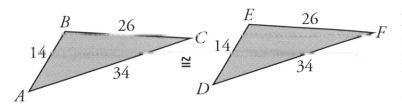

It's a good habit to check off each letter of the postulate as you find the *pair* of equal parts that the letter stands for:

$\overline{AB} \cong \overline{DE}$ S✓ $\overline{BC} \cong \overline{EF}$ S✓ $\overline{CA} \cong \overline{FD}$ S✓

Naming the Congruency: Order Counts! You must name the congruence by pairing up congruent parts in the two triangles. Choose a starting point and a path when naming the first triangle and choose the corresponding starting point and follow the same path when naming the second:

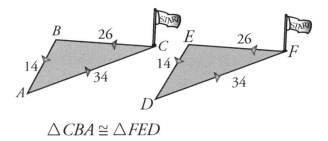

$\triangle CBA \cong \triangle FED$

It also makes sense to "check off the letters" on the other postulates and theorems that prove that triangles are congruent. This is especially true when you are working on proofs.

Adjacent means "right next to". If any three adjacent parts of one triangle are congruent to three adjacent parts of another triangle, the two triangles are congruent. This fact leads to two additional congruence postulates:

(*Included*)

II. SIDE ANGLE SIDE CONGRUENCE POSTULATE
If two sides and the included angle of one triangle are congruent to two sides and the included angle of another triangle, the two triangles are congruent.

SAS $\cong$

Study the figures above. Triangle I has a side, the adjacent angle and the immediate next side equal to the corresponding parts of Triangle II. Whenever you see this relationship in a problem, you know that the triangles are congruent by SAS Congruency Postulate. Although it's not part of the name of the postulate, remember the important word "included". The angle *must* be between the two sides that make up the congruent pairs (of sides) for this postulate to work. Here are some examples of pairs of triangles that are congruent based on SAS$\cong$:

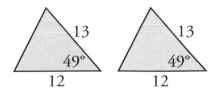

Each letter in SAS$\cong$ stands for one *pair* of congruent corresponding parts.

III. Angle Side Angle Congruence Postulate

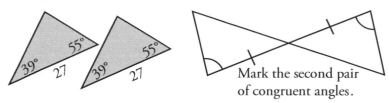

If two angles and the included side of one triangle are congruent to two angles and the included side of another triangle, then the two triangles are congruent.

Here are some examples of triangles that are congruent by ASA ≅:

Each letter in ASA ≅ stands for one *pair* of congruent corresponding parts.

You may have noticed that there is no (Included) in the introduction to this postulate. And yet it is a part of the postulate, so it must be important. Why is it missing?

To find out, we need to re-examine the way triangles work. If two angles of one triangle are equal to two angles of a second triangle, what else must be true? Think of it like this: the interior angles of each triangle total 180°. Subtract the measure of the two angles that are in both triangles and what is left? For example:

$$\triangle \, \text{I} \quad ? = 180° - 43° - 36° = 101°$$

$$\triangle \, \text{II} \quad ? = 180° - 43° - 36° = 101°$$

So the third pair of angles are also equal. This leads to the next theorem.

THEOREM: *If two angles of two triangles are equal, the third angles have no choice but to be equal.*

It might help you to remember this theorem if you think of it as the *third angle* theorem or the *no choice* theorem. Now, re-examining our newest postulate, ASA ≅, the letters ASA stand for: angle, adjacent side, adjacent angle in that order. However, given these conditions, we now know that the third pair of angles are also equal,

which in turn leads to the fourth method for showing that two triangles are congruent.

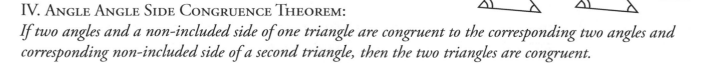

IV. Angle Angle Side Congruence Theorem:
If two angles and a non-included side of one triangle are congruent to the corresponding two angles and corresponding non-included side of a second triangle, then the two triangles are congruent.

Here are some examples of triangles that are congruent by AAS ≅. Notice that the corresponding non-included sides are congruent.

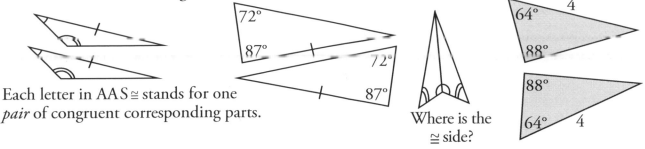

Each letter in AAS ≅ stands for one *pair* of congruent corresponding parts.

Where is the ≅ side?

Matching the Problem to the Correct Theorem: ASA ≅ and AAS ≅ are like first cousins, however, in a proof, you need to name the correct theorem, so we'll compare them:

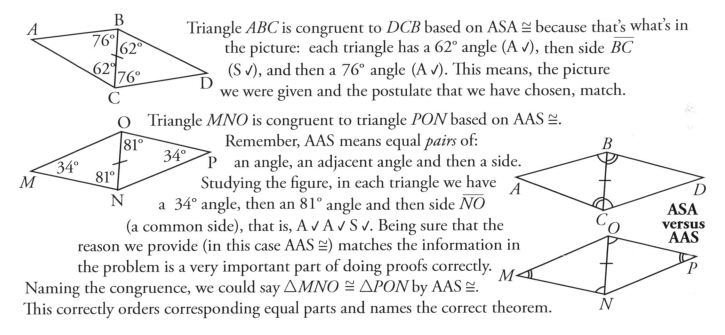

Triangle *ABC* is congruent to *DCB* based on ASA ≅ because that's what's in the picture: each triangle has a 62° angle (A ✓), then side $\overline{BC}$ (S ✓), and then a 76° angle (A ✓). This means, the picture we were given and the postulate that we have chosen, match.

Triangle *MNO* is congruent to triangle *PON* based on AAS ≅. Remember, AAS means equal *pairs* of: an angle, an adjacent angle and then a side. Studying the figure, in each triangle we have a 34° angle, then an 81° angle and then side $\overline{NO}$ (a common side), that is, A ✓ A ✓ S ✓. Being sure that the reason we provide (in this case AAS ≅) matches the information in the problem is a very important part of doing proofs correctly. Naming the congruence, we could say △*MNO* ≅ △*PON* by AAS ≅. This correctly orders corresponding equal parts and names the correct theorem.

ASA versus AAS

And now to answer the question: Why isn't it important to stress that the side must be *included?* Because, if the side is not included, the triangles are still congruent based on AAS ≅ (not ASA ≅). This is not true for SAS ≅. There is no SSA ≅ theorem (because it doesn't work for *all* triangles which a general congruence theorem must do). We will give a counterexample later in the book which disproves SSA ≅.

Getting Ready for Proofs about Congruent Triangles

Example: Decide if the two triangles below are congruent. If the triangles are congruent, name the congruence and state the method which proves they are congruent.

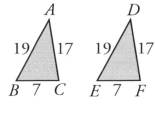

Studying the problem, we see the three pairs of equal sides so we know the triangles are congruent based on SSS ≅:

$$\triangle ABC \cong \triangle DEF$$

Remember, each letter of SSS represents one *pair* of equal corresponding sides: $\overline{AB} \cong \overline{DE}$ SSS $\overline{BC} \cong \overline{EF}$ SSS $\overline{CA} \cong \overline{FD}$ SSS

Proving Triangles are Congruent

To prove triangles are congruent you can use SSS ≅, SAS ≅, ASA ≅, or AAS ≅. But to use any of the 4 methods you must first show that pairs of corresponding parts of the triangles are equal.

Recurring Patterns — To find pairs of equal angles and sides, always look for:

1. Shared sides, 2. Vertical angles, 3. Information from the parallel line methods (especially PAIAC), and 4. Information from definitions. A⁺tip

When you've added as much information as you can find, study what you've shown and pick the right method from the four, the one that matches what you see.

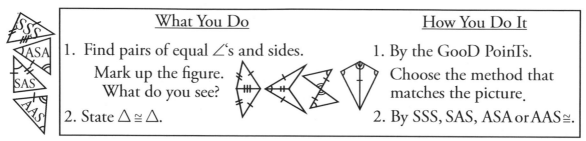

What You Do	How You Do It
1. Find pairs of equal ∠'s and sides. Mark up the figure. What do you see?	1. By the GooD PoinTs. Choose the method that matches the picture.
2. State △ ≅ △.	2. By SSS, SAS, ASA or AAS≅.

Now You Try It — Search the triangles below for additional pairs of equal parts (state your reasons). Then, name the postulate or theorem which proves that the pair of triangles is congruent.

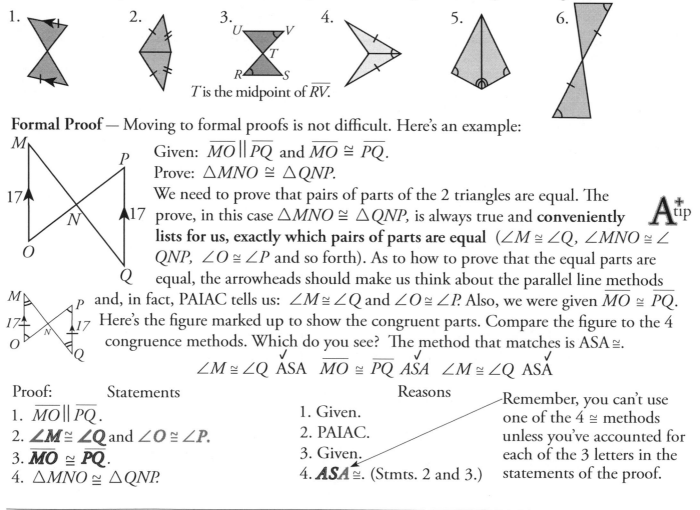

1. 2. 3. *T* is the midpoint of $\overline{RV}$. 4. 5. 6.

Formal Proof — Moving to formal proofs is not difficult. Here's an example:

Given: $\overline{MO} \parallel \overline{PQ}$ and $\overline{MO} \cong \overline{PQ}$.

Prove: $\triangle MNO \cong \triangle QNP$.

We need to prove that pairs of parts of the 2 triangles are equal. The prove, in this case $\triangle MNO \cong \triangle QNP$, is always true and **conveniently lists for us, exactly which pairs of parts are equal** ($\angle M \cong \angle Q$, $\angle MNO \cong \angle QNP$, $\angle O \cong \angle P$ and so forth). As to how to prove that the equal parts are equal, the arrowheads should make us think about the parallel line methods and, in fact, PAIAC tells us: $\angle M \cong \angle Q$ and $\angle O \cong \angle P$. Also, we were given $\overline{MO} \cong \overline{PQ}$. Here's the figure marked up to show the congruent parts. Compare the figure to the 4 congruence methods. Which do you see? The method that matches is ASA≅.

$\angle M \cong \angle Q$ ASA $\overline{MO} \cong \overline{PQ}$ ASA $\angle M \cong \angle Q$ ASA

Proof: Statements

1. $\overline{MO} \parallel \overline{PQ}$.
2. $\angle M \cong \angle Q$ and $\angle O \cong \angle P$.
3. $\overline{MO} \cong \overline{PQ}$.
4. $\triangle MNO \cong \triangle QNP$.

Reasons

1. Given.
2. PAIAC.
3. Given.
4. **ASA**≅. (Stmts. 2 and 3.)

Remember, you can't use one of the 4 ≅ methods unless you've accounted for each of the 3 letters in the statements of the proof.

Now You Try It

Given: $\overline{AB} \parallel \overline{DE}$ and C is the midpoint of $\overline{AE}$.

Prove: $\triangle ACB \cong \triangle ECD$.

Proof:

Statements   Reasons

Hint — When you are given that
C is the midpoint of $\overline{AE}$, you are
supposed to respond with:
$\overline{AC} \cong \overline{CE}$ (Def. of a midpt.)

The 4 methods for establishing congruence work for all triangles. But, for right triangles *only*, there is an additional theorem which proves that two (right) triangles are congruent:

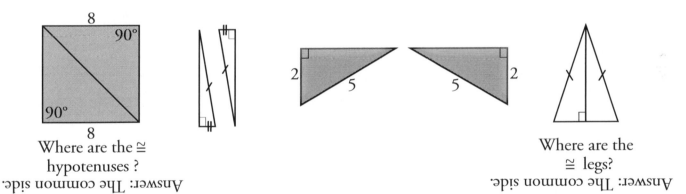

HYPOTENUSE LEG THEOREM:

If the hypotenuse and leg of one right triangle are congruent to the hypotenuse and leg of a second right triangle, the two triangles are congruent.

HL is the usual abbreviation for this theorem. Here are some examples of right triangles that are congruent based on HL:

8
90°
90°
8

Where are the $\cong$
hypotenuses ?
Answer: The common side.

2   5   5   2

Where are the
$\cong$ legs?
Answer: The common side.

Remember what the name HL stands for — **Hypotenuse Leg**.

Rules for using HL:

1. Only use HL for *right* triangles (only right triangles have hypotenuses).
2. In a proof, you must be given or show that the triangles are right, in order to use HL.
3. When a proof involves right triangles, try using HL first.
4. The two hypotenuses must be equal to each other or you can't use HL.

For example: These two triangles are congruent but the
                      congruence is by SAS $\cong$, *not* HL.

Now You Try It

1. Given: $\overline{AB} \cong \overline{CB}$ and $\overline{BD} \perp \overline{AC}$.   Prove: $\triangle ABD \cong \triangle CBD$.

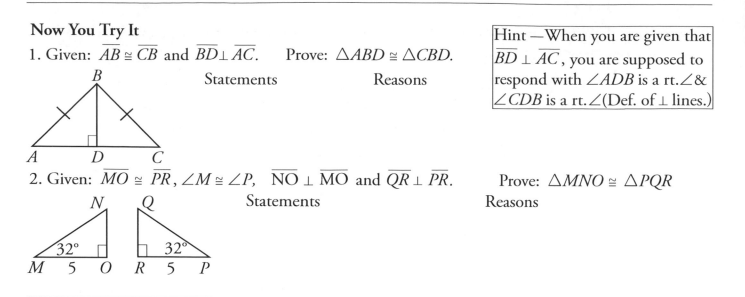

Statements   Reasons

Hint —When you are given that $\overline{BD} \perp \overline{AC}$, you are supposed to respond with $\angle ADB$ is a rt. $\angle$ & $\angle CDB$ is a rt. $\angle$ (Def. of $\perp$ lines.)

2. Given: $\overline{MO} \cong \overline{PR}$, $\angle M \cong \angle P$, $\overline{NO} \perp \overline{MO}$ and $\overline{QR} \perp \overline{PR}$.   Prove: $\triangle MNO \cong \triangle PQR$

Statements   Reasons

Quick Review — To prove that two triangles are congruent:

1. Use one of the 4 Congruency Methods for *all* triangles:

   SSS $\cong$, SAS $\cong$, ASA $\cong$, SAS $\cong$ or a fifth, HL, for *right* triangles only.

2. Use other *right* triangle congruence theorems only if they are given by your teacher or textbook.

Please note: Do not make up theorems. There are no other congruence theorems that work for all triangles.

Ways *Not* to prove Congruence

AAA   Why not?
These two triangles have three pairs of equal corresponding angles and are definitely *not* congruent.

SSA   Why not?
These two triangles have equal pairs of sides, the adjacent sides and the adjacent angles, and yet the triangles are definitely *not* congruent.

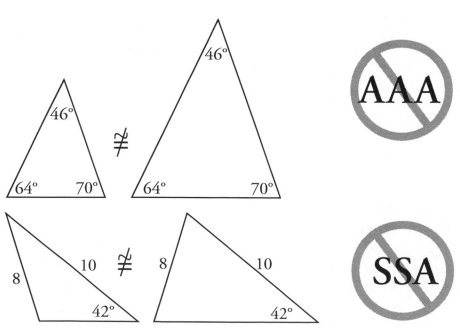

And remember, although we could provide an infinite number of counterexamples in both of these cases, even one counterexample disproves an hypothesis.

Using Congruent Triangles to Prove Other Things — Here are 5 important ways in which congruent triangles are used in theorems, problems and proofs. Pay attention to the *pattern*:

1. Prove triangles are congruent.
2. Use the definition of congruent triangles to prove other facts.

I. A DEFINITION OF CONGRUENT TRIANGLES: **CPCT**

*Corresponding **P**arts of Congruent Triangles are Congruent*

By definition, if 2 triangles are congruent, every corresponding pair of parts of the 2 triangles is congruent. For example:

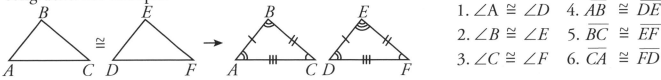

1. $\angle A \cong \angle D$ 4. $\overline{AB} \cong \overline{DE}$
2. $\angle B \cong \angle E$ 5. $\overline{BC} \cong \overline{EF}$
3. $\angle C \cong \angle F$ 6. $\overline{CA} \cong \overline{FD}$

This means that if you can show that 2 triangles are congruent, you can then state that *any* pair of corresponding parts of the triangles are congruent, including those that weren't mentioned in the proof of the congruence of the triangles. Here's an example:

Given: $\overline{AC} \cong \overline{EC}$ and $\overline{BC} \cong \overline{DC}$.

Prove: $\overline{AB} \cong \overline{DE}$.

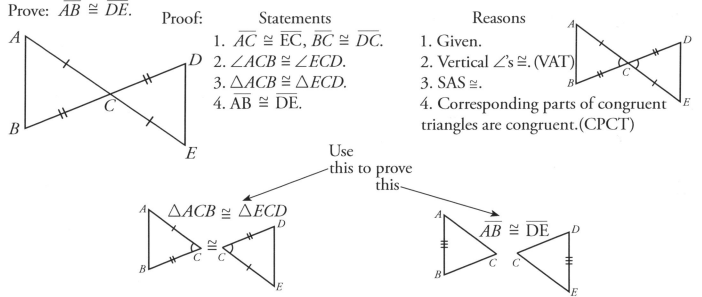

Proof:

Statements	Reasons
1. $\overline{AC} \cong \overline{EC}$, $\overline{BC} \cong \overline{DC}$.	1. Given.
2. $\angle ACB \cong \angle ECD$.	2. Vertical $\angle$'s $\cong$. (VAT)
3. $\triangle ACB \cong \triangle ECD$.	3. SAS $\cong$.
4. $\overline{AB} \cong \overline{DE}$.	4. Corresponding parts of congruent triangles are congruent.(CPCT)

Use this to prove this

$\triangle ACB \cong \triangle ECD$ $\overline{AB} \cong \overline{DE}$

You cannot use CPCT until **AFTER** you have proved that the 2 triangles are congruent. This means that you must first prove $\triangle \cong \triangle$, then you can use CPCT.

Learn the *pattern:*

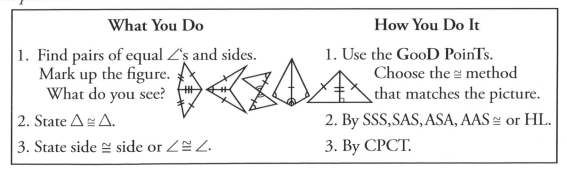

What You Do	How You Do It
1. Find pairs of equal $\angle$'s and sides. Mark up the figure. What do you see?	1. Use the **GooD PoinTs.** Choose the $\cong$ method that matches the picture.
2. State $\triangle \cong \triangle$.	2. By SSS, SAS, ASA, AAS $\cong$ or HL.
3. State side $\cong$ side or $\angle \cong \angle$.	3. By CPCT.

II. Proofs Using More Than One Pair of Congruent Triangles.

Given: $\overline{AB} \cong \overline{AC}$, $\overline{BD} \cong \overline{CD}$.
Prove: $\overline{BE} \cong \overline{CE}$.

First think through the proof:

1. "See" the pairs of triangles that you think are congruent:

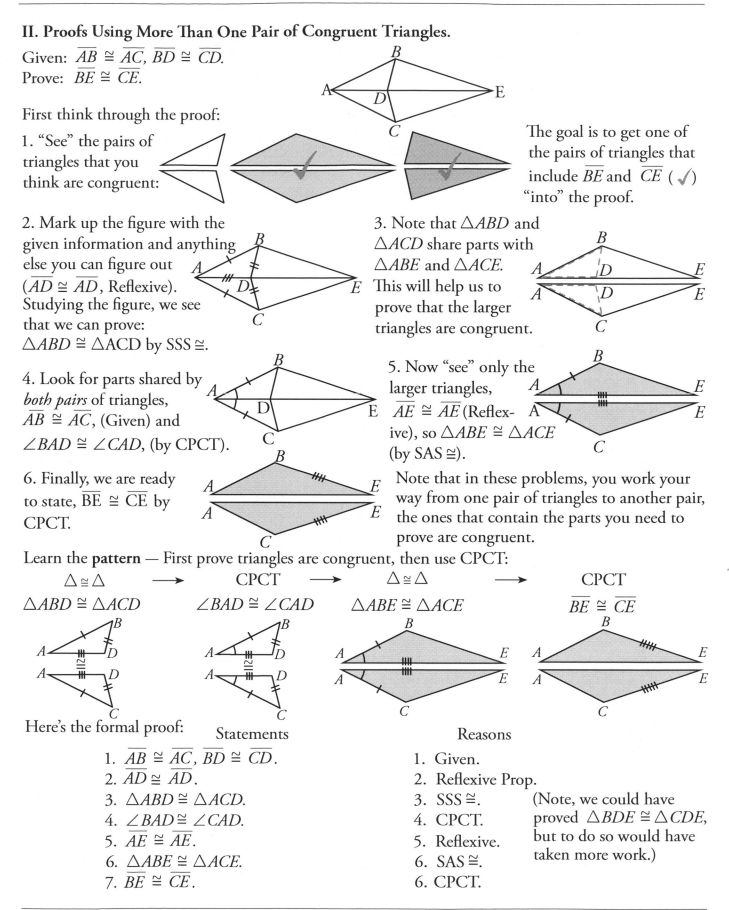

The goal is to get one of the pairs of triangles that include $\overline{BE}$ and $\overline{CE}$ (✓) "into" the proof.

2. Mark up the figure with the given information and anything else you can figure out ($\overline{AD} \cong \overline{AD}$, Reflexive). Studying the figure, we see that we can prove: $\triangle ABD \cong \triangle ACD$ by SSS $\cong$.

3. Note that $\triangle ABD$ and $\triangle ACD$ share parts with $\triangle ABE$ and $\triangle ACE$. This will help us to prove that the larger triangles are congruent.

4. Look for parts shared by *both pairs* of triangles, $\overline{AB} \cong \overline{AC}$, (Given) and $\angle BAD \cong \angle CAD$, (by CPCT).

5. Now "see" only the larger triangles, $\overline{AE} \cong \overline{AE}$ (Reflexive), so $\triangle ABE \cong \triangle ACE$ (by SAS $\cong$).

6. Finally, we are ready to state, $\overline{BE} \cong \overline{CE}$ by CPCT.

Note that in these problems, you work your way from one pair of triangles to another pair, the ones that contain the parts you need to prove are congruent.

Learn the **pattern** — First prove triangles are congruent, then use CPCT:

$\triangle \cong \triangle$ ⟶ CPCT ⟶ $\triangle \cong \triangle$ ⟶ CPCT

$\triangle ABD \cong \triangle ACD$ $\angle BAD \cong \angle CAD$ $\triangle ABE \cong \triangle ACE$ $\overline{BE} \cong \overline{CE}$

Here's the formal proof:

Statements	Reasons
1. $\overline{AB} \cong \overline{AC}$, $\overline{BD} \cong \overline{CD}$.	1. Given.
2. $\overline{AD} \cong \overline{AD}$.	2. Reflexive Prop.
3. $\triangle ABD \cong \triangle ACD$.	3. SSS $\cong$.
4. $\angle BAD \cong \angle CAD$.	4. CPCT.
5. $\overline{AE} \cong \overline{AE}$.	5. Reflexive.
6. $\triangle ABE \cong \triangle ACE$.	6. SAS $\cong$.
7. $\overline{BE} \cong \overline{CE}$.	6. CPCT.

(Note, we could have proved $\triangle BDE \cong \triangle CDE$, but to do so would have taken more work.)

Perpendicular Bisectors and Angle Bisectors

Proofs of these important theorems are exercises on the following page.

III. Perpendicular Bisectors

DEFINITION: *A perpendicular bisector is a segment, ray or line which is perpendicular to a segment at its midpoint.* (Remember, a midpoint divides a segment into 2 congruent pieces.)

There are 2 important theorems about perpendicular bisectors:

THEOREM: <u>*EVERY*</u> *point on the perpendicular bisector of a segment is equidistant from the endpoints of the segment.*

For example, given that $\overleftrightarrow{RS}$ is a perpendicular bisector of $\overline{MN}$, every point on $\overleftrightarrow{RS}$ is equally distant from point *M* and point *N*. For example point C is 25 units from point *M* and 25 units from point *N*.

And the converse:

THEOREM: *Any point that is equidistant from the endpoints of a segment lies on the perpendicular bisector of the segment.*

For example, since *W* is equally distant from points *A* and *B*, *W* lies on the perpendicular bisector of $\overline{AB}$.

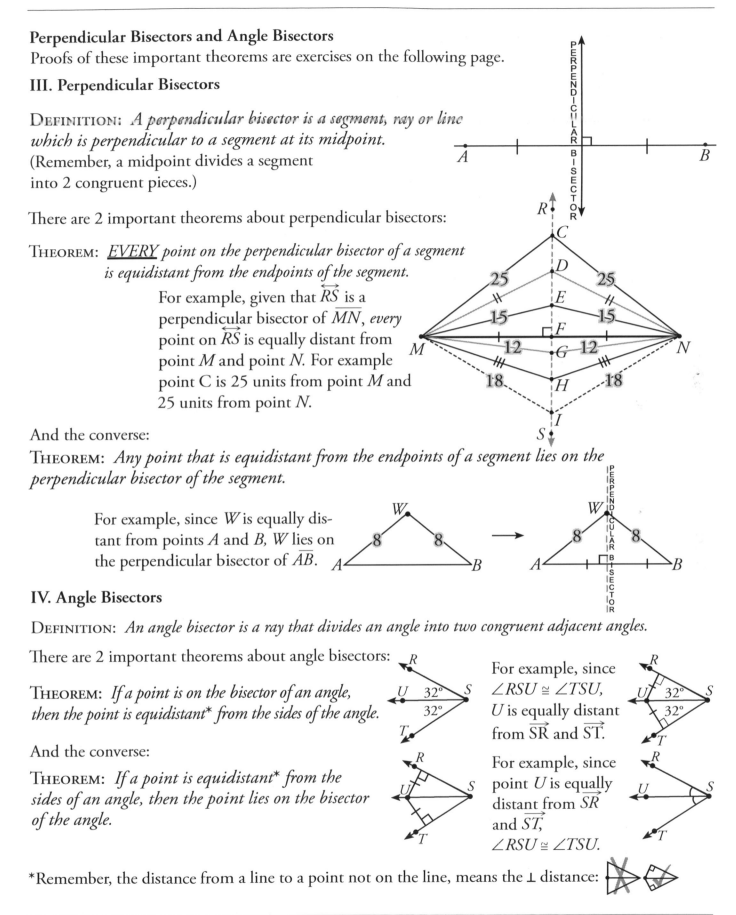

IV. Angle Bisectors

DEFINITION: *An angle bisector is a ray that divides an angle into two congruent adjacent angles.*

There are 2 important theorems about angle bisectors:

THEOREM: *If a point is on the bisector of an angle, then the point is equidistant* from the sides of the angle.*

For example, since $\angle RSU \cong \angle TSU$, *U* is equally distant from $\overrightarrow{SR}$ and $\overrightarrow{ST}$.

And the converse:

THEOREM: *If a point is equidistant* from the sides of an angle, then the point lies on the bisector of the angle.*

For example, since point *U* is equally distant from $\overrightarrow{SR}$ and $\overrightarrow{ST}$, $\angle RSU \cong \angle TSU$.

*Remember, the distance from a line to a point not on the line, means the ⊥ distance:

Now You Try It — Every proof on this page can be done using congruent triangles and/or the definition of congruence.

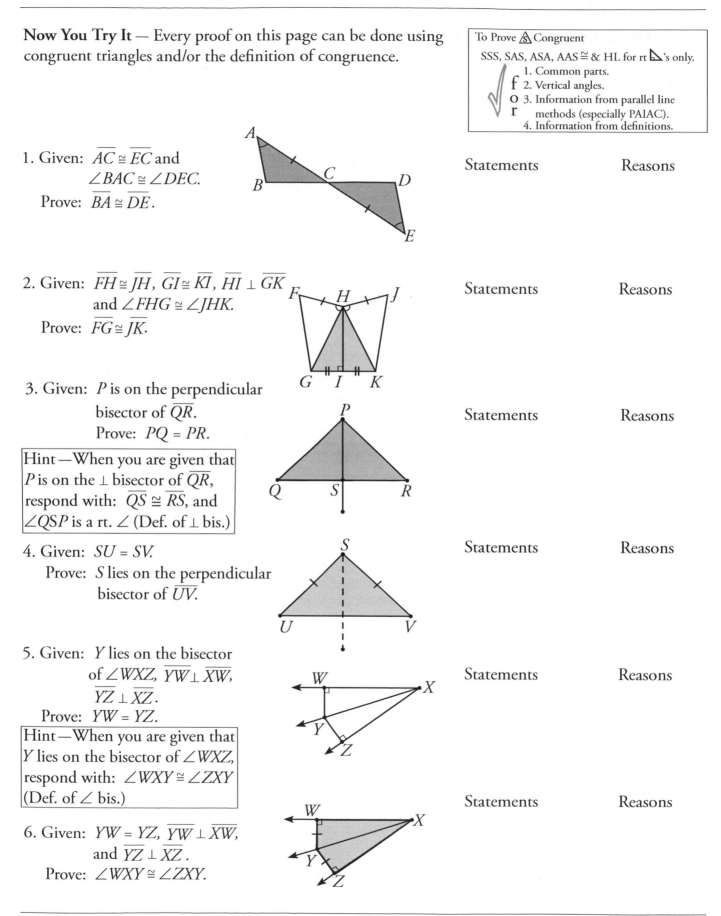

1. Given: $\overline{AC} \cong \overline{EC}$ and
         $\angle BAC \cong \angle DEC$.
   Prove: $\overline{BA} \cong \overline{DE}$.

   Statements   Reasons

2. Given: $\overline{FH} \cong \overline{JH}$, $\overline{GI} \cong \overline{KI}$, $\overline{HI} \perp \overline{GK}$
         and $\angle FHG \cong \angle JHK$.
   Prove: $\overline{FG} \cong \overline{JK}$.

   Statements   Reasons

3. Given: P is on the perpendicular
         bisector of $\overline{QR}$.
         Prove: $PQ = PR$.

   Statements   Reasons

   Hint—When you are given that
   P is on the $\perp$ bisector of $\overline{QR}$,
   respond with: $\overline{QS} \cong \overline{RS}$, and
   $\angle QSP$ is a rt. $\angle$ (Def. of $\perp$ bis.)

4. Given: $SU = SV$.
   Prove: S lies on the perpendicular
         bisector of $\overline{UV}$.

   Statements   Reasons

5. Given: Y lies on the bisector
         of $\angle WXZ$, $\overline{YW} \perp \overline{XW}$,
         $\overline{YZ} \perp \overline{XZ}$.
   Prove: $YW = YZ$.

   Statements   Reasons

   Hint—When you are given that
   Y lies on the bisector of $\angle WXZ$,
   respond with: $\angle WXY \cong \angle ZXY$
   (Def. of $\angle$ bis.)

   Statements   Reasons

6. Given: $YW = YZ$, $\overline{YW} \perp \overline{XW}$,
         and $\overline{YZ} \perp \overline{XZ}$.
   Prove: $\angle WXY \cong \angle ZXY$.

66

Using Congruent Triangles — continued

V. Theorems about Isosceles Triangles

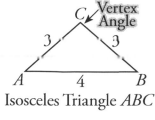

A⁺tip Become an expert with isosceles triangles. They are a powerful tool in geometry!

Isosceles Triangle *ABC*

A triangle with at least two equal sides is an isosceles triangle. The angle between two equal sides is called the *vertex* angle.

Pay special attention to the technique used to prove the next theorem. By drawing the altitude from the vertex angle of the isosceles triangle, we divide the original triangle into 2 *congruent right* triangles.

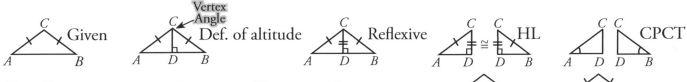

THE ISOSCELES TRIANGLE THEOREM: **ITT**

If a triangle has two congruent sides, the angles opposite those sides are also congruent.

Although we call it ITT, you must call each theorem by the name given by your teacher or textbook.

To prove ITT use congruent triangles:

Given   Def. of altitude   Reflexive   HL   CPCT

THE CONVERSE OF THE ISOSCELES TRIANGLE THEOREM:

If a triangle has two congruent angles, the sides opposite those angles are also congruent.

To prove this theorem use congruent triangles:

Given   Def. of altitude   Reflexive   AAS   CPCT

Name the right theorem!   If s**I**des are congruent, it's **I**TT.

If a**N**gles are congruent, it's the Co**N**verse of ITT.

Working with Isosceles Triangles

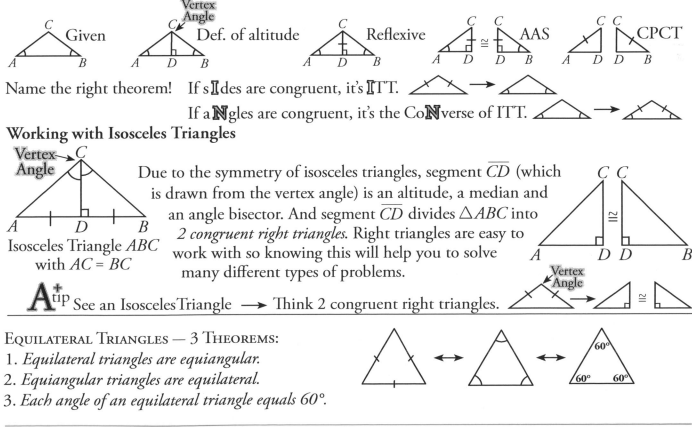

Isosceles Triangle *ABC* with *AC* = *BC*

Due to the symmetry of isosceles triangles, segment $\overline{CD}$ (which is drawn from the vertex angle) is an altitude, a median and an angle bisector. And segment $\overline{CD}$ divides $\triangle ABC$ into *2 congruent right triangles.* Right triangles are easy to work with so knowing this will help you to solve many different types of problems.

A⁺tip See an Isosceles Triangle ⟶ Think 2 congruent right triangles.

EQUILATERAL TRIANGLES — 3 THEOREMS:
1. *Equilateral triangles are equiangular.*
2. *Equiangular triangles are equilateral.*
3. *Each angle of an equilateral triangle equals 60°.*

These are polygons:

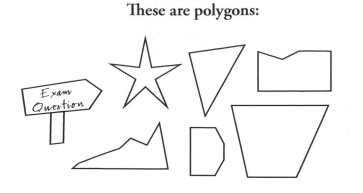

These are not polygons:

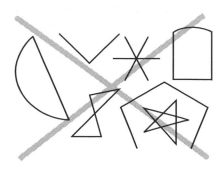

Polygons are: 1. Closed figures.
              2. Made up of (straight, line) segments.
              3. The segments intersect at their endpoints (they are not allowed to cross one another).
              4. Each segment must intersect with exactly two other segments.

When you see a figure, you are expected to know whether or not it is a polygon.

Polygons come in **2** varieties, *convex* and *non-convex* (also called *concave*). You are expected to know the difference between the two types. Here's how:

Convex or Non-Convex? Imagine wrapping a rubber band around a polygon. If the rubber band touches every one of the polygon's vertices, then the polygon is *convex*. To demonstrate, we will test each of the six polygons from above:

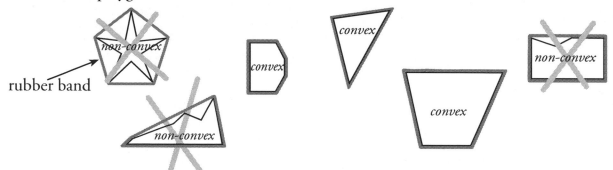

Although you do need to be able to recognize a non-convex polygon, geometry problems are usually about convex polygons. When a problem uses the word polygon, assume it's convex unless the figure or given information tells you it's non-convex.

Naming Polygons

Polygons are named by their consecutive vertices which are points, and points are named by capital letters. The letters are usually named in alphabetical order.

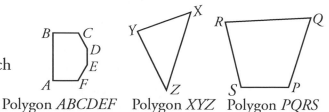

Polygon *ABCDEF*   Polygon *XYZ*   Polygon *PQRS*

Special Names for Special Polygons

You are responsible for memorizing these terms as well as any additional ones that your teacher or textbook think are important:

Number of Sides	Name	Example
3	Triangle	
4	Quadrilateral	
5	Pentagon	
6	Hexagon	
8	Octagon	
10	Decagon	
12	Dodecagon	

n–gons: Another way to name polygons is "number of sides–gon". For example, a triangle is a 3–gon, an octagon is an 8–gon and, speaking in general, we say an "*n*–gon", which means a polygon with an unstated number of sides. Remember, "*n*" stands for the number of sides.

Diagonals of a Polygon: A diagonal of a polygon is a segment that goes from one vertex to a non-adjacent (not next to) vertex.

All the diagonals of a pentagon.

New Term

DEFINITION OF A REGULAR POLYGON: *A polygon with equal sides and equal angles is called a regular polygon.* That is, a regular polygon is both *equilateral* and *equiangular*.

REGULAR
Equal Sides
Equal Angles

Here are some examples of regular polygons:

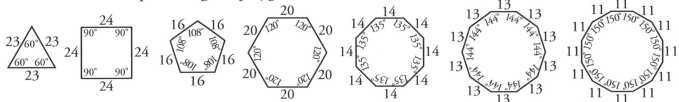

Note that if a triangle is either equiangular or equilateral, then it is *both*. That is, it is regular. This is not true about any other polygon. For example, shown below are polygons that are equiangular *or* equilateral but not *both* (and *not* regular):

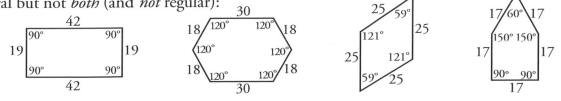

Finding the Sum of the Interior Angles of Polygons

The three interior angles of any triangle add up to 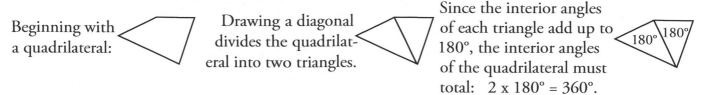 180°. We can use this fact to figure out the formula for the sum of the measures of the interior angles of polygons.

Beginning with a quadrilateral:

Drawing a diagonal divides the quadrilateral into two triangles.

Since the interior angles of each triangle add up to 180°, the interior angles of the quadrilateral must total: 2 x 180° = 360°.

In the example above, we used an oddly shaped quadrilateral, but we would have had the same result with any quadrilateral.

Now we'll try a pentagon:

Drawing the diagonals from a single vertex divides the pentagon into three triangles.

Since the interior angles of each triangle add up to 180°, the interior angles of the pentagon must total: 3 x 180° = 540°.

Now we'll choose a hexagon:

Drawing the diagonals from a single vertex divides the hexagon into four triangles.

Since the interior angles of each triangle add up to 180°, the interior angles of the hexagon must total: 4 x 180° = 720°.

Here's the pattern: Each increase in the number of sides produces one additional triangle, which adds 180° to the total measure of the polygon's interior angles. Here's a table organizing our findings:

Finish the last four rows of the table. It will help you to understand and remember the pattern.

Number of Sides	Number of Triangles	Calculation	Total of Interior Angles
3	1	1 x 180	180
4	2	2 x 180	360
5	3	3 x 180	540
6	4	4 x 180	720
7			
8			
9			
10			

In mathematics, we like to use formulas that are concise and correct for all possible cases. And a formula should include a variable to represent the quantity that is changing from case to case. In our example, it is the number of sides of the polygon that is changing. Notice that the number of triangles is always 2 less than the number of sides. So the correct formula is:

(number of sides − 2) (180)

Letting n stand for the number of sides, the formula becomes:

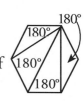

$(n - 2) (180)$ = the sum of the measures of the interior angles of a polygon with n sides.

Here's an example: Find the sum of the measures of the interior angles of a dodecagon.

A dodecagon is 12 sided, so *n* = 12.

(*n* – 2)(180) = the sum of the measure of its interior angles.

$$(\overset{12}{\cancel{n}} - 2)(180) = (10)(180) = \boxed{1800° \checkmark}$$

Here's another example: Find the sum of the interior angles of rectangle *ABCD*:

A rectangle is a quadrilateral (4 sided), so *n* = 4.

(*n* – 2)(180) = the sum of the measure of its interior angles.

$$(n - 2)(180) = (\overset{4}{\cancel{n}} - 2)(180) = (2)(180) = \boxed{360° \checkmark}$$

Since the sides are perpendicular (which means they meet to form 90° angles), we can confirm our solution: 4 x 90 = 360.

Here's an example of how the formula (*n* – 2)(180) might be used in a problem:

Find the measure of ∠*M*, ∠*N*, ∠*O*, ∠*P* and ∠*Q*.

Step 1. Count the number of sides (find *n*).

Step 2. Use the formula (*n* – 2)(180) = ($\overset{5}{\cancel{n}}$ – 2)(180) = (3)(180) = 540.

Step 3. Since 540 is the total sum of the interior angles, set the angles equal to 540:

$6x + (8x + 4) + 6x + (7x + 13) + (7x + 13) = 540$

Step 4. Collect like terms and solve for *x*:

$34x + 30 = 540, \quad 34x = 510, \quad x = 15$

Step 5. Go back and read the question and answer the question you were asked to answer!

∠*M* = 6*x*, = 6(15) = 90 √, ∠*N* = 8*x* + 4 = 8(15) + 4 = 124 √, ∠*O* = 6(15) = 90 √,
∠*P* = 7*x* + 3 = 7(15) + 13 = 118 √, and ∠*Q* = 7*x* + 13 = 7(15) + 13 = 118 √.

Now You Try It

1. Find the total sum of the measure of the interior angles of a 100-gon.

2. Find the measure of the smallest and largest angle of polygon *PQRSTUVW*.

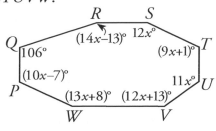

Going Backwards: Finding n When you Know the Sum of the Interior Angles

Example: How many sides does a polygon have if the interior angles total 1440°?

In this problem, the answer to $(n-2)(180)$ is given. So set $(n-2)(180)$ equal to the answer:
$$(n-2)(180) = 1440$$
Carefully distribute each term of the expression $(n-2)$:
$$n(180) - 2(180) = 1440$$
$$180n - 360 = 1440$$

• NOTES
Coefficient means the number in front of the variable.

Balance the equation:
$$\begin{array}{r} +\,360 \quad +\,360 \\ \hline 180n = 1800 \end{array}$$

Divide by the *coefficient* of n: $\dfrac{\cancel{180}n}{\cancel{180}} = \dfrac{\cancel{1800}^{\,10}}{\cancel{180}}$

$$n = 10$$

Finding the Measure of One Interior Angle of a Regular Polygon

• NOTES
Regular means equal sides and equal angles.

Remember that the word regular means equal sides and equal angles. Since the angles are equal, divide the total measure of the interior angles by the number of angles to find the measure of a single angle.

Here's a compact formula to find the value of **one** interior angle of a regular polygon: $\dfrac{(n-2)180}{n}$

Here's an example: Find the measure of each interior angle of a regular pentagon.

First, find the total of the measures of the interior angles: $(5-2)180 = 540$
Then divide the total by the number of angles: $\dfrac{540}{5} = 108°$ ✓

The following example combines two concepts:
How large is one interior angle of a regular polygon if the sum of its interior angles is 1800?

Exam Question

In this problem, the answer to $(n-2)(180)$ is given, so set $(n-2)(180)$ equal to the answer:
$$(n-2)(180) = 1800$$
$$180n - 360 = 1800, \quad 180n = 1800 + 360$$
$$180n = 2160, \quad n = 12$$
Since $n = 12$, and the polygon (which we now know is a dodecagon) is regular, we know that the 12 angles share the 1800° equally, which means the measure of each angle is: $\dfrac{1800}{12} = 150°$ ✓

Now You Try It — If the sum of the interior angles of a regular polygon is 1260, find the measure of one exterior angle of the polygon.

Exterior Angles of Polygons

Quick Review — Exterior angles of a polygon are formed by extending a side of the polygon. Start with one side, extend it and measure back to the polygon. If you need more than one exterior angle, work your way around the polygon in one direction. Draw only one exterior angle for each vertex.

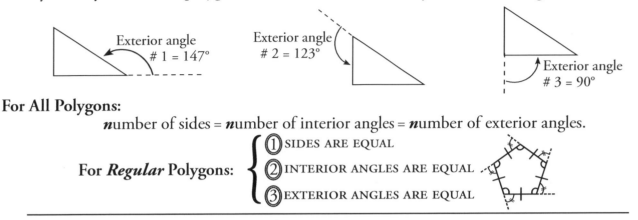

For All Polygons:
number of sides = **n**umber of interior angles = **n**umber of exterior angles.

For *Regular* Polygons:
1. SIDES ARE EQUAL
2. INTERIOR ANGLES ARE EQUAL
3. EXTERIOR ANGLES ARE EQUAL

Finding the Sum of the Exterior Angles

The interior angles of a triangle total 180°. The exterior angles of the triangle above total:

$$147° + 123° + 90° = 360°$$

The interior angles of a quadrilateral total $(n - 2)180$, or $(4 - 2)180 = 360°$. But what about the exterior angles? Here's an example:

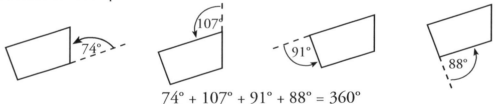

$$74° + 107° + 91° + 88° = 360°$$

So the exterior angles of the above quadrilateral total 360°.

The interior angles of a pentagon total $(5 - 2) \times 180 = 540°$. Checking the exterior angles:

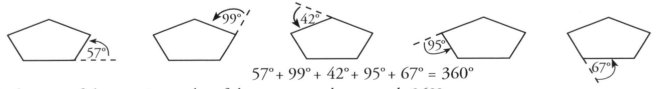

$$57° + 99° + 42° + 95° + 67° = 360°$$

So the sum of the exterior angles of the pentagon above equals 360°.

In fact, **the sum of the exterior angles of any polygon is always 360°!**
Note: As *n* increases, the sum of the interior angles goes up. But, as n increases, the sum of the exterior angles stays the same.

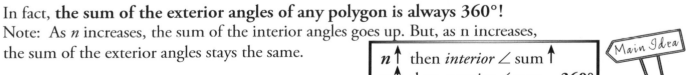

$n \uparrow$ then *interior* $\angle$ sum $\uparrow$
$n \uparrow$ then *exterior* $\angle$ sum = **360°**

Main Idea

THEOREM: *The Sum of the Exterior Angles of a Polygon = 360°*

The above theorem is especially helpful with problems involving regular polygons.

Since the *exterior* angles of a *regular* polygon are equal, they share the 360° equally:

$$\frac{360°}{n} = \text{the measure of one exterior angle of a \textit{regular} polygon} \qquad \frac{360°}{\text{the measure of one exterior angle of a \textit{regular} polygon}} = n$$

An understanding of exterior angles can help solve many polygon problems. Here are some examples:

1. An exterior angle of a *regular* polygon measures 30°. Find the measure of an interior angle.

    A⁺tip →Sketch a *single* vertex of the polygon drawing in the exterior angle.
    Since the interior angle is supplementary to the adjacent exterior angle:

    $$30° + x° = 180°$$
    $$x = 150 \checkmark$$

2. An exterior angle of a regular polygon measures 30°. How many sides does the polygon have?
    Since the sum of the exterior angles of a polygon is always 360°, and since all of the exterior angles of a regular polygon are equal, the 360° is shared equally among the exterior angles. So, if each exterior angle is 30° the polygon has:

    $$\frac{360°}{30°} = 12 \text{ sides} \checkmark$$

3. An interior angle of a regular polygon measures 140°. How many sides does the polygon have?
    Sketch a single vertex of the polygon drawing in the interior angle. Since the interior angle is supplementary to the adjacent exterior angle:

    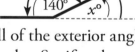

    $$140° + x° = 180°$$
    $$x = 40$$

Since the sum of the exterior angles of a polygon is always 360°, and since all of the exterior angles of a regular polygon are equal, the 360° is shared equally among the exterior angles. So, if each exterior angle is 40° the polygon has:

$$\frac{360°}{40°} = 9 \text{ sides} \checkmark$$

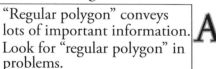

"Regular polygon" conveys lots of important information. **A⁺tip** Look for "regular polygon" in problems.

Now You Try It

1. If an exterior angle of a regular polygon measures 72°, what is the total measure of the interior angles of the polygon?

2. If the interior angles of a regular polygon total 1620°, find the measure of one exterior angle.

3. Polygon *ABCDEFGH* is regular. Find the measure of ∠*I*.

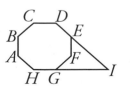

7. Quadrilaterals

Parallelograms

Like all quadrilaterals, parallelograms are named by their vertices. The symbol for a parallelogram is ⧄. The figure at right is ⧄*ABCD*.

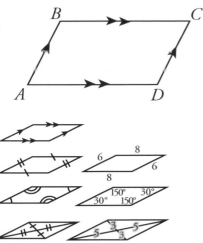

4 Facts You Need To Know About Parallelograms:

1. DEFINITION: *Both pairs of opposite sides are parallel.*

2. THEOREM: *Both pairs of opposite sides are equal.*

3. THEOREM: *Both pairs of opposite angles are equal.*

4. THEOREM: *The diagonals divide each other in half.*

Take a close look at the last theorem. Diagonals of parallelograms aren't necessarily equal. What is true is that the diagonals break each other in half, that is, they bisect each other.

Properties — In mathematics, when things are true about a certain type of object, we say those things are the *properties* of that object. The definition is not a property, the definition is what defines the object. The theorems are the properties of the object. Given a parallelogram (a quadrilateral with two pairs of parallel sides), the other three facts (the properties) are always true.

In a problem, you will know a figure is a parallelogram either from the figure itself, or by the word parallelogram or by the symbol ⧄.

Remember, if a figure is a parallelogram, you know a lot about it. Always think about the **4 facts** and add any information to the figure to show those facts that are connected to a problem. For example:

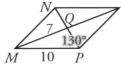

Given ⧄*MNOP*, on the left, we can add the information shown on the right to the original figure.

How did we figure out that ∠*M* and ∠*O* = 50°? There are two ways: 1. Like all quadrilaterals, the interior angles of ⧄ *MNOP* add up to 360° (remember, $(n-2)180$). This tells us that ∠*M* and ∠*O* together share 100° (360° – 130° – 130° = 100°). Since opposite angles of parallelograms are equal, ∠*O* and ∠*M* must both equal 50°.   2. When a transversal cuts two parallel lines, by PSSIS, same side interior angles are supplementary.

So, any two consecutive angles of a parallelogram are supplementary! A⁺tip

Now You Try It

Given ⧄*GHIJ* on the right, add the correct measurements for the following:

1. m∠*GHI*   3. m∠*HGJ*   5. $\overline{HI}$

2. m∠*HIJ*   4. $\overline{HG}$ 6. $\overline{GI}$

Proving the Properties of Parallelograms — A Detailed Look at Proof

If a quadrilateral has two pairs of parallel sides, the other parallelogram properties are always present. And since those properties were given as theorems, we must be able to prove each of them based only on the figure's two pairs of parallel sides. Let's investigate:

We will use congruent triangles to prove all three theorems.

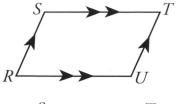

Given: $\square RSTU$.

Prove: The opposite sides of a parallelogram are congruent.

For the given figure we need to prove that $\overline{SR} \cong \overline{TU}$ and $\overline{ST} \cong \overline{RU}$.

We will begin by drawing diagonal $\overline{RT}$ (Fig. 1). Our reason is, two points determine a line.

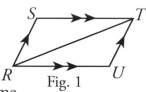
Fig. 1

Since $\overline{ST} \parallel \overline{RU}$ and $\overline{SR} \parallel \overline{TU}$, our transversal theorems should come to mind. There are no angles on the outside of the figure so we should think about alternate interior angles. Studying Figure 1: $m\angle STR = m\angle URT$ and $m\angle SRT = m\angle UTR$ by PAIAC (Fig. 2).

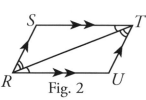
Fig. 2

Studying Fig. 2, we see that the diagonal creates 2 triangles. If the triangles are congruent, we'll have our "prove". That is:

$$\overline{SR} \cong \overline{TU} \text{ and } \overline{ST} \cong \overline{RU} \text{ by CPCT.}$$

But in order to prove congruence we must use one of the four triangle congruence methods. Since $\overline{RT}$ is a side common to both triangles, we can state that it is equal to itself by the Reflexive Property and mark it. Now, comparing Fig. 3 to the four congruence methods, we see that $\triangle RTS \cong \triangle TRU$ by ASA $\cong$.

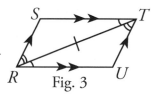
Fig. 3

Having proved that the two triangles are congruent, we can now say that $\overline{SR} \cong \overline{TU}$ and $\overline{ST} \cong \overline{RU}$ because corresponding parts of congruent triangles are congruent (CPCT) (Fig. 4).

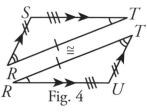
Fig. 4

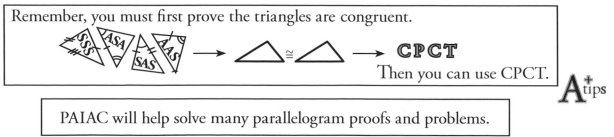

Remember, you must first prove the triangles are congruent.

$\longrightarrow$ CPCT

Then you can use CPCT.

PAIAC will help solve many parallelogram proofs and problems.

Given: ▱RSTU. Prove: Opposite angles of a parallelogram are congruent.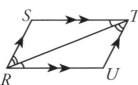

We need to prove $m\angle TSR = m\angle RUT$ and $m\angle SRU = m\angle STU$. Using the same reasoning (the same steps) that we used in the previous proof, we arrive at $\triangle RTS \cong \triangle TRU$, which proves that $m\angle TSR = m\angle RUT$ by CPCT. This completes one-half of the proof. Continuing with the steps of the previous proof we also found that:

$$m\angle \textbf{SRT} = m\angle \textbf{UTR} \text{ and}$$
$$\angle \textbf{URT} = m\angle \textbf{STR}$$

The addition property of equality allows us to add the left sides of two equations and set them equal to the sum of the right sides of the two equations:

$$m\angle \textbf{SRT} + m\angle \textbf{URT} = m\angle \textbf{STR} + m\angle \textbf{UTR}$$

Keeping in mind that our goal is to prove $m\angle SRU = m\angle STU$, we combine the two pieces of each angle to make the whole which gets $\angle SRU$ and $\angle STU$ into the proof:

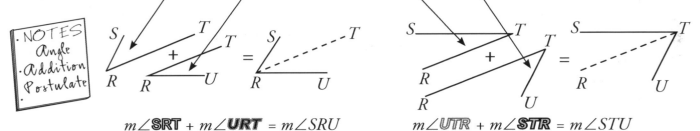

$$m\angle \textbf{SRT} + m\angle \textbf{URT} = m\angle SRU \qquad m\angle \textbf{UTR} + m\angle \textbf{STR} = m\angle STU$$

The reason for the last statement is the Angle Addition Postulate. And finally, since we have already shown that the left sides of the above two equations are equal, we can state that the right sides must also be equal, $m\angle SRU = m\angle STU$, by the Substitution Property. Notice that we also could have completed the last half of the proof by simply arguing that since $m\angle TSR = m\angle RUT$, that is, one pair of opposite angles were proved equal in an arbitrary parallelogram, so any pair of opposite angles are equal.

Now You Try It
Supply the reasons to prove the diagonals of a parallelogram bisect each other.
Given: ▱ABCD.   Prove: $\overline{BD}$ bisects $\overline{AC}$; $\overline{AC}$ bisects $\overline{BD}$.

	Statements		Reasons
	1. ▱ABCD.	1.	
	2. ∠AEB ≅ ∠DEC.	2.	
	3. $\overline{AB} \cong \overline{DC}$.	3.	
	4. $\overline{BC} \parallel \overline{AD}$; $\overline{AB} \parallel \overline{DC}$.	4.	
	5. ∠DBA ≅ ∠CDB.	5.	
	6. △AEB ≅ △CED.	6.	
	7. $\overline{AE} = \overline{EC}$; $\overline{BE} = \overline{ED}$.	7.	
	8. E is the midpoint of $\overline{AC}$; E is the midpoint of $\overline{BD}$.	8.	
	9. $\overline{BD}$ bisects $\overline{AC}$; $\overline{AC}$ bisects $\overline{BD}$.	9.	

Proving a Quadrilateral is a Parallelogram

A quadrilateral with two pairs of parallel sides is, by definition, a parallelogram. But if certain other conditions are met, could a quadrilateral be a parallelogram even though we don't know if the sides are parallel? Notice that we are *going backwards*, we are starting with a quadrilateral and asking, is the figure a parallelogram?

What if we are given a quadrilateral with two pairs of equal opposite sides? Is the figure a parallelogram and, if so, can we prove it? That is, can we prove that both pairs of opposite sides are parallel as well as being equal?

Fig. 1

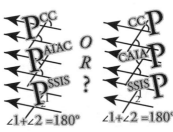

If we are going to prove that lines are parallel, we'll need the transversal theorems since they are our only theorems that deal with parallel lines. But which of the two groups of transversal theorems do we want? We need to get the order right. That is, given certain conditions, we want to be able to prove that pairs of lines are parallel. Recall that the transversal theorems that *end* with a "p" are the ones that do this. Now we need to find out exactly which one of the group we need.

$\angle 1 + \angle 2 = 180°$

The angles in Fig. 1 are in the interior of the figure, which makes CCP unlikely; the problem gives no angle measures which eliminates SSISP. This leaves CAIAP, which says that if two lines are cut by a transversal and alternate interior angles are congruent, then the lines are parallel. Diagonal $\overline{MO}$ (Fig. 2) serves as a transversal. Looking at Fig. 2, we see that we have created two triangles, and that they share $\overline{MO}$. We mark $\overline{MO}$ equal to itself by the Reflexive Property (Fig. 3).

Fig. 2

Figure 3 has two triangles that are congruent (by SSS ≅) and, therefore, three pairs of corresponding angles congruent by CPCT (Fig. 4). $\angle NMO \cong \angle POM$ tells us $MN \parallel PO$ by CAIAP. $\angle NOM \cong \angle PMO$ tells us $NO \parallel MP$, also by CAIAP. These two statements prove that *MNOP* is a parallelogram based on the definition of a parallelogram, a quadrilateral with two pairs of parallel sides.

Fig. 3

Fig. 4

Here is the theorem we just proved:

THEOREM: *A quadrilateral with two pairs of equal opposite sides is a parallelogram.*

Here are two examples:

A_{tip}^+ Remember, once a theorem is given, you can use it in any subsequent (later) proofs.

There are 3 other sets of conditions that guarantee that a quadrilateral is a parallelogram.

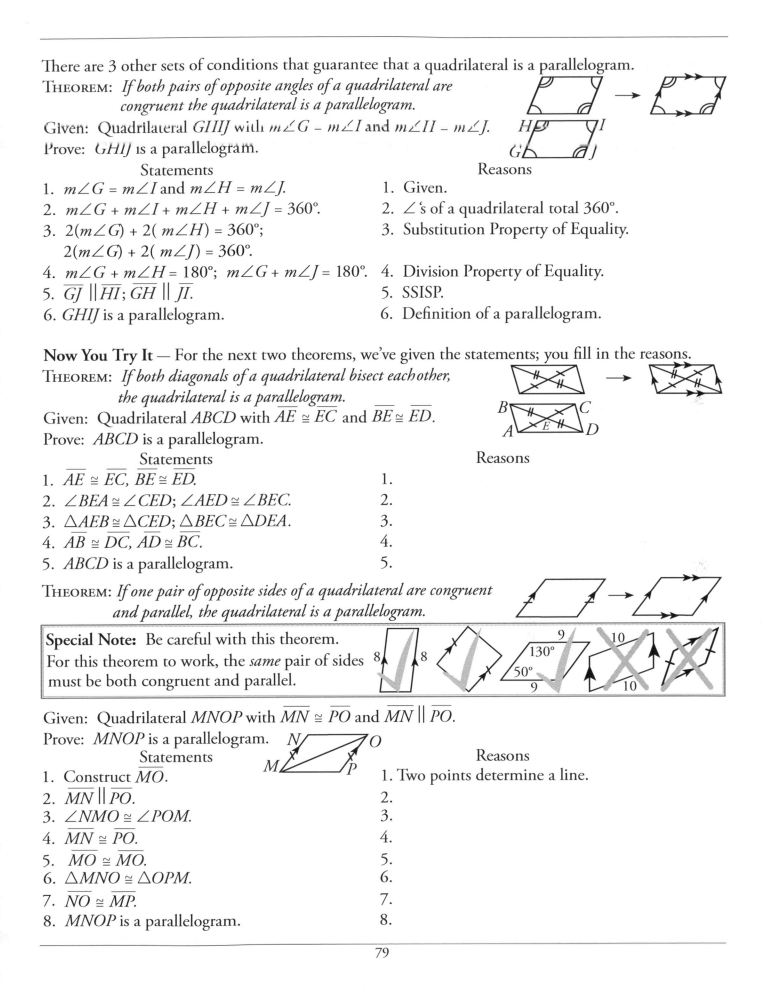

THEOREM: *If both pairs of opposite angles of a quadrilateral are congruent the quadrilateral is a parallelogram.*

Given: Quadrilateral *GHIJ* with $m\angle G = m\angle I$ and $m\angle H = m\angle J$.

Prove: *GHIJ* is a parallelogram.

Statements	Reasons
1. $m\angle G = m\angle I$ and $m\angle H = m\angle J$.	1. Given.
2. $m\angle G + m\angle I + m\angle H + m\angle J = 360°$.	2. ∠'s of a quadrilateral total 360°.
3. $2(m\angle G) + 2(m\angle H) = 360°$; $2(m\angle G) + 2(m\angle J) = 360°$.	3. Substitution Property of Equality.
4. $m\angle G + m\angle H = 180°$; $m\angle G + m\angle J = 180°$.	4. Division Property of Equality.
5. $\overline{GJ} \parallel \overline{HI}$; $\overline{GH} \parallel \overline{JI}$.	5. SSISP.
6. *GHIJ* is a parallelogram.	6. Definition of a parallelogram.

Now You Try It — For the next two theorems, we've given the statements; you fill in the reasons.

THEOREM: *If both diagonals of a quadrilateral bisect each other, the quadrilateral is a parallelogram.*

Given: Quadrilateral *ABCD* with $\overline{AE} \cong \overline{EC}$ and $\overline{BE} \cong \overline{ED}$.

Prove: *ABCD* is a parallelogram.

Statements	Reasons
1. $\overline{AE} \cong \overline{EC}$, $\overline{BE} \cong \overline{ED}$.	1.
2. $\angle BEA \cong \angle CED$; $\angle AED \cong \angle BEC$.	2.
3. $\triangle AEB \cong \triangle CED$; $\triangle BEC \cong \triangle DEA$.	3.
4. $\overline{AB} \cong \overline{DC}$, $\overline{AD} \cong \overline{BC}$.	4.
5. *ABCD* is a parallelogram.	5.

THEOREM: *If one pair of opposite sides of a quadrilateral are congruent and parallel, the quadrilateral is a parallelogram.*

Special Note: Be careful with this theorem. For this theorem to work, the *same* pair of sides must be both congruent and parallel.

Given: Quadrilateral *MNOP* with $\overline{MN} \cong \overline{PO}$ and $\overline{MN} \parallel \overline{PO}$.

Prove: *MNOP* is a parallelogram.

Statements	Reasons
1. Construct $\overline{MO}$.	1. Two points determine a line.
2. $\overline{MN} \parallel \overline{PO}$.	2.
3. $\angle NMO \cong \angle POM$.	3.
4. $\overline{MN} \cong \overline{PO}$.	4.
5. $\overline{MO} \cong \overline{MO}$.	5.
6. $\triangle MNO \cong \triangle OPM$.	6.
7. $\overline{NO} \cong \overline{MP}$.	7.
8. *MNOP* is a parallelogram.	8.

Now You Try It — For problems 1 – 4, find any parallelograms in the drawings. Be sure to thoroughly explain your conclusions.

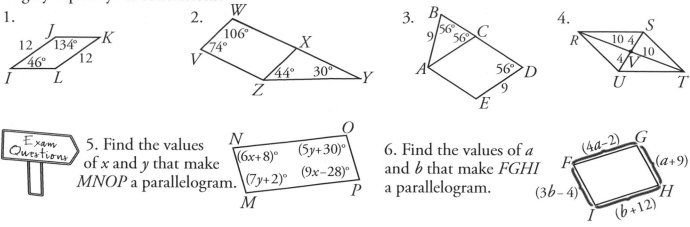

1.

2.

3.

4.

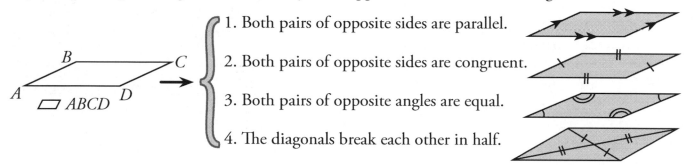

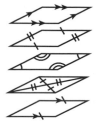

5. Find the values of x and y that make *MNOP* a parallelogram.

6. Find the values of a and b that make *FGHI* a parallelogram.

Quick Review — Properties of Parallelograms

If you're given a parallelogram, here's what you're supposed to know about the figure:

1. Both pairs of opposite sides are parallel.

2. Both pairs of opposite sides are congruent.

3. Both pairs of opposite angles are equal.

4. The diagonals break each other in half.

Remember, whenever you see the ▱ symbol, you know all of the above facts. Another fact about parallelograms, is that same side interior (consecutive) angles always add up to 180°.

When is a Quadrilateral a Parallelogram?

If you can show *even one* of the following conditions, you have proved that the quadrilateral is a parallelogram. If you're given a quadrilateral that satisfies *even one* of the following conditions you are supposed to know that the quadrilateral is a parallelogram

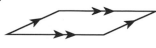

If
1. Both pairs of opposite sides are parallel *or*
2. Both pairs of opposite sides are congruent *or*
3. Both pairs of opposite angles are equal *or*
4. Both diagonals bisect each other *or*
5. One pair of opposite sides are both equal and parallel,
then
the quadrilateral is a parallelogram.

Special Parallelograms

The following figures are parallelograms, and it's important to remember that they have all of the properties of parallelograms, but they also have special properties that are all their own. First we'll define them:

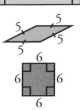

DEFINITION OF A RECTANGLE: *A rectangle is a quadrilateral with four right angles.*

DEFINITION OF A RHOMBUS: *A rhombus is a quadrilateral with four congruent sides.*
Note: The word rhombus has two plural forms, rhombuses and rhombi (rhom'bī).

DEFINITION OF A SQUARE: *A square is a quadrilateral with four congruent sides and four right angles.*

Based on their definitions, it is easy to prove that squares, rectangles and rhombi are parallelograms.

Use (one of) these theorems:

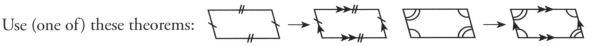

The logic diagram below shows the relationship between the three special parallelograms:

All squares are rectangles.
Not all rectangles are squares.

All squares are rhombi.
Not all rhombi are squares.

Remember, *every* figure in the drawing has *all* of the properties of parallelograms.

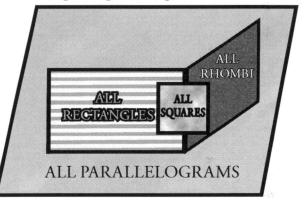

Rectangles

THEOREM: *The diagonals of a rectangle are equal.*

Given: Rectangle *GHIJ*.   Prove: $\overline{GI} \cong \overline{HJ}$

Statements	Reasons
1. Construct $\overline{GI}$ & $\overline{HJ}$.	1. Two points determine a line.
2. $\overline{HG} \cong \overline{IJ}$.	2. Opposite sides of a ▱ ≅.
3. $\overline{GJ} \cong \overline{GJ}$.	3. Reflexive.
4. ∠*JGH* is rt., ∠*GJI* is rt.	4. Definition of a rectangle.
5. $m\angle JGH = m\angle GJI$.	5. All right angles are =.
6. $\triangle JGH \cong \triangle GJI$.	6 SAS ≅.
7. $\overline{GI} \cong \overline{HJ}$.	7. CPCT.

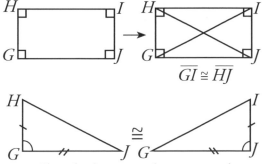

$\overline{GI} \cong \overline{HJ}$

Sketch the triangles separately

Here is an important theorem about right triangles, which follows from the previous theorem:

THEOREM: *In a right triangle, the midpoint of the hypotenuse is equidistant from each vertex.*

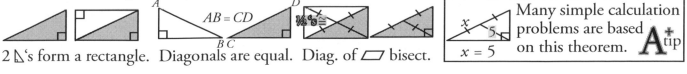

$AB = CD$ ½'s ≅

2 △'s form a rectangle.   Diagonals are equal.   Diag. of ▱ bisect.

$x = 5$

Many simple calculation problems are based on this theorem. **A**⁺tip

Rhombuses

THEOREM: *The diagonals of a rhombus are perpendicular.*

Proof:

The proof of this theorem demonstrates why knowing your definitions and properties is so important in geometry. It is also an example of using congruent triangles and then CPCT in a proof.

Given: Rhombus *ABCD*.
Prove: $\overline{BD} \perp \overline{AC}$.

Fig. 1

Looking at Fig. 1 above, we might not have any idea how to begin, except we see four triangles in the drawing. Now we'll add some knowledge to the picture:

By its definition, a rhombus has four congruent sides:

Fig. 2

Since a rhombus is a parallelogram, the diagonals must bisect each other (Property of a ▱):

Fig. 3

Studying Fig. 3, clearly the four triangles are congruent by SSS≅.

Our goal is to show $\overline{BD} \perp \overline{AC}$, and there are theorems which show two lines are perpendicular. However, it's probably easier, especially on an exam, to use the definition of perpendicular lines: two lines that intersect to form right angles. So we have to figure out a way to show that the center angles are right.

Working on ∠*AEB* and ∠*CEB*, we see straight ∠*AEC*, which we know equals 180°. Since the two triangles are congruent, CPCT tells us that ∠*AEB* ≅ ∠*CEB*. If the measures of two equal angles total 180° (either as a linear pair or using the Angle Addition Postulate, depending on which method your text uses), then each angle must measure 90° by the division property of equality. If an angle equals 90°, it is a right angle (definition of right angles), which then proves that the diagonals are perpendicular (the definition of perpendicular lines). QED (which means what was to have been proven is proved)!

THEOREM: *The angles of a rhombus are bisected by its diagonals.*

Here's the proof for two pairs of angles. The same argument proves the other pairs of angles are equal.
Given: Rhombus *MNOP*. Prove: ∠*QMN* = ∠*QMP* and ∠*MNQ* = ∠*ONQ*.

Statements	Reasons
1. *MNOP* is a rhombus.	1. Given.
2. $\overline{MN} \cong \overline{NO} \cong \overline{OP} \cong \overline{PM}$.	2. Definition of a rhombus.
3. $\overline{MQ} \cong \overline{OQ}$, $\overline{NQ} \cong \overline{PQ}$.	3. Diagonals of a ▱ bisect each other.
4. △*MNQ* ≅ △*MPQ* ≅ △*ONQ*.	4. SSS≅.
5. ∠*QMN* ≅ ∠*QMP*, ∠*MNQ* = ∠*ONQ*.	5. CPCT.

Due to the symmetry of rhombuses, there are other facts that your textbook may introduce as theorems. For example:

A rhombus is divided into **4 congruent *right* triangles** by its diagonals.

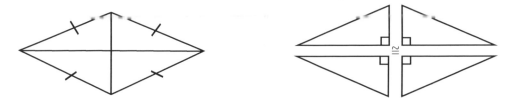

If **2** *consecutive* (right next to each other) sides of a parallelogram are congruent, the figure is a rhombus. (Which property of parallelograms proves this?)

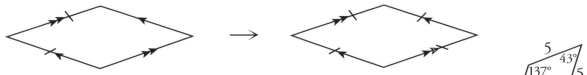

Notice that the 4 angles of a rhombus aren't necessarily equal, it's the *sides* that are equal: If the angles are equal as well, then the figure is a square.

Squares — The definition of a square is a quadrilateral with 4 congruent sides and 4 right angles. So a square is both a rectangle *and* a rhombus! Everything that is true about a rectangle is true about a square *and* everything that is true about a rhombus is true about a square. The square is the ultimate parallelogram!

A⁺tip **Be sure you know the information in the chart below:**

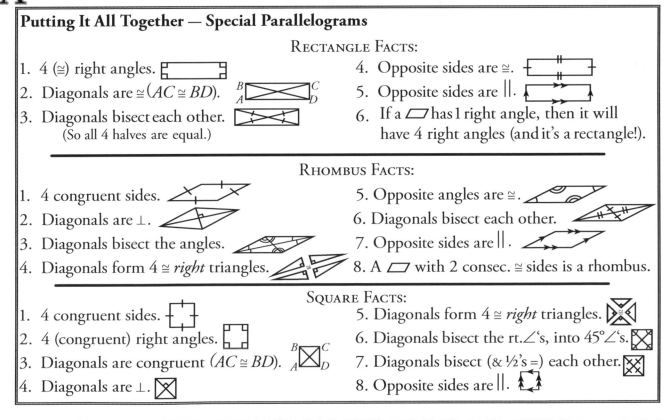

Putting It All Together — Special Parallelograms

RECTANGLE FACTS:
1. 4 (≅) right angles.
2. Diagonals are ≅ ($AC \cong BD$).
3. Diagonals bisect each other. (So all 4 halves are equal.)
4. Opposite sides are ≅.
5. Opposite sides are ‖.
6. If a ▱ has 1 right angle, then it will have 4 right angles (and it's a rectangle!).

RHOMBUS FACTS:
1. 4 congruent sides.
2. Diagonals are ⊥.
3. Diagonals bisect the angles.
4. Diagonals form 4 ≅ *right* triangles.
5. Opposite angles are ≅.
6. Diagonals bisect each other.
7. Opposite sides are ‖.
8. A ▱ with 2 consec. ≅ sides is a rhombus.

SQUARE FACTS:
1. 4 congruent sides.
2. 4 (congruent) right angles.
3. Diagonals are congruent ($AC \cong BD$).
4. Diagonals are ⊥.
5. Diagonals form 4 ≅ *right* triangles.
6. Diagonals bisect the rt.∠'s, into 45°∠'s.
7. Diagonals bisect (& ½'s =) each other.
8. Opposite sides are ‖.

Now You Try It

1. Find the perimeter of ▱ABCD, m∠AED and m∠EAD.

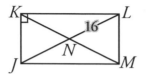

2. a. Find *x*, *y* and *z* in ▱ *FGHI*.
   b. △*GEF* ≅ which triangle?

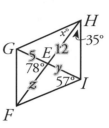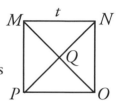

3. Given: ▱ *JKLM*, find *NM*.

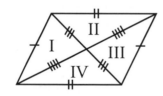

4. Find the perimeter of square *MNOP*, m∠MQN, m∠POQ and name any triangles that are congruent to △OQP.

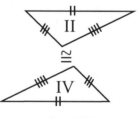

The diagonals divide *any parallelogram* into **2 pairs of congruent triangles**. Here's why:

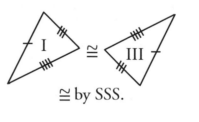

Opposite sides of a ▱ are ≅ and, diagonals of a ▱ bisect each other.

≅ by SSS.

≅ by SSS.

The diagonals divide a *rhombus* into **4 congruent right triangles**. Here's why:

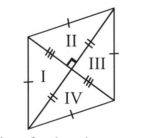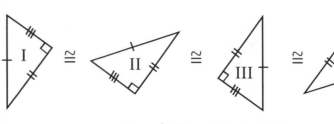

All sides of a rhombus are ≅ . Diagonals of any ▱ bisect each other. Diagonals of a rhombus are ⊥.

4 rt. △ ≅ by SSS (or HL).

Trapezoids

Quick Review — A quadrilateral is a four-sided polygon. Quad means four and lateral means side, so it makes sense that a quadrilateral is a four (straight) sided figure. It's easy to figure out that the interior angles of a quadrilateral total 360°. To do this we can use the formula for polygons, $(n - 2)180°$ or $(4 - 2)180° = 360°$. However, it's better just to memorize this important fact:

The interior angles of a quadrilateral total 360°.

DEFINITION: *A trapezoid is a quadrilateral with **exactly one** pair of parallel sides.*

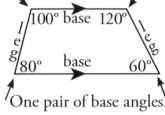

The parts of a trapezoid:

Bases of Trapezoids — A base of a trapezoid is one of its parallel sides. It doesn't matter whether the base is on the side or on the top. If it's a base, it's one of the parallel sides, if it's a parallel side, it's a base.

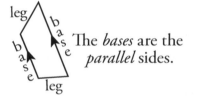

The *bases* are the *parallel* sides.

Like all polygons, a trapezoid is named by its vertices. Many texts use the small trapezoid figure ⌂ to indicate a trapezoid. Here are some examples of the different ways trapezoids may be presented in problems:

Trapezoid *ABCD* with $\overline{BC} \parallel \overline{AD}$. | The figure on the right. | ⌂ *MNOP* on the right. | The figure on the right.

In the fourth figure, the same side interior angles between the top and bottom sides of the figure are supplementary, which means (by SSISP) that the top and bottom sides are parallel. Since the ∥ sides are unequal (which means the figure is not a parallelogram), the figure is a trapezoid.

DEFINITION: *An isosceles trapezoid is a trapezoid with both legs congruent.*
Many of the trapezoids you work with in geometry are isosceles trapezoids.
Note: The *bases* of a trapezoid can never be equal. Why not?
(Because then it would be a parallelogram!)

THEOREM: *The base angles of an isosceles trapezoid are congruent.*

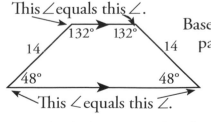

Base angles come in *pairs*. Each trapezoid has 2 pairs of base angles. Each pair shares one of the bases. Notice how the top side of the trapezoid, a base, is a side of both of the 132° angles. Therefore, the 132° angles make up one pair of base angles of this trapezoid. The pair of 48° angles are the other pair of base angles.

Median of a Trapezoid

A trapezoid is a quadrilateral with exactly 2 parallel sides called the bases. The two non-parallel sides of a trapezoid are called the legs.

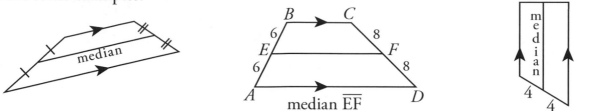

DEFINITION: *The median of a trapezoid is the segment that connects the midpoints of the legs.*

Here are some examples:

median $\overline{EF}$

THEOREM: *The length of the median of a trapezoid is equal to the average length of its bases.*

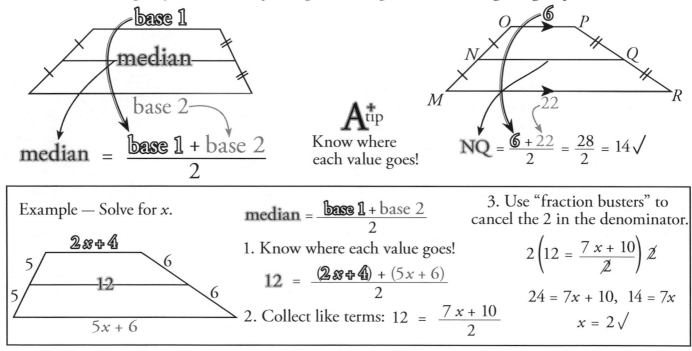

$$\text{median} = \frac{\text{base 1} + \text{base 2}}{2}$$

A^{+}_{tip} Know where each value goes!

$$NQ = \frac{6 + 22}{2} = \frac{28}{2} = 14 \checkmark$$

Example — Solve for x.

$$\text{median} = \frac{\text{base 1} + \text{base 2}}{2}$$

1. Know where each value goes!

$$12 = \frac{(2x + 4) + (5x + 6)}{2}$$

2. Collect like terms: $12 = \dfrac{7x + 10}{2}$

3. Use "fraction busters" to cancel the 2 in the denominator.

$$2\left(12 = \frac{7x + 10}{2}\right)2$$

$$24 = 7x + 10, \quad 14 = 7x$$

$$x = 2 \checkmark$$

THEOREM: *The median of a trapezoid is parallel to the bases.*

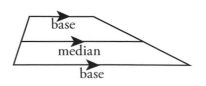

Example — Given isosceles trapezoid *MNOP* with median $\overline{QR}$, find $m\angle NMP$ and x.

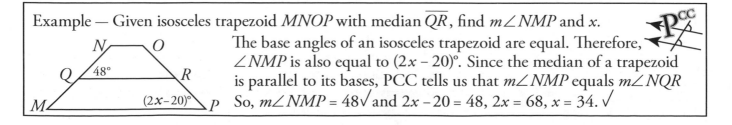

The base angles of an isosceles trapezoid are equal. Therefore, $\angle NMP$ is also equal to $(2x - 20)°$. Since the median of a trapezoid is parallel to its bases, PCC tells us that $m\angle NMP$ equals $m\angle NQR$. So, $m\angle NMP = 48 \checkmark$ and $2x - 20 = 48$, $2x = 68$, $x = 34$. $\checkmark$

Now You Try It

1. Is *ABCD* a trapezoid?

2. Given ▱ *MNOP* with bases $\overline{NO}$ and $\overline{MP}$, find *x* and *y*.

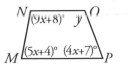

3. Given *GHIJ*, find *u* and *v*.

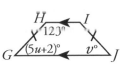

4. Given ▱ *UVWX* with median $\overline{YZ}$, find *a*, *b* and *c*.

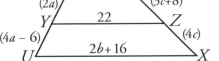

5. Given ▱ *QRST*, find *x*.

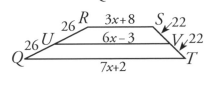

6. Given ▱ *IJKL* with median $\overline{MN}$, find *z*.

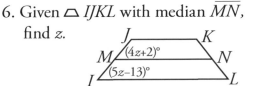

Other Parallel Line Theorems

The median of a *trapezoid* equals $\dfrac{\text{base 1} + \text{base 2}}{2}$

But what happens to the median when the top base of a trapezoid gets smaller and smaller?

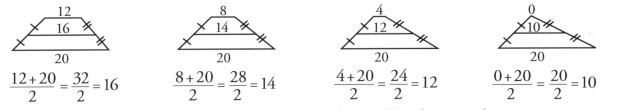

$$\frac{12+20}{2}=\frac{32}{2}=16 \qquad \frac{8+20}{2}=\frac{28}{2}=14 \qquad \frac{4+20}{2}=\frac{24}{2}=12 \qquad \frac{0+20}{2}=\frac{20}{2}=10$$

THEOREM: *The segment connecting the midpoints of two sides of a triangle is:*
    1. Half as long as the third side, and
    2. Parallel to the third side.

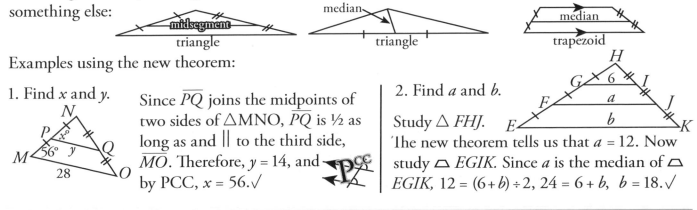

$DE = \frac{1}{2}AC$
$\overline{DE} \parallel \overline{AC}$

For triangles, this special segment is called a *midsegment* because a median of a triangle is something else:

midsegment — triangle   median — triangle   median — trapezoid

Examples using the new theorem:

1. Find *x* and *y*.

Since $\overline{PQ}$ joins the midpoints of two sides of △MNO, $\overline{PQ}$ is ½ as long as and ∥ to the third side, $\overline{MO}$. Therefore, *y* = 14, and by PCC, *x* = 56. ✓

2. Find *a* and *b*.

Study △ *FHJ*. The new theorem tells us that *a* = 12. Now study ▱ *EGIK*. Since *a* is the median of ▱ *EGIK*, 12 = (6 + *b*) ÷ 2, 24 = 6 + *b*, *b* = 18. ✓

Now You Try It

1. Find z and $m\angle DEC$.

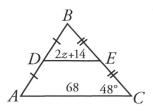

2. Solve for x, y and $m\angle HJI$.

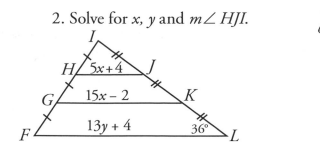

Segment Lengths and Parallel Lines

The *distance* between two parallel lines is the perpendicular distance.

THEOREM: *If 3 or more parallel lines cut off equal segments on a transversal, they are the same distance apart. If 3 or more parallel lines are the same distance apart, they cut off equal segments on all transversals.*

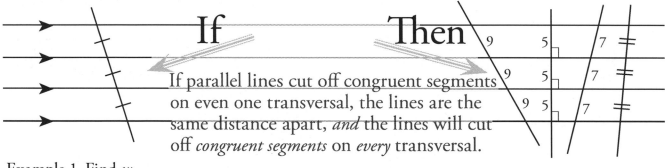

If parallel lines cut off congruent segments on even one transversal, the lines are the same distance apart, *and* the lines will cut off *congruent segments* on *every* transversal.

Example 1. Find w.

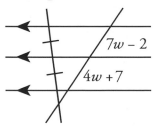

Since the 3 ∥ lines cut off equal parts on one transversal, they cut off equal parts on the other transversal, so set the two expressions equal:
$7w - 2 = 4w + 7$,
$3w = 9$, $w = 3$. √

A⁺tip When you see 3 or more parallel lines in a problem, think of this theorem.

Example 2. Find x.

Since the 3 ∥ lines cut off equal parts on line m, they cut off equal parts on line n, so $3x = 6$, and $x = 2$. √

Now You Try It

1. Given $a = b$, find z.

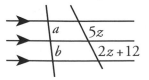

2. The three lines in the figure are equally far apart. Solve for x and y. (Hint: Think about which segment equals which segment.)

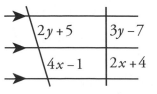

KITES

If your geometry course doesn't include kites, you may want to skip this page.

A kite is a 4-sided polygon with exactly two pairs of consecutive congruent sides. Here are some examples:

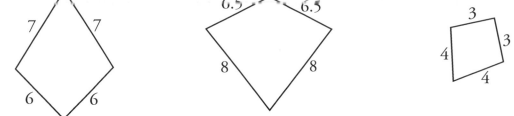

Kites have interesting properties that we can discover by using what we know about congruent and isosceles triangles. We'll investigate using the first figure:

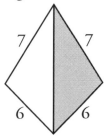

Draw the longer diagonal (which will always be between the pairs of *equal* sides).

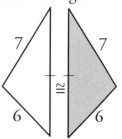

The longer diagonal divides the kite into two congruent triangles by SSS ≅.

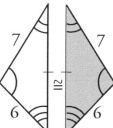

By CPCT, the angles between the unequal sides are equal and the angles between the equal sides are bisected.

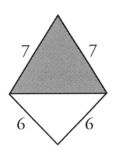

Then draw the shorter diagonal, which will always be between the pairs of *unequal* sides.

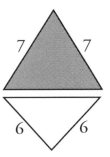

The shorter diagonal divides the kite into two unequal isosceles triangles.

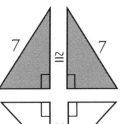

The longer diagonal divides each isosceles triangle into two congruent right triangles.

Facts about kites: 1. The opposite sides are never equal.
2. The angles between the equal sides are never equal.
3. The diagonals are never equal.
4. The diagonals are perpendicular.
5. The longer diagonal bisects the shorter diagonal.

8. INEQUALITIES

In most problems, we work with equalities, that is, expressions that are equal to each other or figures that are congruent to each other. This section deals with geometric objects that are not equal to each other. The goal is to compare two (or more) objects and to figure out which is larger.

Properties of Inequalities

Here are the *Properties of Inequalities* that are used most often in geometry:

I. **If $a > b$ and $c \geq d$, then $a + c > b + d$.**

> Here are some examples:
>
> 1) $8 > 3$ and $2 = 2$, so $8 + 2 > 3 + 2$.
> 2) $5 > 2$ and $4 > 3$, so $5 + 4 > 2 + 3$.
> 3) $AB > CD$, $BC = BC$ so $AB + BC > CD + BC$.

> 4) $MN > PN$ and $NO > NQ$ so $MN + NO > PN + NQ$.

> 5) $m\angle 1 > m\angle 3$ and $m\angle 2 = m\angle 2$, so $m\angle 1 + m\angle 2 > m\angle 3 + m\angle 2$.

II. **If $a > b$ and $b > c$, then $a > c$.**

> Here's an example:
>
> If $EF > FG$ and $FG > GH$, then $EF > GH$.

III. **If $a = b + c$ and $b > 0$ and $c > 0$, then $a > b$ and $a > c$.**
(The whole thing is bigger than either one of its parts.)

> Here are some examples:
>
> 1) $10 = 2 + 8$ so $10 > 2$ and $10 > 8$.
> 2) $IK = IJ + JK$ (Seg. Add. Post.), so $IK > IJ$ and $IK > JK$.

> 3) If $m\angle ABD = m\angle ABC + m\angle CBD$ (Angle Add. Post.), then $m\angle ABD > m\angle ABC$ and $m\angle ABD > m\angle CBD$.

An Important Theorem Used in Many Inequality Problems

Quick Review — An exterior angle of a triangle is formed by extending a side of a triangle and measuring back to the triangle.

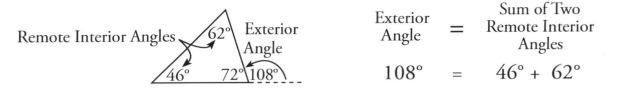

EXTERIOR ANGLE THEOREM: *The measure of an exterior angle of a triangle is equal to the sum of the measures of the two remote interior angles.*

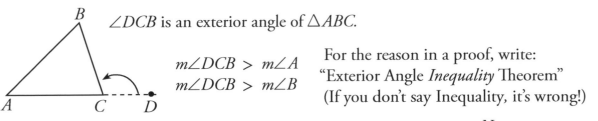

Exterior Angle	=	Sum of Two Remote Interior Angles
108°	=	46° + 62°

Now, recall Property III of Inequalities:

If $a = b + c$, and $b > 0$ and $c > 0$, then $a > b$ and $a > c$.
(The whole thing is bigger than either one of its parts.)

Using the previous problem, $108° = 46° + 62°$, so $108° > 46°$ and $108° > 62°$.

The above property leads to a theorem, which is the key to solving many inequality problems:

EXTERIOR ANGLE **INEQUALITY** THEOREM: *The measure of an exterior angle of a triangle is larger than the measure of either of the two remote interior angles.*

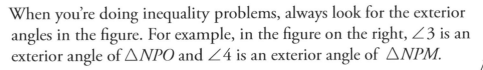

$\angle DCB$ is an exterior angle of $\triangle ABC$.

$m\angle DCB > m\angle A$
$m\angle DCB > m\angle B$

For the reason in a proof, write:
"Exterior Angle *Inequality* Theorem"
(If you don't say Inequality, it's wrong!)

When you're doing inequality problems, always look for the exterior angles in the figure. For example, in the figure on the right, $\angle 3$ is an exterior angle of $\triangle NPO$ and $\angle 4$ is an exterior angle of $\triangle NPM$.

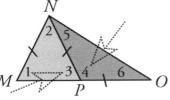

Now You Try It

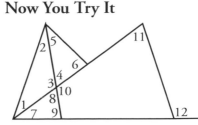

Problems 1-3

1. (a) How many exterior angles are named in the figure? (b) List them. (c) Can you find any exterior angles that are shown but not named?

2. Using the Exterior Angle Inequality Theorem, list as many inequalities as you can.

3. If $m\angle 3 > m\angle 4$ (never assume that it is based on the drawing) what else can you conclude?

Triangle Inequality Theorem

You can't make a triangle with sides of just any lengths. Here's an example:

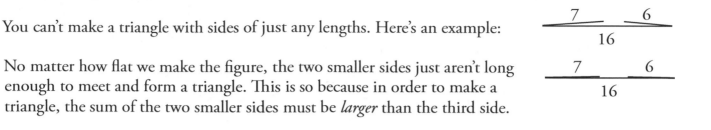

No matter how flat we make the figure, the two smaller sides just aren't long enough to meet and form a triangle. This is so because in order to make a triangle, the sum of the two smaller sides must be *larger* than the third side.

TRIANGLE INEQUALITY THEOREM: *In a triangle, the sum of the lengths of the two smaller sides is larger than the length of the third side.*

There are 2 types of Triangle Inequality problems:

1st Type of Problem — Given three lengths, can you make a triangle?
To check: 1. Add the two *smaller* lengths together.
2. Is their sum *larger* than the largest length?
Yes   No

Examples — Can you make a triangle with the given 3 lengths?

·NOTES-
Add the
2 Smaller
numbers
together

6, 3 and 2 ?   No. 2+3 = 5 and 5 is not larger than 6.
2, 6 and 4 ?   No. 2+4 ≯ 6. "Equal to" isn't good enough. The sum must be larger!
2, 4.01 and 6? Yes. 2+4.01 = 6.01 which is larger than 6. (Even .01 larger is enough.)
3, 3 and 5.8 ? Yes. 3+3 = 6 and 6 is larger than 5.8.

Isosceles ⚠ { 2, 5 and 5 ?   Yes. 2+5 = 7 and 7 is larger than 5.
Equilateral △→ { 4, 4 and 4 ?   Yes. 4+4 = 8 and 8 is larger than 4. (Any equilateral triangle "works".)

2nd Type of Problem — Given the lengths of 2 sides of a triangle, find the lower and upper limits of the length of the third side. } $\underline{?} <$ length of third side $< \underline{?}$

Example — If the lengths of two sides of a triangle are 9 and 4, find the lower and upper limits of the length of the third side: 9 – 4 < length of third side < 9 + 4, so 5 < length of third side < 13 √

A⁺tip Here's the formula★

Big Number − Small Number	< Length of Third Side <	Big Number + Small Number

Example — If 3 sides of a triangle are 6.3, 8 and x, find numbers a and b such that: $a < x < b$

8 – 6.3 < x < 8 + 6.3   1.7 < x < 14.3   a = 1.7 √, b = 14.3 √

Now You Try It

1. Can you make a triangle with sides equal to a) 4, 4 and 8?   b) 4.01, 1 and 3? c) 2, 7.$\overline{9}$ and 6?

2. Three sides of a triangle are 5, 5 and x. Find numbers a and b such that $a < x < b$.

★Why does the formula work? See page 256 in the Answer Section for a complete explanation.

Inequalities for ONE Triangle

As the size of an angle increases, the sides of the angle move further apart. And, as the sides of an angle move further **apart**, the size of the angle increases.

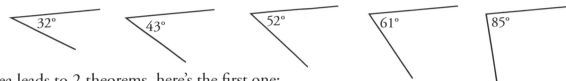

This idea leads to 2 theorems, here's the first one:

THEOREM: *Within a **single** triangle, if one side is larger than a second side, then the angle opposite the first side is larger than the angle opposite the second side.*

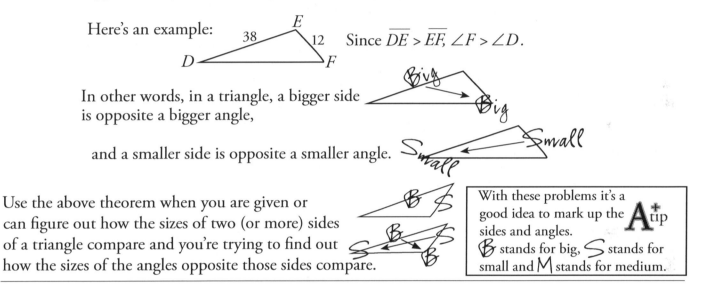

Here's an example: Since $\overline{DE} > \overline{EF}$, $\angle F > \angle D$.

In other words, in a triangle, a bigger side is opposite a bigger angle,

and a smaller side is opposite a smaller angle.

Use the above theorem when you are given or can figure out how the sizes of two (or more) sides of a triangle compare and you're trying to find out how the sizes of the angles opposite those sides compare.

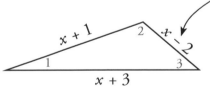

> With these problems it's a good idea to mark up the sides and angles.
> B stands for big, S stands for small and M stands for medium.

Example: Find the largest angle in the triangle below.

Important Mathematics Note:
Physical quantities (like the side of a triangle, for example) are positive, so the expression $x - 2$ must be positive. This tells us that x is positive and in fact that $x > 2$.

Step 1. Label the sides of the triangle.

Step 2. Label the angles opposite the sides.

Step 3. Studying the figure in Step 2, we see that:

$\angle 2$ is the largest angle in the triangle.

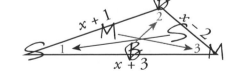

The converse of this theorem is also true: As the sides of an angle move further apart, the larger the angle becomes. This idea leads to the second theorem.

THEOREM: *Within a **single** triangle, if one angle is larger than a second angle, then the side opposite the first angle is larger than the side opposite the second angle.*

Since $\angle D > \angle F$, $EF > DE$.

Notice that $\overline{EF}$ is not the larg*est* side in the triangle but it is larg*er* than $\overline{DE}$. This is all that the theorem is claiming.

Use this theorem when you can compare the sizes of angles in a triangle and you're asked to compare the lengths of its sides.

Remember — Use the 1st theorem when you have information about the sides.
Use the 2nd theorem when you have information about the angles.

Example: Put the side lengths of the two triangles at right in order from smallest to largest.

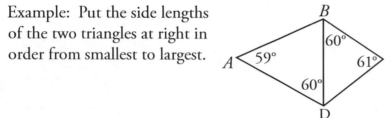

Since we are given the sizes of the angles and we are asked to compare the lengths of the sides, we should use the second theorem.

Step 1. Figure out how big the third angles are in each triangle and add that information to the figure.

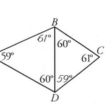

Step 2. Within each triangle, mark the angles to indicate their relative sizes, then mark the sides opposite each angle with the corresponding letter.

Step 3. Now study the drawing. Since the biggest side of $\triangle DCB$, is the same size as the smallest side of $\triangle DAB$, we are able to put the side lengths of the two triangles in order :

$$\overline{BC} < \overline{DC} < \overline{BD} < \overline{AB} < \overline{AD} \checkmark$$

Confused? Think of it like this: The biggest side in $\triangle DCB$ is $\overline{BD}$, but $\overline{BD}$ is the smallest side of $\triangle DAB$. This means that the other sides of $\triangle DAB$ are all larger than every side of $\triangle DCB$.

Example: Find the largest side of triangle *TUV*.

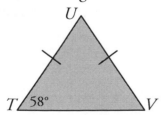

Step 1. This is an isosceles triangle, so mark $\angle T$ and $\angle V$ equal (by ITT). This means $\angle V$ is also equal to 58° and $\angle U$ is equal to:

$$180° - 58° - 58° = 64°$$

Step 2. Add the angle measures and markings showing $\angle U$ is the largest angle. This means that $\overline{TV}$ must be the largest side of the triangle. $\checkmark$

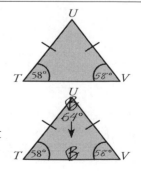

Here is another example of the use of these theorems:

Given: $MN = PN = OP$.

Prove: ON is the largest side of $\triangle PNO$ and $ON > MN$.

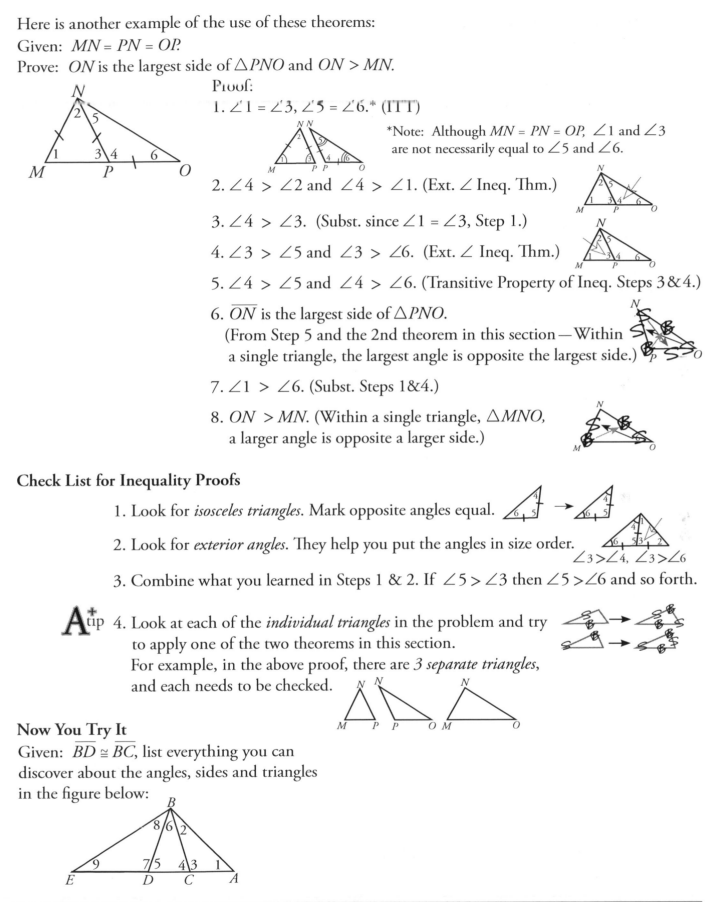

Proof:

1. $\angle 1 = \angle 3$, $\angle 5 = \angle 6$.* (ITT)

   *Note: Although $MN = PN = OP$, $\angle 1$ and $\angle 3$ are not necessarily equal to $\angle 5$ and $\angle 6$.

2. $\angle 4 > \angle 2$ and $\angle 4 > \angle 1$. (Ext. $\angle$ Ineq. Thm.)

3. $\angle 4 > \angle 3$. (Subst. since $\angle 1 = \angle 3$, Step 1.)

4. $\angle 3 > \angle 5$ and $\angle 3 > \angle 6$. (Ext. $\angle$ Ineq. Thm.)

5. $\angle 4 > \angle 5$ and $\angle 4 > \angle 6$. (Transitive Property of Ineq. Steps 3 & 4.)

6. $\overline{ON}$ is the largest side of $\triangle PNO$.
   (From Step 5 and the 2nd theorem in this section—Within a single triangle, the largest angle is opposite the largest side.)

7. $\angle 1 > \angle 6$. (Subst. Steps 1 & 4.)

8. $ON > MN$. (Within a single triangle, $\triangle MNO$, a larger angle is opposite a larger side.)

Check List for Inequality Proofs

1. Look for *isosceles triangles*. Mark opposite angles equal.

2. Look for *exterior angles*. They help you put the angles in size order.

   $\angle 3 > \angle 4$, $\angle 3 > \angle 6$

3. Combine what you learned in Steps 1 & 2. If $\angle 5 > \angle 3$ then $\angle 5 > \angle 6$ and so forth.

4. Look at each of the *individual triangles* in the problem and try to apply one of the two theorems in this section.
   For example, in the above proof, there are *3 separate triangles*, and each needs to be checked.

Now You Try It

Given: $\overline{BD} \cong \overline{BC}$, list everything you can discover about the angles, sides and triangles in the figure below:

Inequalities for 2 Triangles

Even when two triangles are not congruent, if they have certain parts in common, we can make comparisons of some of their other parts.

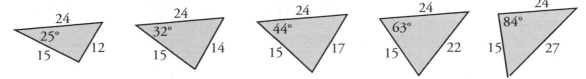

The triangles above have two equal pairs of sides and an unequal included angle. (The included angle in this case, is the angle between the sides measuring 15 and 24.) As the size of the included angle increases, the length of the opposite side also increases. This idea leads to the following theorem:

*The word "**Ineq.**" has to be there or it's not correct.*

Included

SIDE ANGLE SIDE INEQUALITY THEOREM (THE HINGE THEOREM):   **SAS Ineq.**

If two pairs of sides of two triangles are equal and the included angle of the first triangle is smaller than the included angle of the second triangle, then the third side of the first triangle is smaller than the third side of the second triangle.

Sound Confusing? Read on.

What we know: Each of the two S's in SAS Ineq. stands for one *pair* of congruent sides. The A stands for one *pair* of **unequal** included angles.

What we find out (the conclusion): Which of the sides opposite the unequal angles is larger and which is smaller.

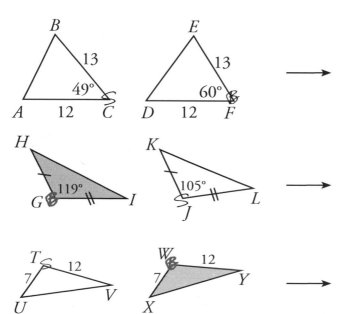

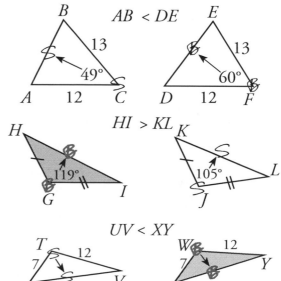

Look at the previous theorem from another point of view: the triangles below have two equal sides but their third *sides* are *unequal*. As the length of the third side increases, the size of the opposite angle also increases.

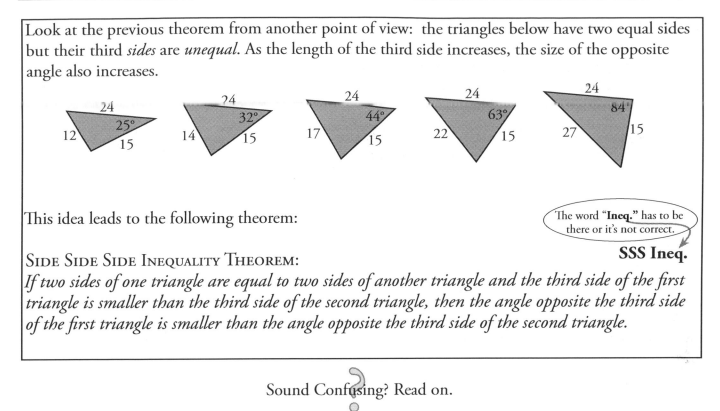

This idea leads to the following theorem:

The word "**Ineq.**" has to be there or it's not correct.

SSS Ineq.

SIDE SIDE SIDE INEQUALITY THEOREM:
If two sides of one triangle are equal to two sides of another triangle and the third side of the first triangle is smaller than the third side of the second triangle, then the angle opposite the third side of the first triangle is smaller than the angle opposite the third side of the second triangle.

Sound Confusing? Read on.

$\overset{\cong}{S}\overset{\cong}{S}\overset{\lt}{S}$ **Ineq.**

What we know: Each of the first two S's in SSS Ineq. stands for one pair of congruent sides. The third S stands for one pair of *unequal* sides.

What we find out (the conclusion): Which of the angles opposite the unequal sides is larger and which is smaller.

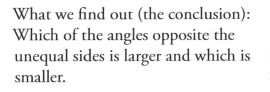

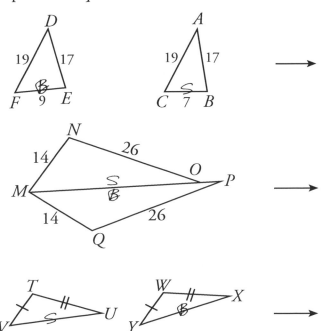

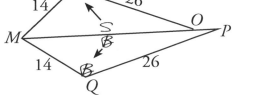

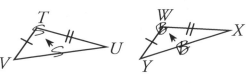

Which Theorem To Use — SAS Ineq. or SSS Ineq. ???

In proofs and on some exam problems you need to know which of these two theorems is the 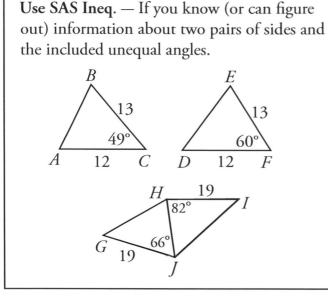 right one for the problem you're doing. You decide based on the information that is coming into the problem (the given), together with any information that you are able to add. Here's how.

Use SAS Ineq. — If you know (or can figure out) information about two pairs of sides and the included unequal angles.

Use SSS Ineq. — If you know (or can figure out) information about three pairs of sides, two pairs equal, one pair unequal.

What information comes out of each theorem — ???

SAS Ineq. — Tells us which of the sides opposite the unequal angles is larger and which is smaller.

SSS Ineq. — Tells us which of the angles opposite the unequal sides is larger and which is smaller.

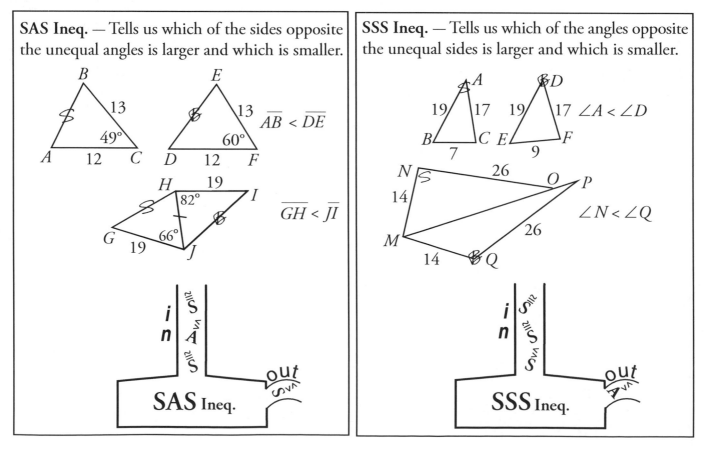

Example:

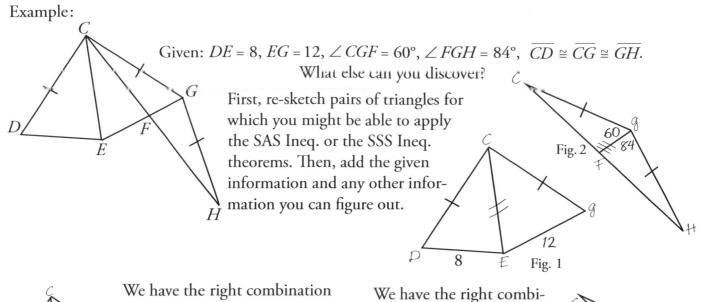

Given: $DE = 8$, $EG = 12$, $\angle CGF = 60°$, $\angle FGH = 84°$, $\overline{CD} \cong \overline{CG} \cong \overline{GH}$.
What else can you discover?

First, re-sketch pairs of triangles for which you might be able to apply the SAS Ineq. or the SSS Ineq. theorems. Then, add the given information and any other information you can figure out.

We have the right combination of information about the two triangles in Fig. 1, to use the SSS Ineq. theorem. And remember, SSS Ineq. tells us which of the angles opposite the unequal sides is larger and which is smaller:

Since $DE < EG$, $\angle DCE < \angle ECG$. √

We have the right combination of information about the two triangles in Fig. 2, to use SAS Ineq. theorem. And remember, SAS Ineq. tells us which of the sides opposite the unequal angles is larger and which is smaller:

Since $\angle CGF < \angle FGH$, $CF < FH$. √

Now You Try It

For problems 1-3, use the figure at left and state what you can conclude:

1. If $\angle NPO = 93°$, and P is the midpoint of $\overline{OM}$.

2. If $\overline{NP}$ is a segment bisector of $\overline{OM}$ and $\angle MPN = 91°$.

3. If $\overline{NO} \cong \overline{NM}$ and $OP < PM$.

Prob. 1-3

4. Given: $WV < XY$ and $\overline{WX} \cong \overline{VY}$.
Prove: $\angle WXV < \angle XYZ$.

Prob. 4

9. SIMILAR FIGURES

An understanding of ratios and the properties of proportions is required for the study of similar figures.

Ratios

We create a ratio when we compare two numbers using division. For example: $1 \div 4$

Although you might not realize it, you think in terms of ratios everyday. For example, when you notice that for every \$4 you have, you are spending \$1 on fast food, you are realizing that the cost ratio of fast food to all of your purchases is 1 to 4, that is, you are spending one fourth of your money on fast food.

The ways in which your text will indicate a ratio are:

$1 \div 4$ $\dfrac{1}{4}$ $1/4$ "one to four"   "one is to four" $1:4$ $\Big\{$ These all mean a ratio of 1 to 4

Usually ratios are written as fractions and reduced to "simplest form" which means that we have removed all common factors:

$$\frac{\overset{1}{\cancel{4}}}{\underset{4}{\cancel{16}}} = \frac{1}{4}$$

A ratio can have *more than two* terms. When this is the case, we use the "colon" form, for example:

1:2:7 which means the three ratios, 1:2, 2:7 and 1:7

Proportions

If two ratios are equal, we can create an equation called a *proportion*: $\dfrac{a}{b} = \dfrac{c}{d}$ $\dfrac{1}{2} = \dfrac{2}{4}$ $\dfrac{2}{3} = \dfrac{4}{6}$

We have special vocabulary for the terms of a proportion:

$$\frac{\text{extreme}}{\text{mean}} = \frac{\text{mean}}{\text{extreme}} \qquad \overset{extreme}{\underset{mean}{\frac{a}{b}}} = \overset{mean}{\underset{extreme}{\frac{c}{d}}} \qquad \overset{extreme}{\underset{mean}{\frac{1}{2}}} = \overset{mean}{\underset{extreme}{\frac{2}{4}}} \qquad \overset{extreme}{\underset{mean}{\frac{2}{3}}} = \overset{mean}{\underset{extreme}{\frac{4}{6}}}$$

You can do lots of interesting things with proportions, some of which will surprise you.

Note: For each transformation, we've gone back and started out with the boxed given form of the proportion.

Given:	$\boxed{\dfrac{a}{b} = \dfrac{c}{d}}$	$\boxed{\dfrac{1}{2} = \dfrac{3}{6}}$	$\boxed{\dfrac{2}{3} = \dfrac{4}{6}}$
Means Extremes Property:	$ad = bc$	$1(6) = 2(3)$	$2(6) = 3(4)$
Flip (invert) the ratios:	$\dfrac{b}{a} = \dfrac{d}{c}$	$\dfrac{2}{1} = \dfrac{6}{3}$	$\dfrac{3}{2} = \dfrac{6}{4}$
Swap the means:	$\dfrac{a}{c} = \dfrac{b}{d}$	$\dfrac{1}{3} = \dfrac{2}{6}$	$\dfrac{2}{4} = \dfrac{3}{6}$
Swap the extremes:	$\dfrac{d}{b} = \dfrac{c}{a}$	$\dfrac{6}{2} = \dfrac{3}{1}$	$\dfrac{6}{3} = \dfrac{4}{2}$
Bring up the denominator:	$\dfrac{a+b}{b} = \dfrac{c+d}{d}$	$\dfrac{1+2}{2} = \dfrac{3+6}{6}$	$\dfrac{2+3}{3} = \dfrac{4+6}{6}$
Add them up:	$\dfrac{a+c}{b+d} = \dfrac{a}{b}$	$\dfrac{1+3}{2+6} = \dfrac{4}{8} = \dfrac{1}{2}$	$\dfrac{2+4}{3+6} = \dfrac{6}{9} = \dfrac{2}{3}$

You learned this originally as "Cross Multiplying".

In problems in which you are asked to decide if proportions are equivalent, compare the given proportion to the original, and figure out what was changed. Then ask yourself if the properties of proportion were followed.

Example: If $\boxed{\dfrac{w}{x} = \dfrac{y}{z}}$ ⟵ The original proportion. which of the following are equivalent proportions?

a.) $\dfrac{w}{z} = \dfrac{y}{x}$ What happened? The denominators got swapped. None of the properties of proportion include swapping denominators. Conclusion: **No**, this is not an equivalent proportion.

b.) $\dfrac{w}{y} = \dfrac{x}{z}$ What happened? (Remember to go back and compare to the original proportion.) The means got swapped. Swapping the means *is* a property of proportions. Conclusion: **Yes**, this is an equivalent proportion.

Here's another type of proportion problem:

If $\dfrac{w}{x} = \dfrac{y}{z}$ then $\dfrac{w+x}{x} = ?$ What happened? The denominator of the left side was "brought up" so you have to do the same thing to the right side. $\dfrac{w+x}{x} = \dfrac{y+z}{z}$ ✓

Now You Try It

1. Name the 6 properties of proportion and give an example of each.

(1)   (2)

(3)   (4)

(5)   (6)

If one ratio changed and the other ratio stayed the same, the answer is: **No**, it is not an equivalent propotion.

2. If $\dfrac{g}{h} = \dfrac{i}{j}$ which of the following are equivalent proportions?

a.) $\dfrac{j}{h} = \dfrac{i}{g}$ b.) $\dfrac{h}{g} = \dfrac{j}{i}$ c.) $\dfrac{g}{h} = \dfrac{j}{i}$

3. If $\dfrac{w}{x} = \dfrac{y}{z}$ then $\dfrac{z}{y} = ?$

4.) If $\dfrac{a}{b} = \dfrac{c}{d}$ then, $\dfrac{b}{b+a} = ?$

Similarity

Set Designers for rock shows spend a lot of time and money drawing and building scale models before construction of the actual set is ever begun. Models help the designers make sure that the set will have more than enough room for the artists to perform as well as room for the correct placement of the acoustical, lighting and special effects equipment.

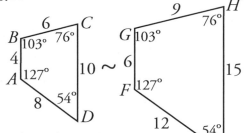

Scale models are smaller but are otherwise exact copies of the full sized project which they represent. Since the measurements of every part of the model are "in proportion" to every corresponding measurement of the real project, the designers can improve and adjust the model and know that when they build the actual set, it will look and work as planned.

Scale models and their life size counterparts are perfect examples of similar objects.

Similar objects have the same shape but are not necessarily the same size. What is it that gives objects the same shape? The answer to this question leads to the definition of similar objects:

DEFINITION OF SIMILARITY: *1. Corresponding angles are congruent and*
2. Corresponding sides are "in proportion".

The symbol for similarity is: ~

Polygon *ABCD* is similar to polygon *FGHI*.

In symbols write: *ABCD ~ FGHI*

Similar Objects have
Exactly the Same Shape!

Note that it is *not* the segments and *not* the angles that are similar. A^tip
It is the *pair of figures* that are similar to *each other*.

A Close Look:	The *figures* are *similar* ◁~◁ when:
3 Ideas in 1 Definition	Corresponding *angles* are *congruent, and*
	Corresponding *sides* are *in proportion*.

What you Need to Know about Similarity Problems

1. In Proportion:

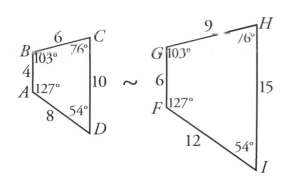

The corresponding sides of similar figures are *in proportion* which means the ratios of the lengths of corresponding sides are the same. Each ratio reduces to the same fraction.

$$\frac{4}{6} = \frac{6}{9} = \frac{10}{15} = \frac{8}{12} = \frac{2}{3}$$

2. Scale Factor:

— The *scale factor* of two similar figures is the (reduced) ratio of the corresponding sides.

— The scale factor of the figures above is: $\frac{2}{3}$ or 2:3 (say "2 to 3" when reading either one).

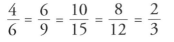**tip** Once you've figured out the scale factor, write it on the figures:

— Remember, only figures that are similar have a scale factor.

3. The Order of the Scale Factor — "Small to Big" or "Big to Small":

— In the example above, the first figure (read left to right or up to down) was the smaller figure and the second figure, the bigger; therefore, the ratio for this particular problem is "small to big", so the scale factor is "small to big" that is $\frac{2}{3}$ or 2:3.

Note: If you say $\frac{3}{2}$ or "3 to 2" it's wrong! $\cancel{\frac{3}{2}}$ $\frac{2}{3}$ ✓

— When you first start the problem, notice, is it "small to big" or "big to small", and then, follow the same order in your answer.

4. The Order of the Similarity — Order Counts:

$ABCD \sim FGHI$ Can you tell what information this similarity is giving?

Four angle congruencies: $\angle A \cong \angle F$ $\angle B \cong \angle G$ $\angle C \cong \angle H$ $\angle D \cong \angle I$.

&

An extended proportion: $\frac{AB}{FG} = \frac{BC}{GH} = \frac{CD}{HI} = \frac{DA}{IF}$

New Term

Segments Divided Proportionally
This means 2 or more segments that are divided so that the ratio of the lengths of their parts is the same.

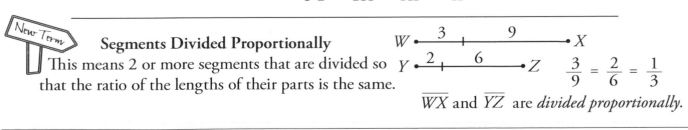

$\frac{3}{9} = \frac{2}{6} = \frac{1}{3}$

$\overline{WX}$ and $\overline{YZ}$ are *divided proportionally.*

Find the Scale Factor — Solve the Problem

Example 1. Given $QRST \sim MNOP$ find x, y, and z.

Step 1. In this problem, the bigger figure was given first, so the order of the ratio will be big to small, that is, $\dfrac{\text{big}}{\text{small}}$.

Step 2. Study the pair of figures and find 2 *corresponding sides for which lengths are given*. In this case, $\overline{RS}$ and $\overline{NO}$.

Step 3. Using the given lengths, $RS = 8$ and $NO = 4$, find the *scale factor* of the two figures. That means the ratio of the corresponding sides, being careful to keep the correct order (big to small):

$$\frac{RS}{NO} = \frac{8}{4} = \frac{2}{1}$$

Step 4. Because the figures are similar, the scale factor equals the ratio of each (and every) pair of corresponding sides. Therefore, we can use the scale factor to solve for all of the missing lengths:

Think "big is to small as 14 is to x".
$$\frac{2}{1} = \frac{14}{x}$$
Cross multiply: $2(x) = (1)(14)$
$$\frac{\cancel{2}x}{\cancel{2}} = \frac{\cancel{14}^{7}}{\cancel{2}}$$
$$x = 7\ \checkmark$$

Think "big is to small as y is to 5".
$$\frac{2}{1} = \frac{y}{5}$$
Cross multiply: $2(5) = (1)(y)$
$$10 = y\ \checkmark$$

Think "big is to small as z is to 3".
$$\frac{2}{1} = \frac{z}{3}$$
Cross multiply: $2(3) = (1)(z)$
$$6 = z\ \checkmark$$

In the next example, the key to the solution is knowing that corresponding sides are in proportion.

Example 2. Given $JKLM \sim NOPQ$ find OP.

Since the two figures are similar, we are able to form the following proportion:

$$\frac{JK}{NO} = \frac{KL}{OP}$$

$$\frac{3x-3}{x+3} = \frac{3x+3}{2x+2}$$

Now cross multiply:
$(3x - 3)(2x + 2) = (x+3)(3x+3)$
$6x^2 - 6 = 3x^2 + 12x + 9$
$3x^2 - 12x - 15 = 0$ Now factor.
$(3x + 3)(x - 5) = 0$

This would make KJ
negative! (& OP & $KL = 0$)

$(3x + 3) = 0$,
$x = \cancel{-1}$

$(x - 5) = 0$
$x = 5$, so $OP = 2(5)+2 = 12\ \checkmark$

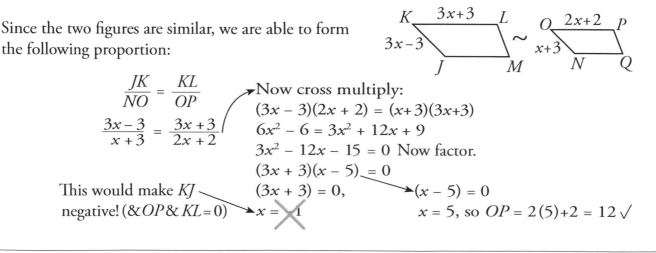

104

Example 3. Given *ABCD ~ EFGH* find *x, y* and *z*.

This example is similar to Example 1, (the scale factor is the key to solving) but these two figures are oriented differently. Therefore, it's a good idea to re-sketch one of them.

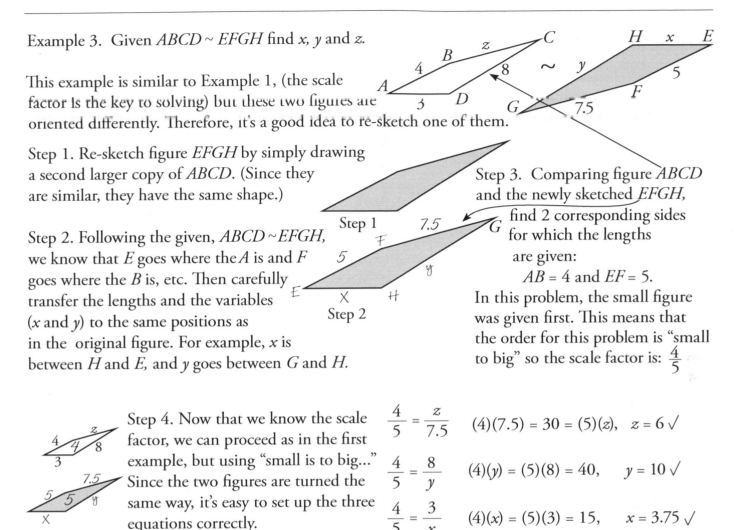

Step 1. Re-sketch figure *EFGH* by simply drawing a second larger copy of *ABCD*. (Since they are similar, they have the same shape.)

Step 2. Following the given, *ABCD ~ EFGH*, we know that *E* goes where the *A* is and *F* goes where the *B* is, etc. Then carefully transfer the lengths and the variables (*x* and *y*) to the same positions as in the original figure. For example, *x* is between *H* and *E*, and *y* goes between *G* and *H*.

Step 3. Comparing figure *ABCD* and the newly sketched *EFGH*, find 2 corresponding sides for which the lengths are given:
$$AB = 4 \text{ and } EF = 5.$$
In this problem, the small figure was given first. This means that the order for this problem is "small to big" so the scale factor is: $\frac{4}{5}$

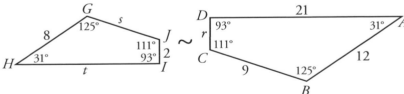

Step 4. Now that we know the scale factor, we can proceed as in the first example, but using "small is to big..." Since the two figures are turned the same way, it's easy to set up the three equations correctly.

$\frac{4}{5} = \frac{z}{7.5}$ $(4)(7.5) = 30 = (5)(z),\; z = 6 \checkmark$

$\frac{4}{5} = \frac{8}{y}$ $(4)(y) = (5)(8) = 40,\quad y = 10 \checkmark$

$\frac{4}{5} = \frac{3}{x}$ $(4)(x) = (5)(3) = 15,\quad x = 3.75 \checkmark$

Now You Try It

1. Write four congruencies and the extended proportion given by this similarity: *MNOP ~ QRST.*

2. The two quadrilaterals shown below are similar.

Remember to re-sketch figure *GHIJ* by simply drawing a second smaller copy of *ABCD*.

a. Name two similar quadrilaterals.

b. *CBAD* is similar to what quadrilateral?

c. What is the scale factor of the two figures?

d. Find *r, s,* and *t.*

Proving Triangles are Similar — 3 Shortcuts

How do we know if two triangles are similar? One way is to use the definition of similar polygons; we can check to see if corresponding angles are equal and corresponding sides are in proportion. But that means finding 12 measurements. Here are three "shortcuts" that you can use to find out if two triangles are similar. Remember, there aren't any other shortcuts. Besides finding and comparing all 12 measurements, these are the only ways to show that two triangles are similar:

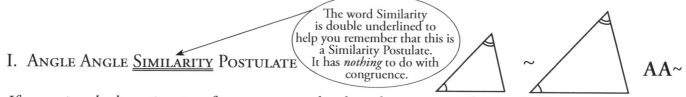

I. ANGLE ANGLE <u><u>SIMILARITY</u></u> POSTULATE

The word Similarity is double underlined to help you remember that this is a Similarity Postulate. It has *nothing* to do with congruence.

AA~

If two triangles have 2 pairs of congruent angles then the two triangles are similar.
Notice that by the definition of similarity, this means that each pair of corresponding sides of the two triangles are *in proportion*.

Remember, each A in AA~ stands for one *pair* of congruent corresponding angles, $\overset{\cong}{\mathbf{A}}\overset{\cong}{\mathbf{A}}$~.

Here's an example of how to use AA~.

Given $\triangle MNO$ and $\triangle ABC$, name a similarity and find n and c.

Starting with $\triangle MNO$, we are given two of the angles so we should solve for the third: $180° - 80° - 40° = 60°$. With this information, we see that the two triangles have two pairs of congruent angles which means they are similar by the AA~ Postulate.

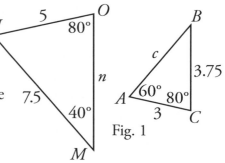

Fig. 1

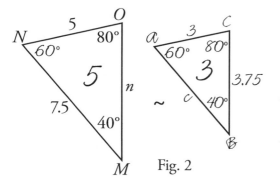

Fig. 2

Here's how to re-sketch $\triangle ABC$ so that it's oriented like $\triangle MNO$:
1. Draw a smaller copy of $\triangle MNO$. (Since the two triangles are similar they have the same shape, but $\triangle ABC$ is smaller.)
2. First include the angle measures (copy them from $\triangle MNO$) and then the vertex letters and finally the lengths.

Now it's easy to write down or "name" a similarity. Here's one correct answer: $\triangle MNO \sim \triangle BAC$.

Studying the two figures, find the two corresponding sides for which lengths are given, $\overline{NO}$ and $\overline{AC}$. Now form the scale factor of the two figures $\frac{5}{3}$, noting that the order is "big to small" and writing the scale factor on the figures. Since the scale factor equals the ratio of each pair of corresponding sides, we can use the scale factor to create correct proportions and solve for the missing lengths:

$$\frac{5}{3} = \frac{7.5}{c}, \ (5)(c) = (3)(7.5) = 22.5, \ c = 4.5 \checkmark \quad \text{and} \quad \frac{5}{3} = \frac{n}{3.75}, \ (5)(3.75) = 18.75 = 3n, \ n = 6.25 \checkmark$$

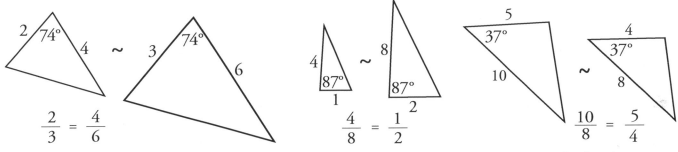

II. Side Angle Side Similarity Theorem

Included

If two triangles have 2 pairs of sides which are in proportion and the included angles are congruent, then the two triangles are similar.

$$\frac{a}{d} = \frac{b}{e}, \quad \angle C = \angle F, \quad \triangle BCA \sim \triangle EFD$$

Notice that each S in SAS~ stands for a *ratio* of two corresponding sides which must equal the *ratio* of a second pair (the second S) of corresponding sides.

$$\overset{ratio \ ratio}{\mathbf{SAS}} \sim \atop \underset{\angle \cong \angle}{}$$

Here are some examples of pairs of triangles that are similar by SAS~.

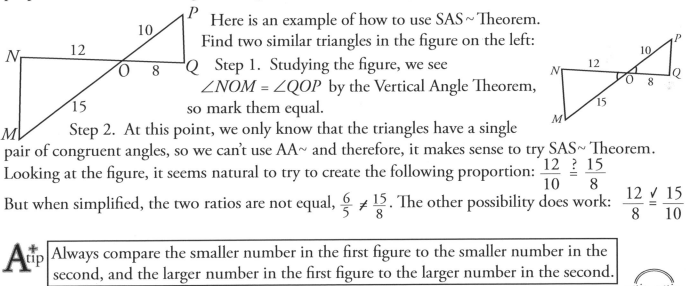

$$\frac{2}{3} = \frac{4}{6} \qquad\qquad \frac{4}{8} = \frac{1}{2} \qquad\qquad \frac{10}{8} = \frac{5}{4}$$

The congruent angles must be the ones that are "wedged in between" the two pairs of sides that are in proportion. That is, the congruent angles are the *included* angles.

Here is an example of how to use SAS~ Theorem. Find two similar triangles in the figure on the left:

Step 1. Studying the figure, we see $\angle NOM = \angle QOP$ by the Vertical Angle Theorem, so mark them equal.

Step 2. At this point, we only know that the triangles have a single pair of congruent angles, so we can't use AA~ and therefore, it makes sense to try SAS~ Theorem. Looking at the figure, it seems natural to try to create the following proportion: $\frac{12}{10} \overset{?}{=} \frac{15}{8}$

But when simplified, the two ratios are not equal, $\frac{6}{5} \neq \frac{15}{8}$. The other possibility does work: $\frac{12}{8} \overset{\checkmark}{=} \frac{15}{10}$

$$\mathbf{A^+_{tip}} \boxed{\text{Always compare the smaller number in the first figure to the smaller number in the second, and the larger number in the first figure to the larger number in the second.}}$$

Since the ratios are equal (which means they form a proportion), the two triangles are similar.

$$\overset{ratio \ ratio}{\mathbf{SAS}} \sim \atop \underset{\angle \cong \angle}{}$$

Step 3. To write down or "name" the similarity, follow the *order* of the correct proportion,

$$\frac{12}{8} = \frac{15}{10} \quad \text{so} \quad \frac{NO}{QO} = \frac{MO}{PO} \quad \text{and} \quad \triangle NOM \sim \triangle QOP.$$

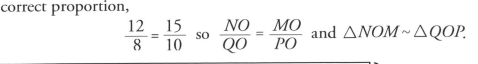

$$\mathbf{A^+_{tip}} \boxed{\text{When you use SAS~ the vertices of the included angles will appear in the names of the segments that form the proportion.}} \Big\} \begin{array}{l} \text{Included } \angle O, \text{ segments:} \\ \overline{NO} \ \overline{QO} \ \overline{MO} \ \overline{PO} \end{array}$$

III. Side Side Side <u>Similarity</u> Theorem

If two triangles have 3 pairs of sides which are in proportion, then the two triangles are similar.

$$\frac{AB}{DE} = \frac{BC}{EF} = \frac{CA}{FD}$$

SSS~

Notice that each S in SSS~ stands for a *ratio* of two corresponding sides: $\overset{ratio=ratio=ratio}{\textbf{S S S}\sim}$

Remember, **all 3** of the ratios must be equal!

The scale factor of the two triangles is the ratio of a pair of corresponding sides, reduced to its lowest terms. Here are two examples of pairs of triangles that are similar by SSS~.

$$\frac{15}{9} = \frac{17.5}{10.5} = \frac{20}{12}$$

$$\text{scale factor} = \frac{5}{3}$$

$$\frac{6}{9} = \frac{10}{15} = \frac{12}{18}$$

$$\text{scale factor} = \frac{2}{3}$$

To test if triangles are similar by SSS~ form the following ratios and reduce. **All 3** must be equal:

$$\frac{\text{smallest side } \triangle 1}{\text{smallest side } \triangle 2} \overset{?}{=} \frac{\text{medium side } \triangle 1}{\text{medium side } \triangle 2} \overset{?}{=} \frac{\text{longest side } \triangle 1}{\text{longest side } \triangle 2}$$

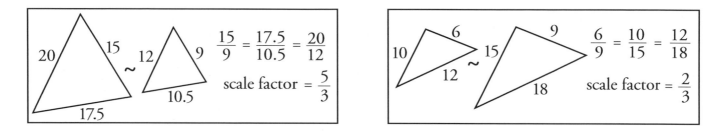

Example.

1. Are the two triangles at right similar?

Form the ratios and test to see if they form a proportion (i.e. are the ratios equal?) $\frac{6}{8} \overset{?}{=} \frac{15}{20} \overset{?}{=} \frac{21}{28}$

Since each ratio reduces to $\frac{3}{4}$, the sides are proportional and the triangles are similar by $\overset{ratio=ratio=ratio}{\textbf{S S S}}\sim$

2. Write down the similarity in the problem above.

Re-sketch the second triangle, orienting it (turning it) like the first. To do this, simply draw a larger copy of the first triangle, then correctly transfer the lengths and then the letters to the new sketch.

Now it's easy to name the similarity. Start with one triangle, choose a starting point and path; then choose the corresponding starting point and path in the second triangle. One correct answer is: $\triangle ACB \sim \triangle FED$

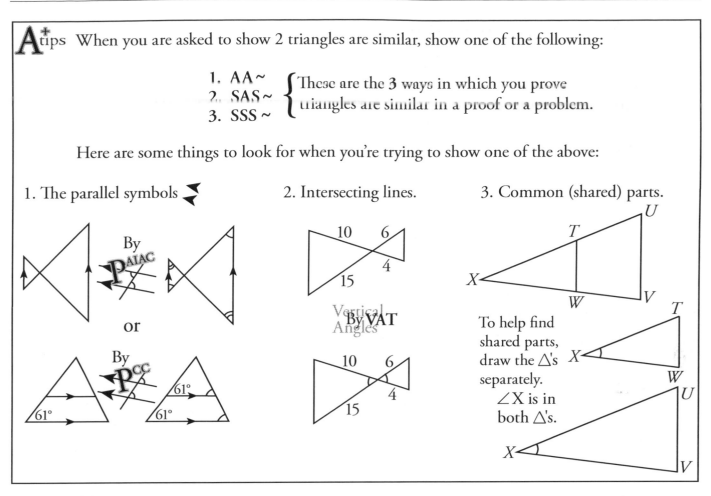

A^{+}tips When you are asked to show 2 triangles are similar, show one of the following:

    1. AA ~   ⎰ These are the **3** ways in which you prove
    2. SAS ~   ⎱ triangles are similar in a proof or a problem.
    3. SSS ~

Here are some things to look for when you're trying to show one of the above:

1. The parallel symbols

2. Intersecting lines.

3. Common (shared) parts.

By PAIAC

or

By PCC

By VAT

Vertical Angles

To help find shared parts, draw the △'s separately.
∠X is in both △'s.

Now You Try It

In problems 1–4, decide if the two triangles are similar. If they are similar, explain why and name the similarity. If not, explain why not.

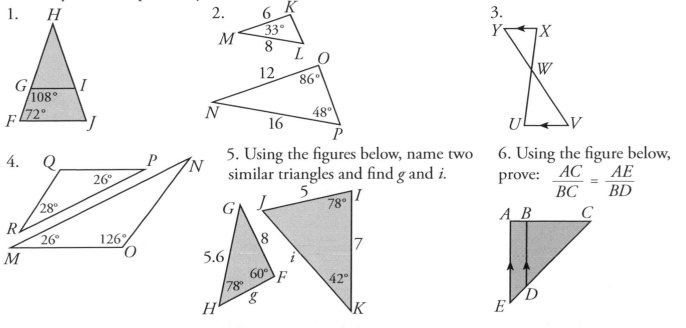

1.

2.

3.

4.

5. Using the figures below, name two similar triangles and find g and i.

6. Using the figure below, prove: $\dfrac{AC}{BC} = \dfrac{AE}{BD}$

Similarity Versus Congruence— Know the Difference

Similarity and Congruency Theory sound the same....but they are completely **different!**

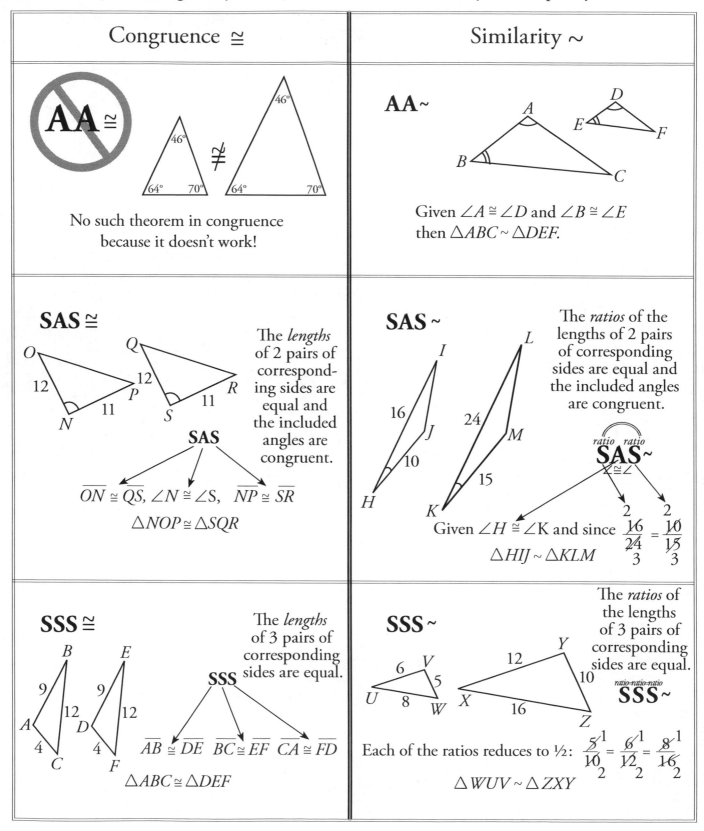

Congruence ≅	Similarity ~
AA ≅ (crossed out) No such theorem in congruence because it doesn't work!	**AA ~** Given ∠A ≅ ∠D and ∠B ≅ ∠E then △ABC ~ △DEF.
SAS ≅ The *lengths* of 2 pairs of corresponding sides are equal and the included angles are congruent. $\overline{ON} \cong \overline{QS}$, ∠N ≅ ∠S, $\overline{NP} \cong \overline{SR}$ △NOP ≅ △SQR	**SAS ~** The *ratios* of the lengths of 2 pairs of corresponding sides are equal and the included angles are congruent. Given ∠H ≅ ∠K and since $\frac{\cancel{16}^2}{\cancel{24}_3} = \frac{\cancel{10}^2}{\cancel{15}_3}$ △HIJ ~ △KLM
SSS ≅ The *lengths* of 3 pairs of corresponding sides are equal. $\overline{AB} \cong \overline{DE}$ $\overline{BC} \cong \overline{EF}$ $\overline{CA} \cong \overline{FD}$ △ABC ≅ △DEF	**SSS ~** The *ratios* of the lengths of 3 pairs of corresponding sides are equal. Each of the ratios reduces to ½: $\frac{\cancel{5}^1}{\cancel{10}_2} = \frac{\cancel{6}^1}{\cancel{12}_2} = \frac{\cancel{8}^1}{\cancel{16}_2}$ △WUV ~ △ZXY

Getting Ready for Right Triangles – Simplifying Radical Expressions

> Check with your teacher to see if you are required to simplify radicals.

What You Do	A^+ tips **Why You Do It**
Example 1. $(4)(5) = 20$	Number times the number.

Example 2. $(\sqrt{3})(\sqrt{6}) = \sqrt{18}$

        Radical times the radical. $(\sqrt{x})(\sqrt{y}) = \sqrt{xy})$
Since 18 (the number under the radical sign) is not prime, break it down into its factors.

$\qquad = \sqrt{(2)(3)(3)}$

Study the factors. Two of a kind? Yes.

$\qquad = \sqrt{(2)(3)(3)}$

$\qquad = 3\sqrt{(2)(3)(3)}$

Cross out both twins, *one* escapes outside as a *factor*.

$\qquad = 3\sqrt{2}$

Done!

Example 3. $(4\sqrt{3})(5) = 20\sqrt{3}$

Number times the number. Since 3, the number under the radical sign, can't be broken down into smaller factors, we're done!

Example 4. $(4)(5\sqrt{6}) = 20\sqrt{6}$

Number times the number.

$\qquad = 20\sqrt{(2)(3)}$

Since 6, the number under the radical sign, is not prime, break it down into its factors. Since no factor appears two times (no twins), the radical can't be reduced.

$\qquad = 20\sqrt{6}$

Re-multiply the "orphans" together. We're done!

Example 5. $(4\sqrt{3})(5\sqrt{6}) = 20\sqrt{18}$

Number times the number, radical times the radical.

$\qquad = 20(3)\sqrt{(2)(3)(3)}$

Now reduce 18, the number under the radical sign. (See Example 2, above.)

$\qquad = 20(3)\sqrt{2} = 60\sqrt{2}$

Done! Be sure to notice that 3 is a factor (so multiply).

Example 6. $(4)(2\sqrt{54}) = 8\sqrt{54}$

Number times the number.

$\qquad = 8\sqrt{(2)(3)(3)(3)}$

Since 54 (the number under the radical sign) is not prime, break it down into its factors.

$\qquad = 8\sqrt{(2)(3)(3)(3)}$

Study the factors. Two of a kind? Yes! Note that we are looking for *pairs* of factors (twins). The third 3 is an orphan.

$\qquad = (8)(3)\sqrt{(2)(3)(3)(3)}$

Cross out both twins, one escapes outside as a factor.

$\qquad = 24\sqrt{6}$

Multiply the factors together, re-multiply the "orphans". Done!

★*Important Mathematics Note:* The rules on these two pages work for positive numbers only. In second year algebra, you will learn different rules for negative numbers under the radical sign.

Radical Rules — You are not finished with a radical until you have obeyed these **3** Rules:

Rule 1. No perfect square factors (that is, equal factors,) left under the radical sign.

We were working on this rule on the previous page.
Example: $\sqrt{32}$ is not finished. Why not?
Because $\sqrt{32} = \sqrt{(2)(16)}$ and 16 is a perfect square.
$$\sqrt{32} = \sqrt{(2)(16)} = \sqrt{(2)(4)(4)} = 4\sqrt{2} \text{ Done!}$$

Or, break 4 down: $\sqrt{32} = \sqrt{(2)(2)(2)(2)(2)} = (2)(2)\sqrt{2} = 4\sqrt{2}$ Either method works.

Rule 2. No radicals in the denominator.

Example: $\dfrac{1}{2\sqrt{5}}$ is not finished. Why not?

Because $\dfrac{1}{2\sqrt{5}}$ has a radical, $\sqrt{5}$, in the denominator.

Any radical times itself equals the number under the radical. $(\sqrt{x})(\sqrt{x}) = x)$.
For example, $(\sqrt{5})(\sqrt{5}) = 5$. Use this idea to "rationalize the denominator":

Rationalizing the denominator
$$\frac{1}{2\sqrt{5}} = \frac{1}{2\sqrt{5}}\left(\frac{\sqrt{5}}{\sqrt{5}}\right) = \frac{1\sqrt{5}}{2\,(5)} = \frac{\sqrt{5}}{10} \quad \text{Done! (It's okay to have a radical in the numerator.)}$$

Note: A number divided by itself equals 1; this is why we can multiply by $\left(\dfrac{\sqrt{5}}{\sqrt{5}}\right)$

Rule 3. No fractions under the radical sign.

Example: $\sqrt{\dfrac{3}{15}}$ is not finished. Why not?

Because $\sqrt{\dfrac{3}{15}}$ is a fraction under a radical sign.

A fraction under a radical sign can be broken up into two parts, the numerator and the denominator:
$$\sqrt{\frac{3}{15}} = \frac{\sqrt{3}}{\sqrt{15}} \quad \text{Now rationalize the denominator.}$$

Rationalizing the denominator
$$\frac{\sqrt{3}}{\sqrt{15}}\left(\frac{\sqrt{15}}{\sqrt{15}}\right) = \frac{\sqrt{45}}{15} = \frac{\sqrt{(5)(9)(3)}}{15} = \frac{3\sqrt{5}}{15} = \frac{\sqrt{5}}{5} \quad \text{Done! (It's okay to have a radical in the numerator.)}$$

Radical Talk — "Radical 2", "Rad 2", "Root 2", "The square root of 2" all mean the same thing: $\sqrt{2}$

Now You Try It — Complete the indicated operation and if necessary simplify (finish) the
following radical expressions.

1. $(3)(2\sqrt{27})$ 　　　 2. $(4\sqrt{8})(2\sqrt{6})$ 　　　 3. $\dfrac{3}{4\sqrt{3}}$ 　　　 4. $\sqrt{\dfrac{4}{15}}$

Geometric Mean

In order to study right triangles, you need to know about a special number which lies between two positive numbers called the geometric mean.

For example, 8 is the geometric mean between 4 and 16.

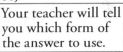

The geometric mean is easy to find, let x = the geometric mean between 4 and 16, then form the proportion:

$$\frac{4}{x} = \frac{x}{16}$$ Now cross-multiply, $x^2 = 64$, and solve: $x = 8$.

$\overset{+}{tip}$ The **Geometric Mean** always appears **2 times** in the equation! 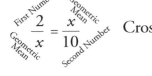 $\frac{4}{x} = \frac{x}{16}$ Learn the pattern.

Examples: 1.) Find the geometric mean between 2 and 10.

 $\frac{2}{x} = \frac{x}{10}$ Cross multiplying $x^2 = (2)(10)$, taking the square root,
$\sqrt{x^2} = \sqrt{20}$ so $x = 2\sqrt{5} \approx 4.47$.
Note: "$\approx$" means "is approximately".

> Your teacher will tell you which form of the answer to use.

2.) Find x if 9 is the geometric mean between x and 27.

$\frac{x}{9} = \frac{9}{27}$ Cross multiplying, $27x = (9)(9) = 81$.
Dividing both sides by 27, $x = 3$.

Quick Review — In a right triangle, the side opposite the right angle is called the *hypotenuse*. The hypotenuse is always the longest side of the triangle.

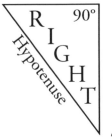

> ALTITUDE ⟶
> PERPENDICULAR

An *altitude* of a triangle is the *perpendicular* line segment that goes from a vertex to the line that includes the opposite side.

Here are the three altitudes of a right triangle:

The altitude we are interested in is ②, the perpendicular segment *from the right angle to the hypotenuse.* This altitude divides the triangle in interesting ways:

THEOREM: *When the altitude is drawn from the right angle to the hypotenuse of a right triangle, the two triangles that are formed are similar to the original triangle and similar to each other.*

$\triangle MNO \sim \triangle MPN$

$\triangle MNO \sim \triangle NPO$

$\triangle MPN \sim \triangle NPO$

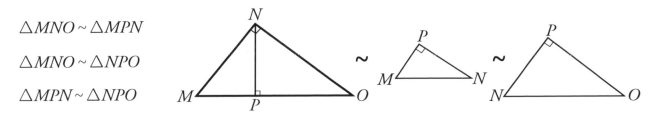

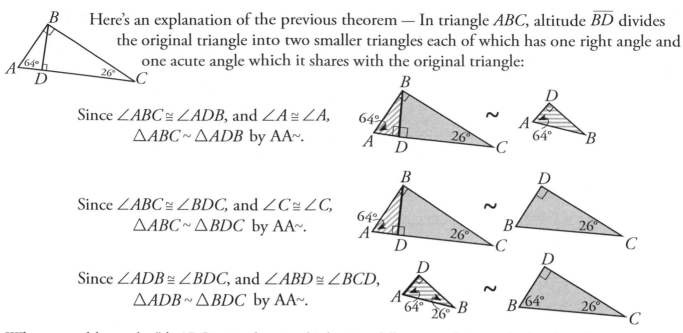

Here's an explanation of the previous theorem — In triangle *ABC*, altitude $\overline{BD}$ divides the original triangle into two smaller triangles each of which has one right angle and one acute angle which it shares with the original triangle:

Since ∠*ABC* ≅ ∠*ADB*, and ∠*A* ≅ ∠*A*, △*ABC* ~ △*ADB* by AA~.

Since ∠*ABC* ≅ ∠*BDC*, and ∠*C* ≅ ∠*C*, △*ABC* ~ △*BDC* by AA~.

Since ∠*ADB* ≅ ∠*BDC*, and ∠*ABD* ≅ ∠*BCD*, △*ADB* ~ △*BDC* by AA~.

When a problem asks "△*ABC* is similar to which triangle" you are being asked to list the vertices of the other triangle in the correct (corresponding) order. For example, using the triangles above, if asked, "△*DBC* is similar to what triangle?" the correct answer would be either △*BAC or* △*DAB*. If asked to "Name 3 similar triangles", one correct answer would be △*ABC* ~ △*ADB* ~ △*BDC*, another would be △*BAC* ~ △*DAB* ~ △*DBC*.

Since similar figures have the same shape, you can draw two smaller versions of the original $\textbf{A}^{+}_{\text{tip}}$ triangle oriented (turned) like the original triangle and carefully place each letter on the correct vertex. Then it's easy to "name three similar triangles."

Now You Try It — Given right △*GHL* with altitude $\overline{HJ}$ drawn from the right angle, draw the 2 smaller triangles oriented in the same way as △*GHL*, carefully label the vertices of the two smaller triangles, and name three pairs of similar triangles. Remember, order counts!

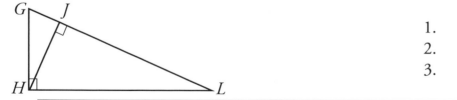

1.

2.

3.

Similar triangles have three pairs of corresponding congruent angles. So it makes sense that:

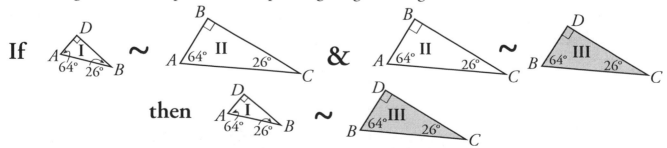

Similarity *transfers*. Similarity is a perfect example of the Transitive Property!

THEOREM: *In a right triangle, the length of the altitude from the right angle to the hypotenuse is the geometric mean between the two segments of the hypotenuse.*

Here's an example:

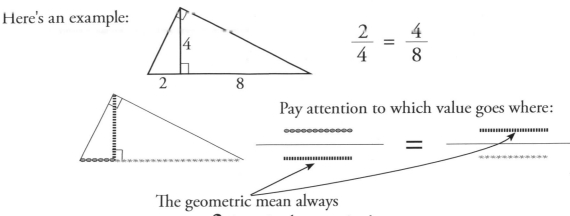

$$\frac{2}{4} = \frac{4}{8}$$

Pay attention to which value goes where:

$$\frac{\text{⊙⊙⊙⊙⊙⊙⊙⊙⊙⊙⊙}}{\text{▪▪▪▪▪▪▪▪▪▪▪}} = \frac{\text{▪▪▪▪▪▪▪▪▪▪▪}}{\text{✳✳✳✳✳✳✳✳✳✳✳}}$$

The geometric mean always appears **2** times in the equation!

Here are some of the ways you can use this theorem:

1. Find the length of altitude $\overline{NP}$.

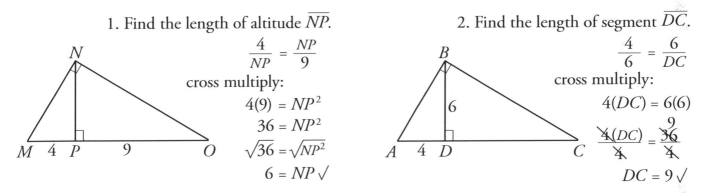

$$\frac{4}{NP} = \frac{NP}{9}$$

cross multiply:

$$4(9) = NP^2$$
$$36 = NP^2$$
$$\sqrt{36} = \sqrt{NP^2}$$
$$6 = NP \checkmark$$

2. Find the length of segment $\overline{DC}$.

$$\frac{4}{6} = \frac{6}{DC}$$

cross multiply:

$$4(DC) = 6(6)$$
$$\frac{\cancel{4}(DC)}{\cancel{4}} = \frac{\cancel{36}^{\,9}}{\cancel{4}}$$
$$DC = 9 \checkmark$$

Pay close attention to:
which part goes where in the equation.

$$\frac{\text{1 Part of Hypot.}}{\text{Altitude to Hypot.}} = \frac{\text{Altitude to Hypot.}}{\text{Other Part of Hypot.}}$$

3. Find x.

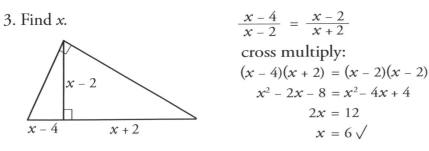

$$\frac{x-4}{x-2} = \frac{x-2}{x+2}$$

cross multiply:

$$(x-4)(x+2) = (x-2)(x-2)$$
$$x^2 - 2x - 8 = x^2 - 4x + 4$$
$$2x = 12$$
$$x = 6 \checkmark$$

And remember, the geometric mean always appears **2** times in the equation!

THEOREM: *In a right triangle, when you draw the altitude from the right angle to the hypotenuse, the length of each leg is the geometric mean between the entire hypotenuse and the part of the hypotenuse nearest the leg.*

This theorem has two parts, the left leg case and the right leg case:

The Left Leg Case

Pay close attention to:
which part goes where in the equation:

$$\frac{\text{Left Part of Hypot.}}{\text{Left Leg}} \quad = \quad \frac{\text{Left Leg}}{\text{(Whole) Hypotenuse}}$$

Example:

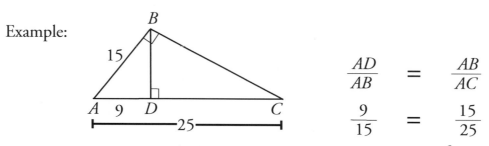

$$\frac{AD}{AB} = \frac{AB}{AC}$$

$$\frac{9}{15} = \frac{15}{25}$$

Since both sides reduce to $\frac{3}{5}$, the fractions are equal.

Here are some of the ways you can use this theorem:

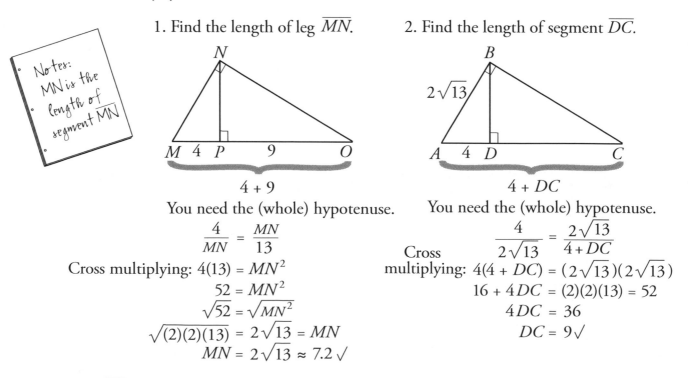

Notes:
MN is the length of segment $\overline{MN}$

1. Find the length of leg $\overline{MN}$.

$4 + 9$

You need the (whole) hypotenuse.

$$\frac{4}{MN} = \frac{MN}{13}$$

Cross multiplying: $4(13) = MN^2$

$$52 = MN^2$$

$$\sqrt{52} = \sqrt{MN^2}$$

$$\sqrt{(2)(2)(13)} = 2\sqrt{13} = MN$$

$$MN = 2\sqrt{13} \approx 7.2 \checkmark$$

2. Find the length of segment $\overline{DC}$.

$4 + DC$

You need the (whole) hypotenuse.

$$\frac{4}{2\sqrt{13}} = \frac{2\sqrt{13}}{4 + DC}$$

Cross multiplying: $4(4 + DC) = (2\sqrt{13})(2\sqrt{13})$

$$16 + 4DC = (2)(2)(13) = 52$$

$$4DC = 36$$

$$DC = 9 \checkmark$$

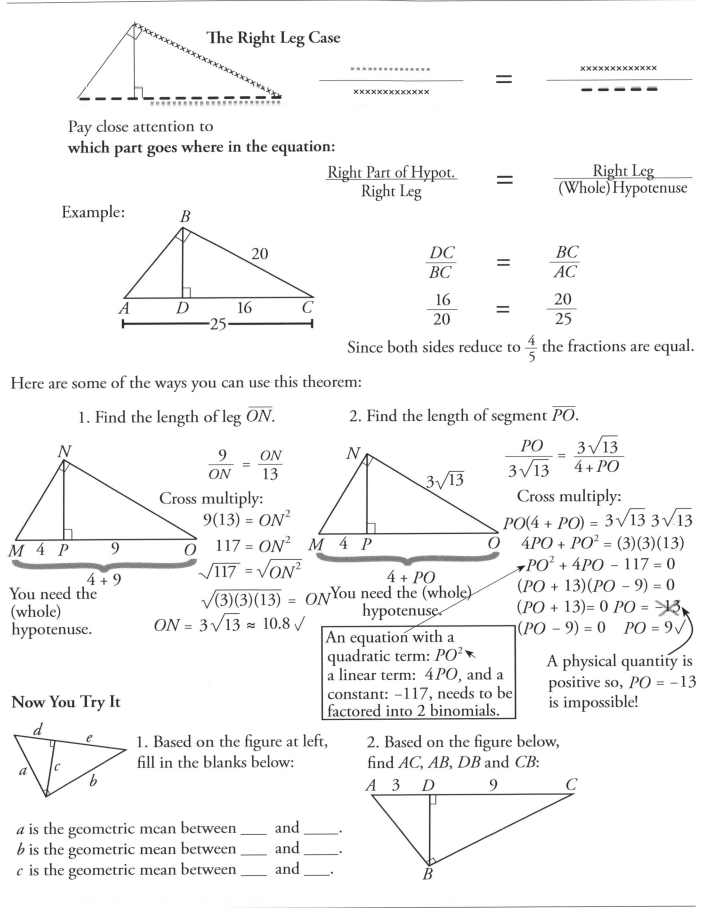

The Right Leg Case

$$\frac{\cdots\cdots\cdots}{\text{xxxxxxxxxxx}} = \frac{\text{xxxxxxxxxxx}}{\text{-----}}$$

Pay close attention to
which part goes where in the equation:

$$\frac{\text{Right Part of Hypot.}}{\text{Right Leg}} = \frac{\text{Right Leg}}{\text{(Whole) Hypotenuse}}$$

Example:

$$\frac{DC}{BC} = \frac{BC}{AC}$$

$$\frac{16}{20} = \frac{20}{25}$$

Since both sides reduce to $\frac{4}{5}$ the fractions are equal.

Here are some of the ways you can use this theorem:

1. Find the length of leg $\overline{ON}$.

$$\frac{9}{ON} = \frac{ON}{13}$$

Cross multiply:
$$9(13) = ON^2$$
$$117 = ON^2$$
$$\sqrt{117} = \sqrt{ON^2}$$
$$\sqrt{(3)(3)(13)} = ON$$
$$ON = 3\sqrt{13} \approx 10.8 \checkmark$$

4 + 9

You need the (whole) hypotenuse.

2. Find the length of segment $\overline{PO}$.

$$\frac{PO}{3\sqrt{13}} = \frac{3\sqrt{13}}{4+PO}$$

Cross multiply:
$$PO(4 + PO) = 3\sqrt{13}\ 3\sqrt{13}$$
$$4PO + PO^2 = (3)(3)(13)$$
$$PO^2 + 4PO - 117 = 0$$
$$(PO + 13)(PO - 9) = 0$$
$$(PO + 13) = 0 \quad PO = -13$$
$$(PO - 9) = 0 \quad PO = 9\checkmark$$

4 + PO

You need the (whole) hypotenuse.

An equation with a quadratic term: PO^2 a linear term: $4PO$, and a constant: -117, needs to be factored into 2 binomials.

A physical quantity is positive so, $PO = -13$ is impossible!

Now You Try It

1. Based on the figure at left, fill in the blanks below:

a is the geometric mean between ___ and ___.
b is the geometric mean between ___ and ___.
c is the geometric mean between ___ and ___.

2. Based on the figure below, find AC, AB, DB and CB:

THE PYTHAGOREAN THEOREM

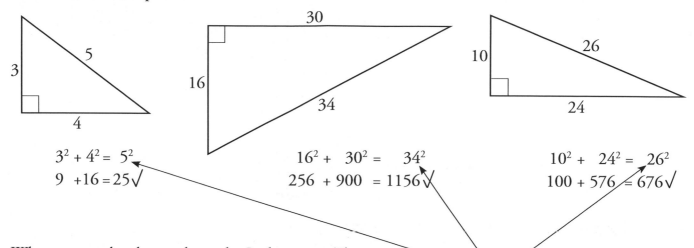

If **Triangle is Right** then $(Leg1)^2 + (Leg2)^2 = (Hypotenuse)^2$

In a right triangle, the square of the hypotenuse is equal to the sum of the squares of the two legs.

Here are some examples:

$3^2 + 4^2 = 5^2$
$9 + 16 = 25 \checkmark$

$16^2 + 30^2 = 34^2$
$256 + 900 = 1156 \checkmark$

$10^2 + 24^2 = 26^2$
$100 + 576 = 676 \checkmark$

What you need to know about the Pythagorean Theorem:

1. The length of the hypotenuse always sits all by itself on one side of the equation.
2. The hypotenuse is opposite the right angle and is always the longest side.
3. This theorem only works for right triangles. (Only right triangles have hypotenuses.)
4. You must square each length individually:

$$(Leg1)^2 + (Leg2)^2 = (Hypotenuse)^2$$

$3 + 4 = 5$ but $3^2 + 4^2 = 5^2$, $9 + 16 = 25 \checkmark$

Example, find y:

$(2y)^2 + y^2 = (2\sqrt{30})^2$ In each case, the entire term is squared.

$2^2 y^2 + y^2 = (2)^2 (\sqrt{30})^2$ The exponent 2 is applied to each member.

$4y^2 + y^2 = (4)(30)$ $(\sqrt{30})^2 = 30$ since $(\sqrt{x})^2 = x$, is true for any number.

$\dfrac{5y^2}{5} = \dfrac{120}{5} \; 24$ Divide both sides by 5, the coefficient of y^2.

$\sqrt{y^2} = \sqrt{24}$ Take the square root of each side.

$y = \sqrt{(2)(2)(2)(3)}$ Break 24 down into its factors looking for two of a kind.

$y = 2\sqrt{(2)(2)(2)(3)}$ Cancel both twins, one escapes as a factor.

$y = 2\sqrt{6} \checkmark$ Re-multiplying the "orphans", 2 and 3 together, equals 6.

> Why didn't we need to factor? Because $y^2 - 24 = 0$ has a quadratic term (y^2) and a constant (-24), but no linear term, for example: $7y$.

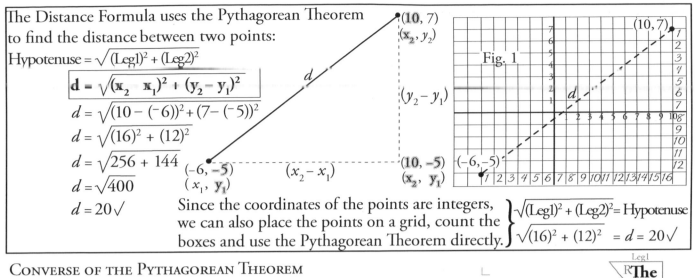

The Distance Formula uses the Pythagorean Theorem to find the distance between two points:

Hypotenuse $= \sqrt{(\text{Leg1})^2 + (\text{Leg2})^2}$

$$\boxed{d = \sqrt{(x_2 - x_1)^2 + (y_2 - y_1)^2}}$$

$d = \sqrt{(10 - (^-6))^2 + (7 - (^-5))^2}$

$d = \sqrt{(16)^2 + (12)^2}$

$d = \sqrt{256 + 144}$

$d = \sqrt{400}$

$d = 20\ \checkmark$

Since the coordinates of the points are integers, we can also place the points on a grid, count the boxes and use the Pythagorean Theorem directly.

$\left.\begin{array}{l} \sqrt{(\text{Leg1})^2 + (\text{Leg2})^2} = \text{Hypotenuse} \\ \sqrt{(16)^2 + (12)^2} = d = 20\ \checkmark \end{array}\right\}$

Fig. 1

CONVERSE OF THE PYTHAGOREAN THEOREM

If (Side 1)^2 + (Side 2)^2 = (LongestSide)^2 then Triangle is Right

In a triangle, if the sum of the squares of the two smaller sides is equal to the square of the largest side, the triangle is right.

Use this theorem to decide if a triangle *is* a right triangle. Remember, only right triangles have a hypotenuse. That's why we can't call the longest side the hypotenuse until we're sure that the triangle is right. Remember, always test! It "looks like" or "doesn't look like" a right triangle isn't good enough.

Triangle *QRS* has sides *QR* = 6, *RS* = 10, and *SQ* = 8. Is triangle *QRS* a right triangle?

1. Put the square of the largest number on the right side of a *test* equation. A test equation means we are not sure if the two sides are equal, so we test to see if it *is* an equation.

$\overset{?}{=}\ 10^2$

2. Put the sum of the squares of the two smaller numbers on the left side of the test equation.

$6^2 + 8^2 \overset{?}{=} 10^2$

3. Now do the math and check:

$36 + 64 = 100\ \checkmark$

Since (Side)^2 + (Side)^2 = (LongestSide)^2 we know that *QRS* is a right triangle.

Here are some more examples:

Is triangle *ABC* right?

Does

$9^2 + 12^2 \overset{?}{=} 15$

$81 + 144 = 225\ \checkmark$

So triangle *ABC* must be a right triangle.

Is triangle *DEF* right?

Does

$6^2 + 9^2 \overset{?}{=} 10^2$

$36 + 81 \neq 100$

So triangle *DEF* can't be a right triangle.

Now You Try It

1. Find *z* in the right triangle below.

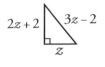

2. Is triangle *PQR* a right triangle? Explain.

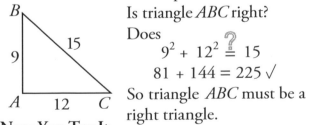

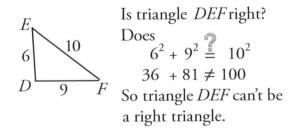

In the Inequality Chapter, we learned that in a triangle, the bigger the angle, the bigger the side opposite the angle. This is true because as the size of an angle increases, the sides of the angle move further apart.

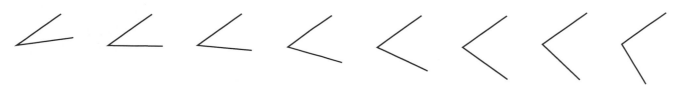

And conversely, the bigger the side opposite an angle, the bigger the angle.

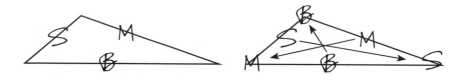

Putting these ideas together with the converse of the Pythagorean Theorem leads to some interesting conclusions about all triangles. Here's an example:

Given triangle *EFG* with sides equal to 4, 6, and 9, what can we conclude? We know that $\angle F$ is the largest angle, followed in size by $\angle G$ and then $\angle E$. $\angle F$ certainly looks obtuse but can we be sure that it is? Let's compare the lengths of the sides using the idea of the converse of the Pythagorean Theorem.

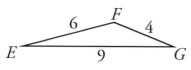

$$4^2 + 6^2 \stackrel{?}{=} 9^2 \quad \text{No,} \quad 4^2 + 6^2 = 52, \quad 9^2 = 81 \text{ so } 4^2 + 6^2 < 9^2$$

What does this mean? If $\angle F$ were a right angle then the length of the side opposite it would have to be $\sqrt{52} \approx 7.2$. But $EG = 9$, which is larger than 7.2. Therefore $\angle F$ must be larger than a right angle; that is, $\angle F$ is obtuse.

THEOREM: $(\text{Side1})^2 + (\text{Side2})^2 < (\text{Longest Side})^2 \longrightarrow$

If the sum of the squares of the lengths of the two smaller sides of a triangle is smaller than the square of the length of the largest side, then the triangle is obtuse.

Here's another example: Is the triangle below a right triangle?

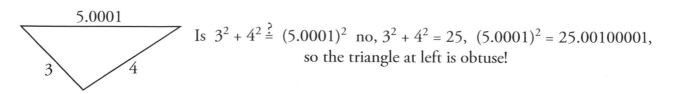

Is $3^2 + 4^2 \stackrel{?}{=} (5.0001)^2$ no, $3^2 + 4^2 = 25$, $(5.0001)^2 = 25.00100001$, so the triangle at left is obtuse!

Now You Try It — Given: $a^2 + b^2 < c^2$ Prove: $\triangle ABC$ is obtuse.

Hints: Construct $\triangle TUV$, with right $\angle V$, legs a and b. Now recall the Triangle Inequality Theorems!

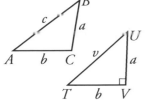

Statements	Reasons

What if the sum of the two smaller sides in a triangle is larger than the square of the largest side?

For example, given triangle HIJ with sides equal to 5, 7, and 8, what can we conclude? We know that $\angle I$ is the largest angle, followed in size by $\angle J$ and then $\angle H$. Let's compare the lengths of the sides using the idea of the converse of the Pythagorean Theorem:

$$5^2 + 7^2 \stackrel{?}{=} 8^2 \quad \text{No,} \quad 5^2 + 7^2 = 74, \quad 8^2 = 64 \quad \text{so} \quad 5^2 + 7^2 > 8^2$$

What does this mean? If $\angle I$ were a right angle then the length of the side opposite it would have to be $\sqrt{74} \approx 8.6$ but $HJ = 8$, which is smaller than 8.6 and therefore, $\angle I$ must be smaller than a right angle, that is, $\angle I$ is acute. But $\angle I$ is the largest angle in the triangle, so if $\angle I$ is acute then the other two (smaller) angles must also be acute, and therefore the triangle is acute.

THEOREM: $(\text{Side1})^2 + (\text{Side2})^2 > (\text{Longest Side})^2$ $\longrightarrow$

If the sum of the squares of the lengths of the two smaller sides of a triangle is larger than the square of the length of the longest side, then the triangle is acute.

Put the previous two theorems together with the converse of the Pythagorean Theorem to decide whether a triangle is acute, right or obtuse, (providing you know the lengths of the sides). Here's how:

1. Put the square of the *longest* length on the right side of a *test* equation.
2. Put the sum of the squares of the two smaller lengths on the left side of the test equation.
3. Do the math and compare:

$$(\text{Side1})^2 + (\text{Side2})^2 \stackrel{?}{\lessgtr} (\text{LargestSide})^2$$

If the left side is smaller, then the triangle is obtuse. $(\text{Side1})^2 + (\text{Side2})^2 < (\text{LongestSide})^2$

If the left side is larger, then the triangle is acute. $(\text{Side1})^2 + (\text{Side2})^2 > (\text{LongestSide})^2$

If the left side equals the right side, then the triangle is right. $(\text{Side1})^2 + (\text{Side2})^2 = (\text{LongestSide})^2$

Here's a memory hint you may find helpful. **<obtuse acute> = right triangle**

Example — Decide whether triangles with sides of the following lengths are acute, right or obtuse. Remember to put the largest number by itself on the right side of a test equation.

a. Triangle *ABC* with sides equal to 6, 12 and 7.

$6^2 + 7^2$? 12^2 $36 + 49 = 85 < 144$

The left side is smaller, triangle *ABC* is obtuse.

b. Triangle *MNO* with sides equal to 8, 9 and 7.

$8^2 + 7^2$? 9^2 $64 + 49 = 113 > 81$

The left side is larger, triangle *MNO* is acute.

Now You Try It — Is triangle *BCD* acute, right or obtuse? Prove it.

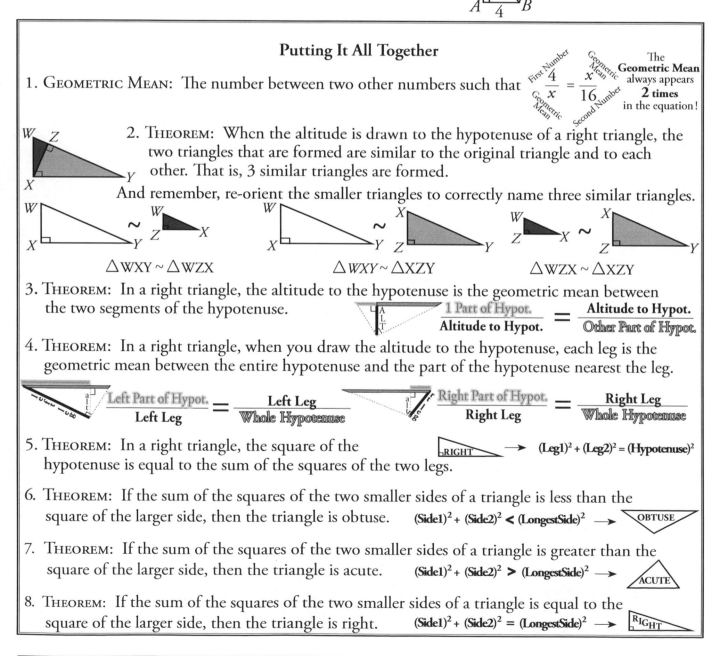

Putting It All Together

1. GEOMETRIC MEAN: The number between two other numbers such that $\frac{4}{x} = \frac{x}{16}$. The **Geometric Mean** always appears **2 times** in the equation!

2. THEOREM: When the altitude is drawn to the hypotenuse of a right triangle, the two triangles that are formed are similar to the original triangle and to each other. That is, 3 similar triangles are formed.

And remember, re-orient the smaller triangles to correctly name three similar triangles.

△WXY ~ △WZX   △WXY ~ △XZY   △WZX ~ △XZY

3. THEOREM: In a right triangle, the altitude to the hypotenuse is the geometric mean between the two segments of the hypotenuse.

$$\frac{\text{1 Part of Hypot.}}{\text{Altitude to Hypot.}} = \frac{\text{Altitude to Hypot.}}{\text{Other Part of Hypot.}}$$

4. THEOREM: In a right triangle, when you draw the altitude to the hypotenuse, each leg is the geometric mean between the entire hypotenuse and the part of the hypotenuse nearest the leg.

$$\frac{\text{Left Part of Hypot.}}{\text{Left Leg}} = \frac{\text{Left Leg}}{\text{Whole Hypotenuse}}$$

$$\frac{\text{Right Part of Hypot.}}{\text{Right Leg}} = \frac{\text{Right Leg}}{\text{Whole Hypotenuse}}$$

5. THEOREM: In a right triangle, the square of the hypotenuse is equal to the sum of the squares of the two legs. $(\text{Leg1})^2 + (\text{Leg2})^2 = (\text{Hypotenuse})^2$

6. THEOREM: If the sum of the squares of the two smaller sides of a triangle is less than the square of the larger side, then the triangle is obtuse. $(\text{Side1})^2 + (\text{Side2})^2 < (\text{LongestSide})^2 \longrightarrow$ OBTUSE

7. THEOREM: If the sum of the squares of the two smaller sides of a triangle is greater than the square of the larger side, then the triangle is acute. $(\text{Side1})^2 + (\text{Side2})^2 > (\text{LongestSide})^2 \longrightarrow$ ACUTE

8. THEOREM: If the sum of the squares of the two smaller sides of a triangle is equal to the square of the larger side, then the triangle is right. $(\text{Side1})^2 + (\text{Side2})^2 = (\text{LongestSide})^2 \longrightarrow$ RIGHT

Special Right Triangles — Two kinds of right triangles occur so often in math problems that they are called "special" right triangles.

I. 45°–45°–90° Triangles

THEOREM: *The length of the hypotenuse of a 45°–45°–90° triangle equals the length of a leg multiplied by $\sqrt{2}$.*

Proof:
Given:

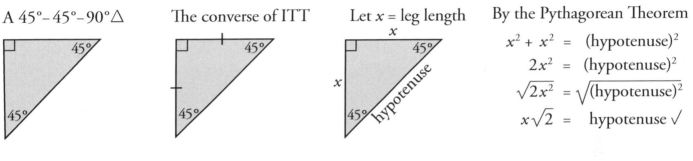

A 45°–45°–90°△   The converse of ITT   Let x = leg length   By the Pythagorean Theorem

$$x^2 + x^2 = (\text{hypotenuse})^2$$
$$2x^2 = (\text{hypotenuse})^2$$
$$\sqrt{2x^2} = \sqrt{(\text{hypotenuse})^2}$$
$$x\sqrt{2} = \text{hypotenuse} \checkmark$$

Here are some examples of 45°–45°–90° triangles.

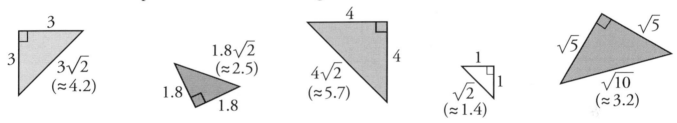

$\sqrt{2}$ is approximately 1.4 so multiplying a number by $\sqrt{2}$, is like multiplying the number by 1.4.

The "$\approx$" symbol means "equals approximately" which means that the number has been rounded off.

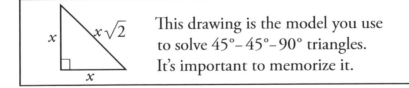

This drawing is the model you use to solve 45°–45°–90° triangles. It's important to memorize it.

Example 1.

Find y:

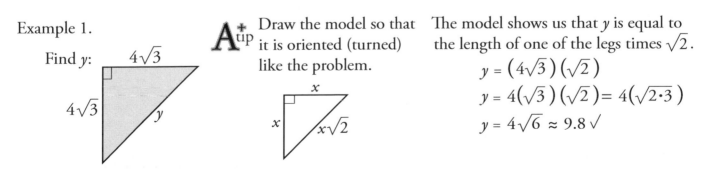

A_{tip}^{+} Draw the model so that it is oriented (turned) like the problem.

The model shows us that y is equal to the length of one of the legs times $\sqrt{2}$.

$$y = \left(4\sqrt{3}\right)\left(\sqrt{2}\right)$$
$$y = 4\left(\sqrt{3}\right)\left(\sqrt{2}\right) = 4\left(\sqrt{2\cdot3}\right)$$
$$y = 4\sqrt{6} \approx 9.8 \checkmark$$

Example 2. Solve for *a*.

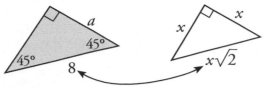

①Draw the model so that it is oriented (turned) like the problem.

② Compare the problem to the model, 8 is in the same position as $x\sqrt{2}$, so **set 8 equal to $x\sqrt{2}$.**

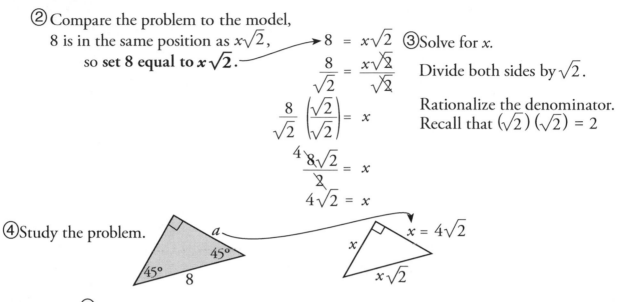

$8 = x\sqrt{2}$ ③Solve for *x*.

$$\frac{8}{\sqrt{2}} = \frac{x\sqrt{2}}{\sqrt{2}}$$ Divide both sides by $\sqrt{2}$.

$$\frac{8}{\sqrt{2}}\left(\frac{\sqrt{2}}{\sqrt{2}}\right) = x$$ Rationalize the denominator. Recall that $(\sqrt{2})(\sqrt{2}) = 2$

$$\frac{\overset{4}{8}\sqrt{2}}{\cancel{2}} = x$$

$$4\sqrt{2} = x$$

④Study the problem.

$x = 4\sqrt{2}$

⑤ Since *a* is in the same position as *x*, $a = x$, that is, $a = 4\sqrt{2} \approx 5.66$ ✓

Example 3.

Find the length of a diagonal of the square below.

$\mathbf{A}^{+}_{tip}$ A diagonal divides a square into two congruent 45°–45°–90° triangles.

Draw the 45°–45° –90° model so that it is oriented like the problem.

Since 5 is the length of a leg (and in the same position as *x*), the diagonal equals $5\sqrt{2}$.

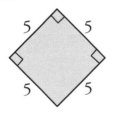 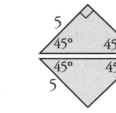 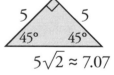 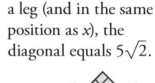

$x\sqrt{2}$

$5\sqrt{2} \approx 7.07$

Now You Try It

1. Find the length of a diagonal of the square below.

2. Find the length of a side and the perimeter of the square below.

II. 30°–60°–90° Triangles

THEOREM: *In a 30°–60°–90° triangle, if x is the length of the side opposite the 30° angle, the side opposite the 60° angle equals $x\sqrt{3}$ and the side opposite the 90° angle equals 2x.*

How do you prove a theorem like this one? Here's one way:

Given: 30°–60°–90° $\triangle$ The midpoint of a hypotenuse is equally distant from each vertex.

By ITT $\triangle$ have 180°   Equiangular $\triangle$ are equilateral

Since ½c = a,
c = 2a √

Since ½c = a

By ITT   The side opposite the 30° angle is ½ the hypotenuse

By Pythagorean Theorem:
$(\frac{1}{2}b)^2 = a^2 - (\frac{1}{2}a)^2$
$\frac{1}{4}b^2 = a^2 - \frac{1}{4}a^2$
$\frac{1}{4}b^2 = \frac{3}{4}a^2$
$b^2 = 3a^2$
$b = a\sqrt{3}$ √

Here are some examples of 30°–60°–90° triangles and remember, $\sqrt{3}$ is approximately 1.73, so multiplying a number by $\sqrt{3}$ is like multiplying a number by 1.73.

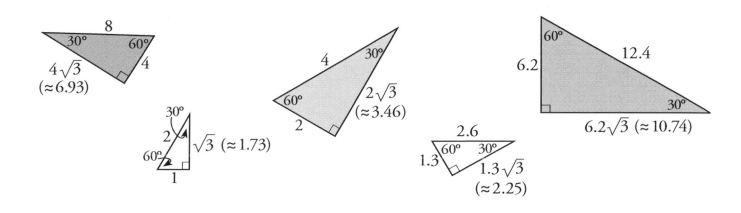

125

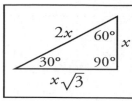 This drawing is the model triangle you use to solve 30°–60°–90° problems. Remember the **3**0°–60°–90° triangle has the $\sqrt{3}$ in the model. It's important to carefully memorize the model triangle.

You need to draw the model correctly. But how do you remember which length is across from which angle?

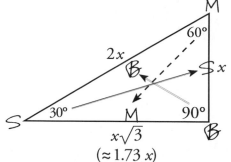

The *smallest* side is across from the *smallest* angle.
The *middle* sized side is across from the *middle* sized angle.
The *biggest* side is across from the *biggest* angle.

Size Order	Angle	Side Opposite
Small	30°	x
Medium	60°	$\approx 1.73\, x$
Big	90°	$2x$

Example 1.

Solve for a.

① Draw the 30°–60°–90° model so that it is oriented (turned) like the problem.

② 7 is in the same position as $x\sqrt{3}$ so
set 7 equal to $x\sqrt{3}$

③ $7 = x\sqrt{3}$ Solve for x.

$\dfrac{7}{\sqrt{3}} = \dfrac{x\sqrt{3}}{\sqrt{3}}$ Divide both sides by $\sqrt{3}$

$\dfrac{7}{\sqrt{3}} \left(\dfrac{\sqrt{3}}{\sqrt{3}} \right) = x$ Rationalize the denominator.
Recall that $(\sqrt{3})(\sqrt{3}) = 3$

$\dfrac{7\sqrt{3}}{3} = x = a \checkmark$ Study the figures, a is in the same position as x, so $a = x = \dfrac{7\sqrt{3}}{3} \approx 4.04$

Your teacher will tell you which form of the answer to use.

126

Example 2. Find x and y.

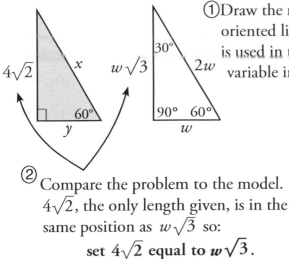

① Draw the model triangle so that it is oriented like the problem and, since x is used in the problem, use a different variable in the model. We used "w".

Size Order	Angle	Side Opp
Small	30°	w
Medium	60°	$w\sqrt{3}$
Big	90°	$2w$

② Compare the problem to the model. $4\sqrt{2}$, the only length given, is in the same position as $w\sqrt{3}$ so:
set $4\sqrt{2}$ equal to $w\sqrt{3}$.

③ $4\sqrt{2} = w\sqrt{3}$ $4\sqrt{2}$ set equal to $w\sqrt{3}$.

$\dfrac{4\sqrt{2}}{\sqrt{3}} = \dfrac{w\sqrt{3}}{\sqrt{3}}$ Divide both sides by $\sqrt{3}$.

$\dfrac{4\sqrt{2}}{\sqrt{3}}\left(\dfrac{\sqrt{3}}{\sqrt{3}}\right) = w$ Rationalize the denominator.

$\dfrac{4\sqrt{6}}{3} = w = y \approx 3.27 \ \checkmark$ You can't divide 3 into $\sqrt{6}$. Think of the "$\sqrt{\ }$" symbol as protecting the number inside.

$2w = 2\left(\dfrac{4\sqrt{6}}{3}\right) = \left(\dfrac{2}{1}\right)\left(\dfrac{4\sqrt{6}}{3}\right) = \dfrac{8\sqrt{6}}{3}$ and $2w = x$, so $x = \dfrac{8\sqrt{6}}{3} \approx 6.53 \ \checkmark$

A_{tip}^{+} To solve a right triangle problem that contains more than one triangle, the key is to begin with the triangle which gives you enough information to get started, and then to work your way over to the other triangle(s). Add any additional values to your sketch, as you find them.

Example. Find y and z.

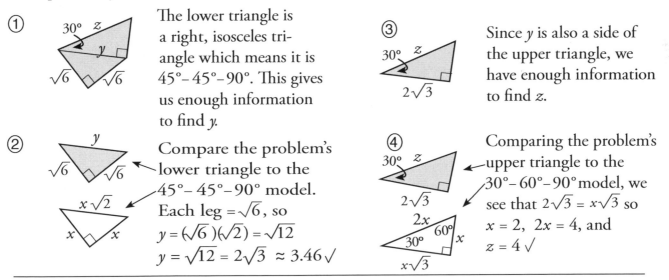

① The lower triangle is a right, isosceles triangle which means it is 45°–45°–90°. This gives us enough information to find y.

② Compare the problem's lower triangle to the 45°–45°–90° model. Each leg $=\sqrt{6}$, so
$y = (\sqrt{6})(\sqrt{2}) = \sqrt{12}$
$y = \sqrt{12} = 2\sqrt{3} \approx 3.46 \checkmark$

③ Since y is also a side of the upper triangle, we have enough information to find z.

④ Comparing the problem's upper triangle to the 30°–60°–90° model, we see that $2\sqrt{3} = x\sqrt{3}$ so $x = 2$, $2x = 4$, and $z = 4 \checkmark$

Now You Try It

Find a.

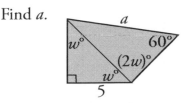

127

TRIGONOMETRY (or "Trig" for short)

Two right triangles are similar if one of the acute angles of the first is equal to one of the acute angles of the second. This is so, because the right angles provide the second pair of congruent angles to satisfy the Angle Angle Similarity Postulate. Then, if the triangles are similar, the ratio of each pair of corresponding sides is the same.

Here's an example using numbers:

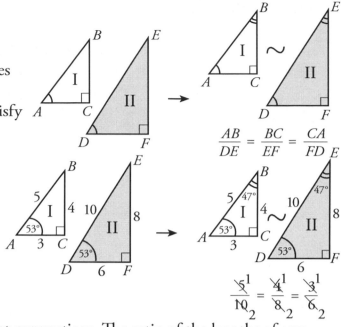

$$\frac{AB}{DE} = \frac{BC}{EF} = \frac{CA}{FD}$$

$$\frac{\cancel{5}^1}{\cancel{10}_2} = \frac{\cancel{4}^1}{\cancel{8}_2} = \frac{\cancel{3}^1}{\cancel{6}_2}$$

The properties of proportion lead to a very important connection: The ratio of the lengths of any two sides of triangle I is the same as the ratio of the lengths of the corresponding sides of triangle II.

$$\frac{BC}{CA} = \frac{4}{3} = \frac{EF}{FD} = \frac{\cancel{8}^4}{\cancel{6}_3} \qquad \frac{CA}{AB} = \frac{3}{5} = \frac{FD}{DE} = \frac{\cancel{6}^3}{\cancel{10}_5} \qquad \frac{BC}{AB} = \frac{4}{5} = \frac{EF}{DE} = \frac{\cancel{8}^4}{\cancel{10}_5}$$

Let's Review — We started out knowing only that angles A and D are equal (both equal 53° in this particular case) and that angles C and F are equal (both are right angles). We ended up knowing that each *ratio* of the lengths of two sides of the first triangle equals the *ratio* of the corresponding sides of the second triangle.

Why is this Important? Trigonometry is based on the fact that the value of the *ratio* of the lengths of two particular sides of a right triangle depends only on the size of the acute angle. The *ratio* does *not* depend on the lengths of the sides. **Your scientific calculator is able to compute all of these ratios!** (More about finding them later.)

Right triangles occur in nature, in our homes and in buildings, and in all sorts of scientific and practical situations. Here's an example of what you can do using trigonometry:

The local mall had a contest offering a gift certificate to whomever could most accurately guess the height of the giant slide at the water slide park. Two geometry students "found the right triangle" in the problem. The angle that the foot of the slide makes with the ground is approximately 53°. The distance from the foot of the slide to a point on the ground beneath the platform is 46 feet. The students knew that the ratio of the length of the leg opposite a 53° angle to the adjacent leg is always $\frac{4}{3}$ (see the triangles above). So they created the following proportion: $\frac{4}{3} = \frac{?}{46}$ Cross multiplying and solving, the students found that ? = 61.3' and won the contest. *And that's trig!*

Trigonometric Functions

The value of a trig function is a ratio determined by the measure of the acute angle and by the specific combination of sides that forms the ratio. We use certain names for the sides of the triangle and when you're first learning trigonometry it's a good idea to label the sides with the correct names.

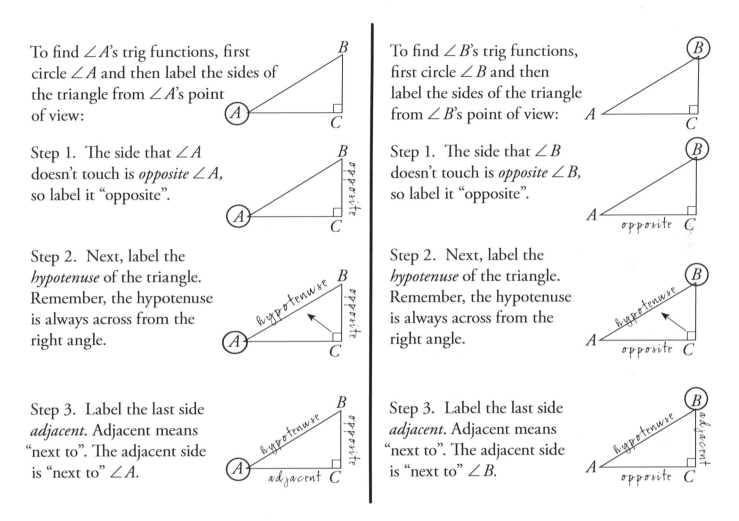

To find ∠A's trig functions, first circle ∠A and then label the sides of the triangle from ∠A's point of view:

Step 1. The side that ∠A doesn't touch is *opposite ∠A,* so label it "opposite".

Step 2. Next, label the *hypotenuse* of the triangle. Remember, the hypotenuse is always across from the right angle.

Step 3. Label the last side *adjacent.* Adjacent means "next to". The adjacent side is "next to" ∠A.

To find ∠B's trig functions, first circle ∠B and then label the sides of the triangle from ∠B's point of view:

Step 1. The side that ∠B doesn't touch is *opposite ∠B,* so label it "opposite".

Step 2. Next, label the *hypotenuse* of the triangle. Remember, the hypotenuse is always across from the right angle.

Step 3. Label the last side *adjacent.* Adjacent means "next to". The adjacent side is "next to" ∠B.

Quick Review — First identify the correct angle and then label the sides from that angle's point of view. Be sure to label the sides in this order: 1. opposite, 2. hypotenuse, 3. adjacent.

Three Trig Functions You Need to Learn:

1. Sine, pronounced sign (like a sign on the wall) and abbreviated sin (also pronounced *sign):*

$$\sin \angle = \frac{\text{length of opposite leg}}{\text{length of hypotenuse}}$$

2. Cosine, abbreviated cos (pronounced like *coas* but without the t):

$$\cos \angle = \frac{\text{length of adjacent leg}}{\text{length of hypotenuse}}$$

3. Tangent, abbreviated tan:

$$\tan \angle = \frac{\text{length of opposite leg}}{\text{length of adjacent leg}}$$

Learning the Trigonometric Functions

A good memory helper is: SOHCAHTOA which is pronounced "sew-cah-toe´-ah".

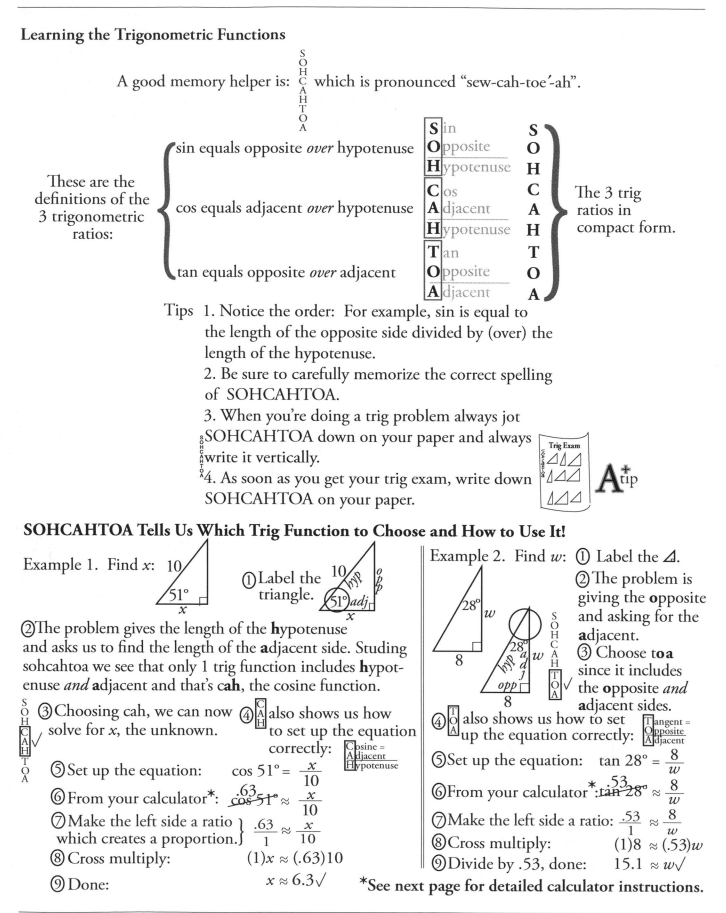

These are the definitions of the 3 trigonometric ratios:

- sin equals opposite *over* hypotenuse
- cos equals adjacent *over* hypotenuse
- tan equals opposite *over* adjacent

Sin	**S**
Opposite	**O**
Hypotenuse	**H**
Cos	**C**
Adjacent	**A**
Hypotenuse	**H**
Tan	**T**
Opposite	**O**
Adjacent	**A**

The 3 trig ratios in compact form.

Tips 1. Notice the order: For example, sin is equal to the length of the opposite side divided by (over) the length of the hypotenuse.

2. Be sure to carefully memorize the correct spelling of SOHCAHTOA.

3. When you're doing a trig problem always jot SOHCAHTOA down on your paper and always write it vertically.

4. As soon as you get your trig exam, write down SOHCAHTOA on your paper.

SOHCAHTOA Tells Us Which Trig Function to Choose and How to Use It!

Example 1. Find x:

① Label the triangle.

② The problem gives the length of the **h**ypotenuse and asks us to find the length of the **a**djacent side. Studing sohcahtoa we see that only 1 trig function includes **h**ypotenuse *and* **a**djacent and that's **cah**, the cosine function.

③ Choosing cah, we can now solve for x, the unknown.

④ **CAH** also shows us how to set up the equation correctly:

Cosine = Adjacent / Hypotenuse

⑤ Set up the equation: $\cos 51° = \dfrac{x}{10}$

⑥ From your calculator*: $\cos 51° \approx \dfrac{x}{10}$ (.63)

⑦ Make the left side a ratio which creates a proportion. $\dfrac{.63}{1} \approx \dfrac{x}{10}$

⑧ Cross multiply: $(1)x \approx (.63)10$

⑨ Done: $x \approx 6.3$ √

Example 2. Find w: ① Label the △.

② The problem is giving the **o**pposite and asking for the **a**djacent.

③ Choose **toa** since it includes the **o**pposite *and* **a**djacent sides.

④ **TOA** also shows us how to set up the equation correctly:

Tangent = Opposite / Adjacent

⑤ Set up the equation: $\tan 28° = \dfrac{8}{w}$

⑥ From your calculator*: $\tan 28° \approx \dfrac{8}{w}$ (.53)

⑦ Make the left side a ratio: $\dfrac{.53}{1} \approx \dfrac{8}{w}$

⑧ Cross multiply: $(1)8 \approx (.53)w$

⑨ Divide by .53, done: $15.1 \approx w$ √

*See next page for detailed calculator instructions.

Finding Trig Ratios with Scientific Calculators

A trig ratio belongs to a particular sized angle. Because of this, people have made tables of the trig ratios and calculator manufacturers have added keys on all scientific calculators which will calculate each trig ratio for any given angle. Your calculator can also find the angle (!) if you enter the value and name of the trig ratio.

 Study your calculator. If the words sin, cos and tan appear on neighboring keys (sin cos tan), you have a scientific calculator. Find the drg or mode key and choose degree mode (it's the default value). Now to find the trigonometric ratios: Most scientific calculators find trigonometric ratios in one of two ways. Test your calculator to see which way it works:

1st Way: Enter 30 (for a 30° angle) and then press the sin key. If your calculator displays .5 then this is how your calculator works: First enter the size of the angle (30 in our example) and then press the key of the particular trig ratio that you wish to find (sin in our example).

2nd Way: Press the sin key, then enter 30 (for a 30° angle) and then press the equal key. If your calculator displays .5 then this is how your calculator works: Press the key of the particular trig ratio you wish to find (sin in our example) and then enter the size of the angle (30 in our example) and then press the equal sign.

> Important Note — How many decimal places of the trig ratio should you use? Use the number that your teacher tells you to use. Some teachers prefer two, some prefer four. Regardless, the methods you use will be the same.

Going Backwards — Finding the Angle When You Know the Trig Ratio

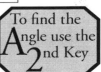

If you know the value of one of the trig ratios of an angle, your calculator can tell you the size of the angle. You have to enter: 1. The name of the trig ratio you have, and 2. Its value. Most scientific calculators work in one of two ways. Test yours to see which way it works:

1st Way: Enter .5, press the 2nd (or Shift or Inv) key and then the sin key. If your calculator displays 30 then this is how your calculator works: First enter the measure of the trig ratio (.5 in our example,) then press the 2nd (or Shift or Inv) key, then the key for the particular trig ratio you have entered (sin in our example). The measure of the angle which corresponds to that information should appear.

2nd Way: Press the 2nd (or Shift or Inv) key, then the sin key and then .5, then press enter. If your calculator displays 30 then this is how your calculator works: First press the 2nd (or Shift or Inv) key, then the key for the trig ratio you are going to enter (sin in our example), then enter the measure of the trig ratio (.5 in our example) and then press equal. The measure of the angle corresponding to that information appears.

If none of the calculator methods shown on page 128 work, you need to find the directions that came with your calculator to see if you have a scientific calculator and if so, how to calculate trig ratios. If you do not have a scientific calculator, your textbook should have a table of trig ratios that you can use.

Why are Your Answers Slightly Different Than the Book's? Most trigonometric ratios are irrational (not simple fractions) and must be rounded off. Also, different calculators use different methods to calculate trigonometric ratios so minor differences in answers are common.

Here are more examples of common trigonometry problems:

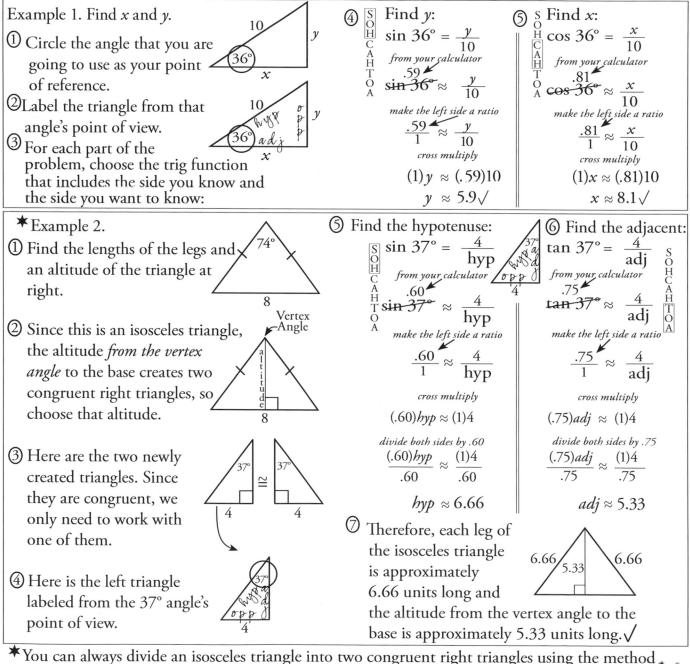

Example 1. Find x and y.

① Circle the angle that you are going to use as your point of reference.

② Label the triangle from that angle's point of view.

③ For each part of the problem, choose the trig function that includes the side you know and the side you want to know:

④ Find y:

$$\sin 36° = \frac{y}{10}$$

from your calculator

$$\sin 36° \approx \frac{y}{10}$$
.59

make the left side a ratio

$$\frac{.59}{1} \approx \frac{y}{10}$$

cross multiply

$$(1)y \approx (.59)10$$

$$y \approx 5.9 \checkmark$$

⑤ Find x:

$$\cos 36° = \frac{x}{10}$$

from your calculator

$$\cos 36° \approx \frac{x}{10}$$
.81

make the left side a ratio

$$\frac{.81}{1} \approx \frac{x}{10}$$

cross multiply

$$(1)x \approx (.81)10$$

$$x \approx 8.1 \checkmark$$

★Example 2.

① Find the lengths of the legs and an altitude of the triangle at right.

② Since this is an isosceles triangle, the altitude *from the vertex angle* to the base creates two congruent right triangles, so choose that altitude.

③ Here are the two newly created triangles. Since they are congruent, we only need to work with one of them.

④ Here is the left triangle labeled from the 37° angle's point of view.

⑤ Find the hypotenuse:

$$\sin 37° = \frac{4}{hyp}$$

from your calculator

$$\sin 37° \approx \frac{4}{hyp}$$
.60

make the left side a ratio

$$\frac{.60}{1} \approx \frac{4}{hyp}$$

cross multiply

$$(.60)hyp \approx (1)4$$

divide both sides by .60

$$\frac{(.60)hyp}{.60} \approx \frac{(1)4}{.60}$$

$$hyp \approx 6.66$$

⑥ Find the adjacent:

$$\tan 37° = \frac{4}{adj}$$

from your calculator

$$\tan 37° \approx \frac{4}{adj}$$
.75

make the left side a ratio

$$\frac{.75}{1} \approx \frac{4}{adj}$$

cross multiply

$$(.75)adj \approx (1)4$$

divide both sides by .75

$$\frac{(.75)adj}{.75} \approx \frac{(1)4}{.75}$$

$$adj \approx 5.33$$

⑦ Therefore, each leg of the isosceles triangle is approximately 6.66 units long and the altitude from the vertex angle to the base is approximately 5.33 units long. $\checkmark$

★You can always divide an isosceles triangle into two congruent right triangles using the method shown in Example 2. Become an *expert* with this method and with isosceles triangles.

A⁺ tip

Example 3. Find the measure of $\angle A$.
This is a "going backwards" problem. Choose the trig ratio that includes the two sides you were given.

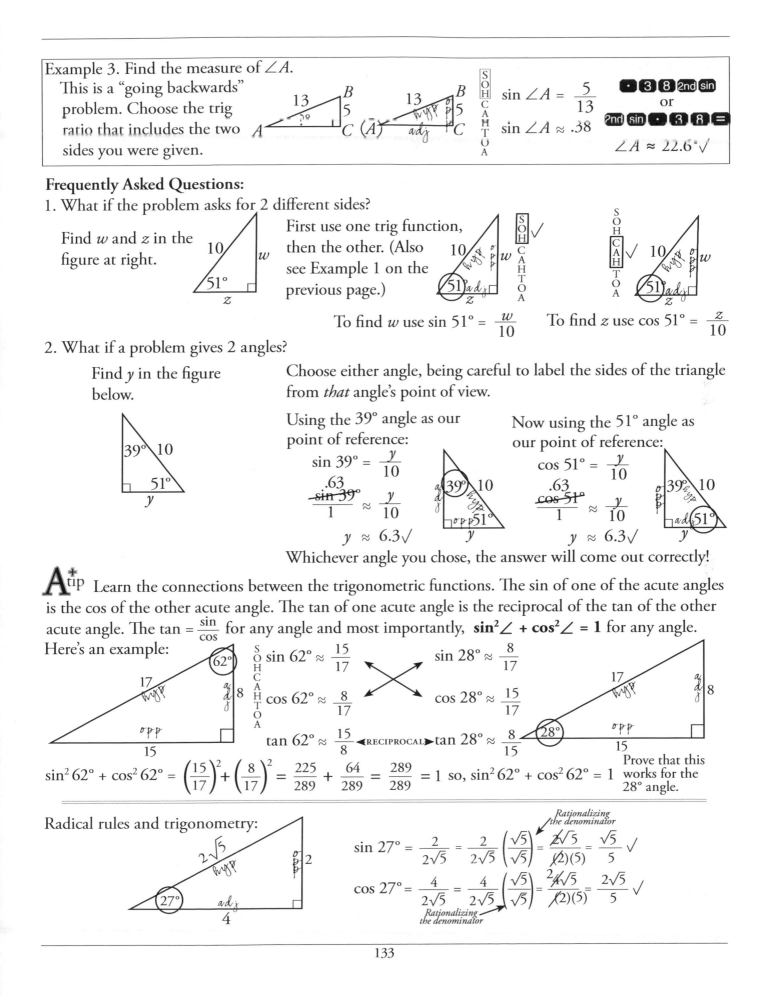

$\sin \angle A = \dfrac{5}{13}$

$\sin \angle A \approx .38$

$\boxed{.}\boxed{3}\boxed{8}\boxed{\text{2nd}}\boxed{\sin}$
or
$\boxed{\text{2nd}}\boxed{\sin}\boxed{.}\boxed{3}\boxed{8}\boxed{=}$

$\angle A \approx 22.6° \checkmark$

Frequently Asked Questions:

1. What if the problem asks for 2 different sides?

Find w and z in the figure at right.

First use one trig function, then the other. (Also see Example 1 on the previous page.)

To find w use $\sin 51° = \dfrac{w}{10}$ To find z use $\cos 51° = \dfrac{z}{10}$

2. What if a problem gives 2 angles?

Find y in the figure below.

Choose either angle, being careful to label the sides of the triangle from *that* angle's point of view.

Using the 39° angle as our point of reference:

$\sin 39° = \dfrac{y}{10}$

$\dfrac{.63}{1} \approx \dfrac{y}{10}$

$y \approx 6.3 \checkmark$

Now using the 51° angle as our point of reference:

$\cos 51° = \dfrac{y}{10}$

$\dfrac{.63}{1} \approx \dfrac{y}{10}$

$y \approx 6.3 \checkmark$

Whichever angle you chose, the answer will come out correctly!

A^+_{tip} Learn the connections between the trigonometric functions. The sin of one of the acute angles is the cos of the other acute angle. The tan of one acute angle is the reciprocal of the tan of the other acute angle. The tan $= \dfrac{\sin}{\cos}$ for any angle and most importantly, $\mathbf{\sin^2 \angle + \cos^2 \angle = 1}$ for any angle. Here's an example:

$\sin 62° \approx \dfrac{15}{17}$ $\sin 28° \approx \dfrac{8}{17}$

$\cos 62° \approx \dfrac{8}{17}$ $\cos 28° \approx \dfrac{15}{17}$

$\tan 62° \approx \dfrac{15}{8}$ ◀RECIPROCAL▶ $\tan 28° \approx \dfrac{8}{15}$

$\sin^2 62° + \cos^2 62° = \left(\dfrac{15}{17}\right)^2 + \left(\dfrac{8}{17}\right)^2 = \dfrac{225}{289} + \dfrac{64}{289} = \dfrac{289}{289} = 1$ so, $\sin^2 62° + \cos^2 62° = 1$

Prove that this works for the 28° angle.

Radical rules and trigonometry:

Rationalizing the denominator

$\sin 27° = \dfrac{2}{2\sqrt{5}} = \dfrac{2}{2\sqrt{5}}\left(\dfrac{\sqrt{5}}{\sqrt{5}}\right) = \dfrac{2\sqrt{5}}{(2)(5)} = \dfrac{\sqrt{5}}{5} \checkmark$

$\cos 27° = \dfrac{4}{2\sqrt{5}} = \dfrac{4}{2\sqrt{5}}\left(\dfrac{\sqrt{5}}{\sqrt{5}}\right) = \dfrac{4\sqrt{5}}{(2)(5)} = \dfrac{2\sqrt{5}}{5} \checkmark$

Rationalizing the denominator

New Term "Solving" a triangle means finding the measurements of all the sides and angles. If you know how to solve a triangle, you can answer any question about the triangle.

How to Solve Right Triangles

GIVEN	FINDING SIDES		FINDING ANGLES
If you know two sides:	Use Pythagorean Theorem to find the third side: $3^2 + x^2 = 5^2$ $x^2 = 16, x = 4$		Use trig (go backwards, use the 2nd key) to find one angle: $\cos \angle A = \dfrac{3}{5}$ $\angle A \approx 53°$ To find the other angle: $\angle B = 90° - 53° \approx 37°$ △ Solved!
If you know one side and one angle:	Use trig to find one missing side: $\cos 53° = \dfrac{3}{AB}$ $\dfrac{.60}{1} \approx \dfrac{3}{AB}$ $AB \approx 5$	Find the other side with trig: $\tan 53° = \dfrac{BC}{3}$ *or* use Pythagorean: $3^2 + (BC)^2 \approx 5^2$ $(BC)^2 \approx 16$ $BC \approx 4$	To find the other angle: $\angle B = 90° - 53° = 37°$ △ Solved!

Remember, if one acute angle is 30°, 45°, or 60°, you can use the special right triangle formulas *or* trig to solve for the missing sides.

Now You Try It

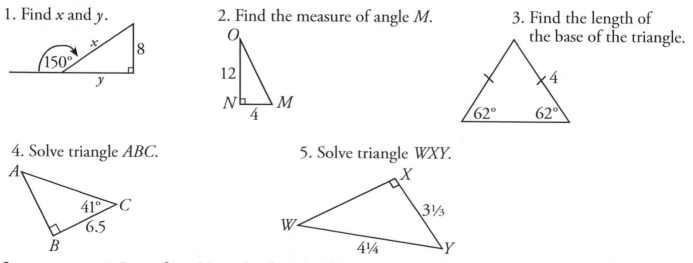

1. Find x and y.

2. Find the measure of angle M.

3. Find the length of the base of the triangle.

4. Solve triangle ABC.

5. Solve triangle WXY.

Learn your trig! Lots of problems for the rest of the school year depend on trigonometry. Think of trigonometry whenever you are given a right triangle or an isosceles triangle. A^+_{tip}

Practical Applications of Trigonometry — Finding the Right Triangle in the Problem

Angle of Elevation and Angle of Depression Problems

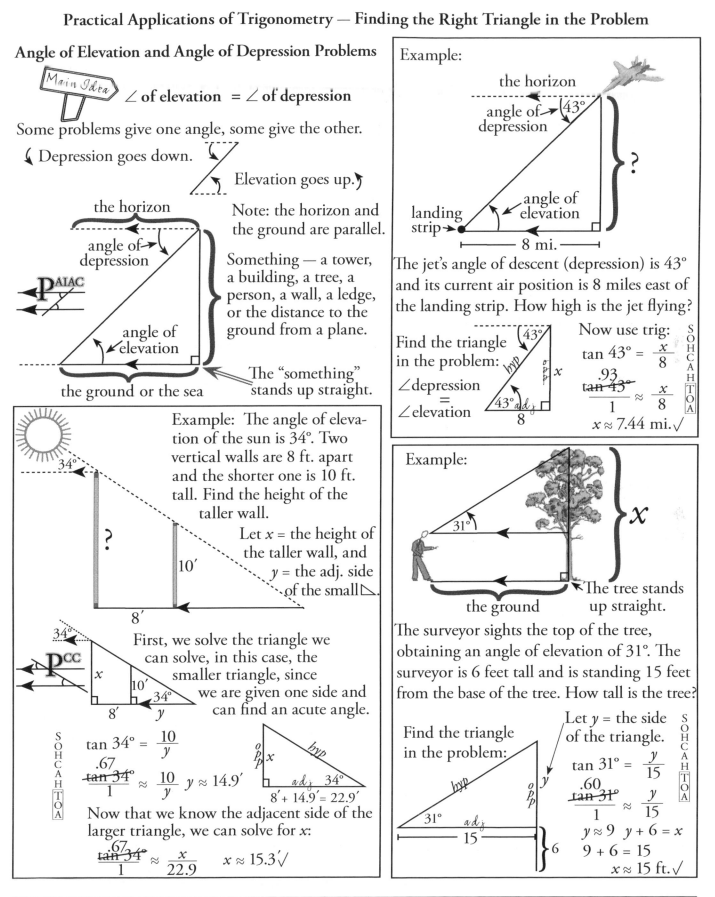

Main Idea ∠ of elevation = ∠ of depression

Some problems give one angle, some give the other.

⟨ Depression goes down. ↘

Elevation goes up. ↗

Note: the horizon and the ground are parallel.

the horizon

angle of depression

P AIAC

angle of elevation

the ground or the sea

Something — a tower, a building, a tree, a person, a wall, a ledge, or the distance to the ground from a plane.

The "something" stands up straight.

Example: The angle of elevation of the sun is 34°. Two vertical walls are 8 ft. apart and the shorter one is 10 ft. tall. Find the height of the taller wall.

34°

?

10′

8′

Let x = the height of the taller wall, and y = the adj. side of the small △.

34° P CC

x

10′

34°

8′   y

First, we solve the triangle we can solve, in this case, the smaller triangle, since we are given one side and can find an acute angle.

$$\tan 34° = \frac{10}{y}$$

$$\frac{.67}{\cancel{\tan 34°} \, 1} \approx \frac{10}{y} \quad y \approx 14.9′$$

o/p/p x   hyp

adj   34°

8′ + 14.9′ = 22.9′

Now that we know the adjacent side of the larger triangle, we can solve for x:

$$\frac{.67}{\cancel{\tan 34°} \, 1} \approx \frac{x}{22.9} \quad x \approx 15.3′ \checkmark$$

Example:

the horizon

angle of depression

(43°)

?

angle of elevation

landing strip →

⊢— 8 mi. —⊣

The jet's angle of descent (depression) is 43° and its current air position is 8 miles east of the landing strip. How high is the jet flying?

Find the triangle in the problem:

(43°)

hyp   o/p/p x

∠depression = ∠elevation

43° adj

8

Now use trig:

$$\tan 43° = \frac{x}{8}$$

$$\frac{.93}{\cancel{\tan 43°} \, 1} \approx \frac{x}{8}$$

$$x \approx 7.44 \text{ mi.} \checkmark$$

S O H C A H T O A

Example:

31°

x

The tree stands up straight.

the ground

The surveyor sights the top of the tree, obtaining an angle of elevation of 31°. The surveyor is 6 feet tall and is standing 15 feet from the base of the tree. How tall is the tree?

Find the triangle in the problem:

hyp   o/p/p y

31°   adj

⊢— 15 —⊣ } 6

Let y = the side of the triangle.

$$\tan 31° = \frac{y}{15}$$

$$\frac{.60}{\cancel{\tan 31°} \, 1} \approx \frac{y}{15}$$

$$y \approx 9 \quad y + 6 = x$$

$$9 + 6 = 15$$

$$x \approx 15 \text{ ft.} \checkmark$$

S O H C A H T O A

135

Solving General Triangles

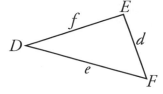
A "general triangle" means any kind of triangle, scalene, isosceles, equilateral, acute, right, obtuse, *any* kind of triangle.

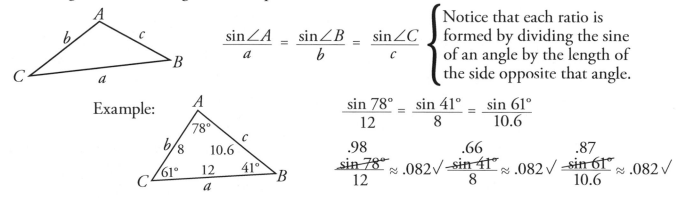

Labeling a General Triangle
A triangle has 6 parts:
3 angles and 3 sides.
Use a capital letter to name each angle.
Use the same letter, but in lower case, for the side opposite each angle.

The Law Of Sines

> ★ Even if your geometry course doesn't include the Law Of Sines, it is an easy method and worth learning.

For *all* triangles, the following relationship works:

$$\frac{\sin\angle A}{a} = \frac{\sin\angle B}{b} = \frac{\sin\angle C}{c}$$

Notice that each ratio is formed by dividing the sine of an angle by the length of the side opposite that angle.

Example:

$$\frac{\sin 78°}{12} = \frac{\sin 41°}{8} = \frac{\sin 61°}{10.6}$$

$$\frac{\overset{.98}{\cancel{\sin 78°}}}{12} \approx .082 \checkmark \quad \frac{\overset{.66}{\cancel{\sin 41°}}}{8} \approx .082 \checkmark \quad \frac{\overset{.87}{\cancel{\sin 61°}}}{10.6} \approx .082 \checkmark$$

Use the Law of Sines to solve a triangle when you have 3 pieces of information and at least 2 pieces are a pair, that is, an angle and the side opposite that angle. (If you have a side and the other 2 angles, then you can find the third angle and make a pair.)

Solve $\triangle RST$.

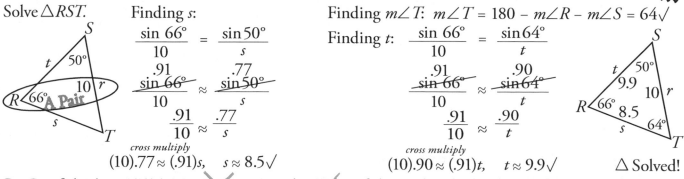

Finding s:

$$\frac{\sin 66°}{10} = \frac{\sin 50°}{s}$$

$$\frac{\overset{.91}{\cancel{\sin 66°}}}{10} \approx \frac{\overset{.77}{\cancel{\sin 50°}}}{s}$$

$$\frac{.91}{10} \approx \frac{.77}{s}$$

cross multiply

$(10).77 \approx (.91)s, \quad s \approx 8.5 \checkmark$

Finding $m\angle T$: $m\angle T = 180 - m\angle R - m\angle S = 64 \checkmark$

Finding t:

$$\frac{\sin 66°}{10} = \frac{\sin 64°}{t}$$

$$\frac{\overset{.91}{\cancel{\sin 66°}}}{10} \approx \frac{\overset{.90}{\cancel{\sin 64°}}}{t}$$

$$\frac{.91}{10} \approx \frac{.90}{t}$$

cross multiply

$(10).90 \approx (.91)t, \quad t \approx 9.9 \checkmark$

$\triangle$ Solved!

Be Careful! It's <u>NOT THE ANGLE</u>, But the SINE of the angle. That's why it's the Law of **Sines.** In some problems you will find the sine of an angle, go backwards (**2nd** **sin**) to find the angle itself.

If an Angle is Obtuse: The meaning of a trig ratio of an *obtuse angle* is explained in Algebra II courses. For now, if you have a sin ratio and know an angle must be obtuse ($a^2 + b^2 < c^2$), take the angle measure your calculator is showing and subtract it from 180. The answer will be the angle's correct measure.

Note: The Law of Sines is a good choice for solving right triangle problems because you can always make a pair. Many students find the Law of Sines easier than trigonometry. $\mathbf{A^{+}_{tip}}$

Law of Cosines

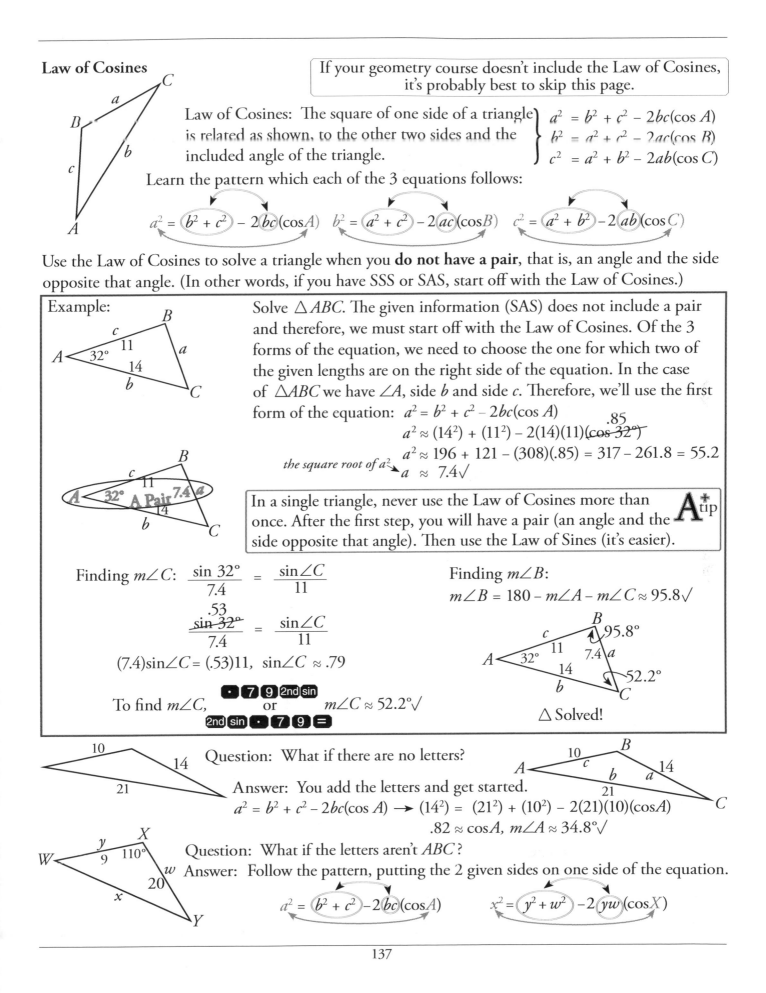

If your geometry course doesn't include the Law of Cosines, it's probably best to skip this page.

Law of Cosines: The square of one side of a triangle is related as shown, to the other two sides and the included angle of the triangle.

$$a^2 = b^2 + c^2 - 2bc(\cos A)$$
$$b^2 = a^2 + c^2 - 2ac(\cos B)$$
$$c^2 = a^2 + b^2 - 2ab(\cos C)$$

Learn the pattern which each of the 3 equations follows:

$$a^2 = (b^2 + c^2) - 2(bc)(\cos A) \qquad b^2 = (a^2 + c^2) - 2(ac)(\cos B) \qquad c^2 = (a^2 + b^2) - 2(ab)(\cos C)$$

Use the Law of Cosines to solve a triangle when you **do not have a pair**, that is, an angle and the side opposite that angle. (In other words, if you have SSS or SAS, start off with the Law of Cosines.)

Example:

Solve $\triangle ABC$. The given information (SAS) does not include a pair and therefore, we must start off with the Law of Cosines. Of the 3 forms of the equation, we need to choose the one for which two of the given lengths are on the right side of the equation. In the case of $\triangle ABC$ we have $\angle A$, side b and side c. Therefore, we'll use the first form of the equation: $a^2 = b^2 + c^2 - 2bc(\cos A)$

$$a^2 \approx (14^2) + (11^2) - 2(14)(11)(\underset{.85}{\cos 32°})$$
$$a^2 \approx 196 + 121 - (308)(.85) = 317 - 261.8 = 55.2$$

the square root of a^2 → $a \approx 7.4 \checkmark$

In a single triangle, never use the Law of Cosines more than once. After the first step, you will have a pair (an angle and the side opposite that angle). Then use the Law of Sines (it's easier). **A$^+_{tip}$**

Finding $m\angle C$: $\dfrac{\sin 32°}{7.4} = \dfrac{\sin\angle C}{11}$

$$\dfrac{\overset{.53}{\sin 32°}}{7.4} = \dfrac{\sin\angle C}{11}$$

$(7.4)\sin\angle C = (.53)11, \quad \sin\angle C \approx .79$

To find $m\angle C$, [·][7][9][2nd sin] or [2nd sin][·][7][9][=] $\qquad m\angle C \approx 52.2° \checkmark$

Finding $m\angle B$:
$$m\angle B = 180 - m\angle A - m\angle C \approx 95.8° \checkmark$$

$\triangle$ Solved!

Question: What if there are no letters?

Answer: You add the letters and get started.
$$a^2 = b^2 + c^2 - 2bc(\cos A) \rightarrow (14^2) = (21^2) + (10^2) - 2(21)(10)(\cos A)$$
$$.82 \approx \cos A, \quad m\angle A \approx 34.8° \checkmark$$

Question: What if the letters aren't ABC?

Answer: Follow the pattern, putting the 2 given sides on one side of the equation.
$$a^2 = (b^2 + c^2) - 2(bc)(\cos A) \qquad x^2 = (y^2 + w^2) - 2(yw)(\cos X)$$

Now You Try It

In problems 1-5, name a method that could be used to solve each triangle, then solve the triangle.

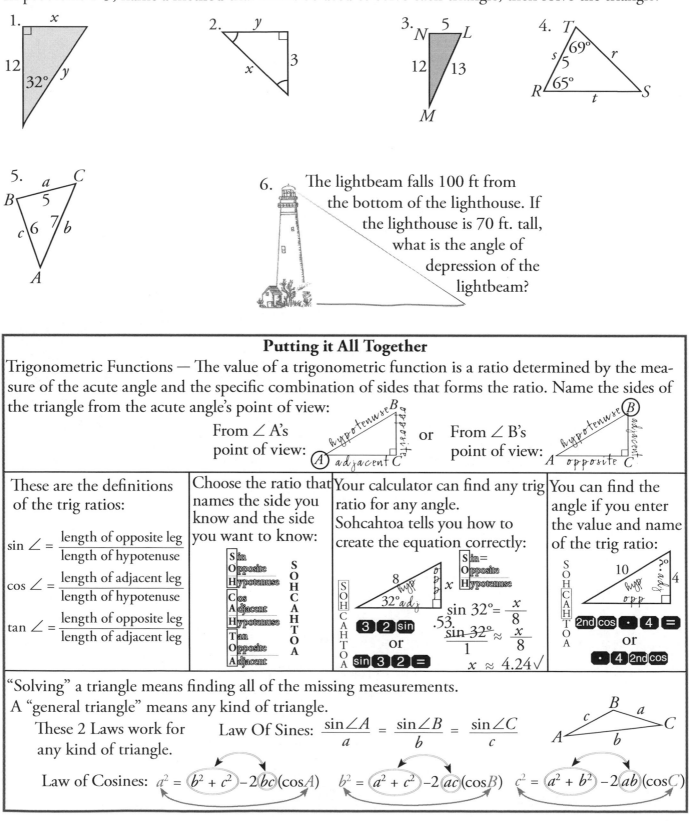

1. x, 12, $32°$, y

2. y, x, 3

3. N 5 L, 12, 13, M

4. T, s $69°$ r, 5, R $65°$ S, t

5. a C, B 5, c 6 7 b, A

6. The lightbeam falls 100 ft from the bottom of the lighthouse. If the lighthouse is 70 ft. tall, what is the angle of depression of the lightbeam?

Putting it All Together

Trigonometric Functions — The value of a trigonometric function is a ratio determined by the measure of the acute angle and the specific combination of sides that forms the ratio. Name the sides of the triangle from the acute angle's point of view:

From ∠ A's point of view: or From ∠ B's point of view:

These are the definitions of the trig ratios:	Choose the ratio that names the side you know and the side you want to know:	Your calculator can find any trig ratio for any angle. Sohcahtoa tells you how to create the equation correctly:	You can find the angle if you enter the value and name of the trig ratio:
$\sin \angle = \dfrac{\text{length of opposite leg}}{\text{length of hypotenuse}}$	**S**in **O**pposite **H**ypotenuse **S O H**		
$\cos \angle = \dfrac{\text{length of adjacent leg}}{\text{length of hypotenuse}}$	**C**os **A**djacent **H**ypotenuse **C A H**	$\sin 32° = \dfrac{x}{8}$ $\dfrac{\sin 32°}{1} \approx \dfrac{x}{8}$ $x \approx 4.24 \checkmark$	2nd cos · 4 = or · 4 2nd cos
$\tan \angle = \dfrac{\text{length of opposite leg}}{\text{length of adjacent leg}}$	**T**an **O**pposite **A**djacent **T O A**		

"Solving" a triangle means finding all of the missing measurements.
A "general triangle" means any kind of triangle.

These 2 Laws work for any kind of triangle.

Law Of Sines: $\dfrac{\sin \angle A}{a} = \dfrac{\sin \angle B}{b} = \dfrac{\sin \angle C}{c}$

Law of Cosines: $a^2 = (b^2 + c^2) - 2(bc)(\cos A)$ $b^2 = (a^2 + c^2) - 2(ac)(\cos B)$ $c^2 = (a^2 + b^2) - 2(ab)(\cos C)$

The Essentials — Be an Expert at These 3 Methods:

1. The Pythagorean Theorem

When you have a right triangle and the lengths of 2 sides are given, use the Pythagorean Theorem to find the length of the third side.

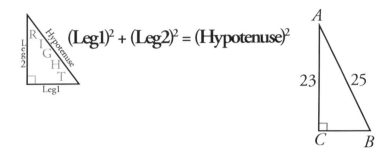

$$(\textbf{Leg1})^2 + (\textbf{Leg2})^2 = (\textbf{Hypotenuse})^2$$

Example: Given $\triangle ABC$, find CB.

Remember, the hypotenuse always goes all by itself on the right side of the equation.

$$23^2 + CB^2 = 25^2$$
$$CB^2 = 625 - 529 = 96$$
$$CB = 4\sqrt{6} \approx 9.8 \checkmark$$

2. Using the Law of Sines for Right Triangles

When you have a right triangle and the measures of one side and either acute angle is given, you can use the Law of Sines to find the missing lengths of the sides of the triangle. Note that the sin of 90° is 1 (which is easy to remember).

$$\frac{\sin\angle A}{a} = \frac{\sin\angle B}{b} = \frac{\sin\angle C}{c}$$

Example: Given $\triangle DEF$, find DF and FE.

$$\frac{\sin 67°}{DF} = \frac{\sin 90°}{25}$$
$$\frac{.92}{DF} \approx \frac{1.0}{25}$$
$$(.92)25 \approx (1.0)DF = DF$$
$$23 \approx DF \checkmark$$

$\angle D = 90° - 67° = 23°$
$$\frac{.39}{FE} \approx \frac{1.0}{25}$$
$$(.39)25 \approx (1.0)FE = FE$$
$$9.8 \approx FE \checkmark$$

3. Solving Isosceles Triangles

It doesn't matter which side(s) of the triangle and/or which angle you are given, always solve an Isosceles Triangle by drawing the altitude from the vertex angle to the base. This will divide the triangle into two congruent right triangles.

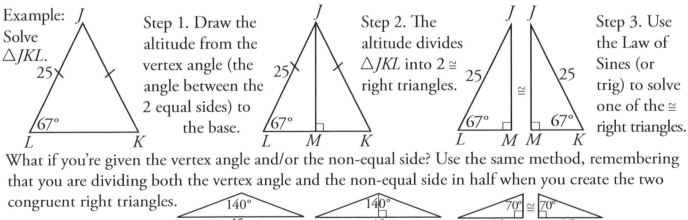

Example: Solve $\triangle JKL$.

Step 1. Draw the altitude from the vertex angle (the angle between the 2 equal sides) to the base.

Step 2. The altitude divides $\triangle JKL$ into 2 ≅ right triangles.

Step 3. Use the Law of Sines (or trig) to solve one of the ≅ right triangles.

What if you're given the vertex angle and/or the non-equal side? Use the same method, remembering that you are dividing both the vertex angle and the non-equal side in half when you create the two congruent right triangles.

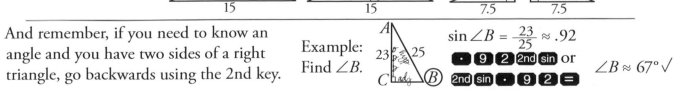

And remember, if you need to know an angle and you have two sides of a right triangle, go backwards using the 2nd key.

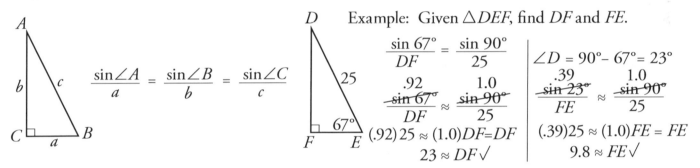

Example: Find $\angle B$.

$$\sin\angle B = \frac{23}{25} \approx .92$$

[.] [9] [2] [2nd] [sin] or
[2nd] [sin] [.] [9] [2] [=]

$$\angle B \approx 67° \checkmark$$

11. CIRCLES

What You Need to Know About Circles

1. A circle is made up of all of the points in a plane that are equally far away from a given point called the center.

   Taken together, the equally distant points form the curved figure that we call a circle.

   The circle
   Its center

2. The distance from the center to the circle is called the *radius* of the circle.

   The radius of this circle is 5.

   The radius of this circle is 7.

   Any segment (a part of a line) that goes from the center to the circle is also called a radius. If you know where the center of a circle is, *you can always add a radius* wherever you need one.

   Each of the segments shown is a radius and each is equal to 5.

   Each of the segments shown is a radius and each is equal to 7.

   The plural of radius is radii (which is pronounced ray-d-i) or radiuses.

   All radii of a circle are equal. ⟵ ★A very important fact about circles★

3. Circles are named by their centers which are points, and points are named by a single capital letter.

   This is circle H.   This is circle O.   This is circle A.

   In each of the figures above, we are being given the location of the center. Always look for the dot (•).

4. Most textbooks use a small circle with a dot in the center as the symbol for a circle: ⊙
   Using the circle symbol we would call the above three circles ⊙ H, ⊙ O, and ⊙ A.

5. Here are terms associated with circles:

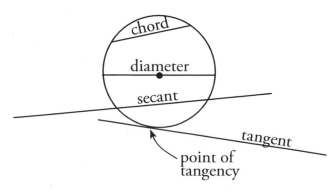

   CHORD — *A line segment whose endpoints are both on the circle.*
   DIAMETER — *A chord that passes through the center of a circle. A diameter is the longest chord in a circle.*
   SECANT — *A line that includes a chord of the circle.*
   TANGENT — *A line which passes through only one point of a circle, or one of its rays or segments which includes that point.*
   POINT OF TANGENCY — *The single point shared by a circle and a tangent.*

What You Need to Know About Circles (continued)

6 All circles have the same shape and therefore *all circles are similar.*

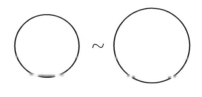

7. DEFINITION: *Congruent circles have congruent radii.*

Of course congruent circles also have congruent diameters and congruent circumferences, but if you're asked to prove that circles are congruent, show that their radii are congruent.

The radius of ⊙M ≅ the radius of ⊙N
so, ⊙M is congruent to ⊙N.

8. DEFINITION: *Concentric circles are circles having the same center.*

Think of the center as the bulls eye of a target.

Four **concentric** circles

Spheres

DEFINITION: *A sphere is the collection of points in space that are equally far away from a given point called its center.*

Spheres are a part of our everyday lives, for example tennis balls, soccer balls, and basketballs. All of the terms that we use with circles can also be used with spheres. Here's an example: If a tennis ball were inside a soccer ball which in turn were inside a basketball and if we arranged them so that all three balls had the same center, then they would be concentric spheres.

Now You Try It

Study the illustration on the right and identify each object. (Hint, some are duplicates and for this exercise, objects that appear tangent are tangent.)

A _____   G _____
B _____   H _____
C _____   I _____
D _____   J _____
E _____   K _____
F _____   L _____

Be sure to check in the answer section at the back of the book to see how you did!

All About Tangents

Tangents pass through a single point of a circle because the tangent is straight and the circle is curved and pulls away from the tangent on each side of the point of tangency.

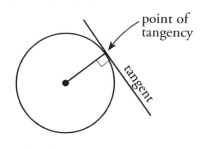

Note: Although tangents are lines, we are usually interested in only a portion of the tangent. Because of this, we refer to tangent *segments* and tangent *rays* as tangents.

THEOREM: *A tangent is perpendicular to a radius or diameter drawn to the point of tangency.*

If you know that a line is a tangent and if there is a radius or diameter **A⁺tip** drawn to the point of tangency, add the perpendicular sign (⌐). If no radius or diameter is drawn but the center is shown, draw in a radius and then add the perpendicular sign.

The above theorem leads to a second theorem which is used in many problems:

THEOREM: *If two tangents to a circle are drawn from the same external point, the tangent segments are congruent.*

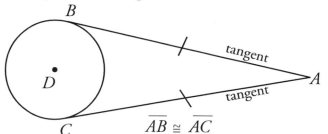

$$\overline{AB} \cong \overline{AC}$$

My students named this the Clown Hat Theorem. Whenever you see two tangents drawn from the same external point, always mark the two tangents as equal. **A⁺tip**

We can prove this theorem using congruent triangles.
Given: $\overline{AB}$ and $\overline{AC}$ are tangent to circle D.
Prove: $\overline{AB} \cong \overline{AC}$.

Statements	Reasons
1. $\overline{AB}$ and $\overline{AC}$ are tangent to circle D.	1. Given.
2. Draw radii from D to B & C.	2. Two points determine a line.
3. $\overline{DB} \perp \overline{AB}$ & $\overline{DC} \perp \overline{AC}$.	3. Tangents are $\perp$ to a radius drawn to the point of tangency.
4. $\angle DBA$ & $\angle DCA$ are rt. angles.	4. Perpendicular lines meet to form rt. angles.
5. $\angle DBA \cong \angle DCA$.	5. All right angles are congruent.
6. $\overline{DB} \cong \overline{DC}$.	6. All radii of a circle are congruent.
7. $\overline{DA} \cong \overline{DA}$.	7. Reflexive Property of Equality.
8. $\triangle DBA$ & $\triangle DCA$ are rt. $\triangle$'s.	8. A triangle with a rt. $\angle$ is a rt. $\triangle$.
9. $\triangle DBA \cong \triangle DCA$.	9. HL.
10. $\overline{AB} \cong \overline{AC}$.	10. CPCT.

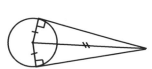

We know that a radius drawn to the point of tangency is perpendicular to the tangent. The converse is also true.

THEOREM: *If a radius is perpendicular at its outer endpoint to a line in the plane of a circle, then the line is tangent to the circle.*

Hypothesis (the "if")   Conclusion (the "then")

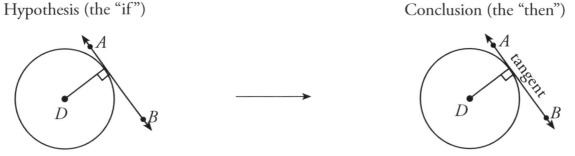

The outer endpoint of the radius is the point of tangency of $\overline{AB}$.

More Tangents

A circle can be tangent to another circle. Tangent circles are tangent to the same line and share the same point of tangency.

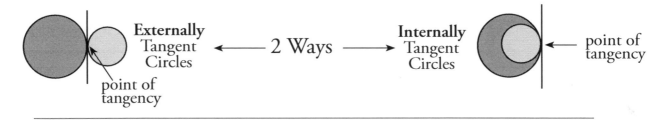

Lines can also be tangent to two non-adjacent (not next to eachother) circles.

Internal or External?? If the common tangent crosses a segment connecting the centers, it's internal.

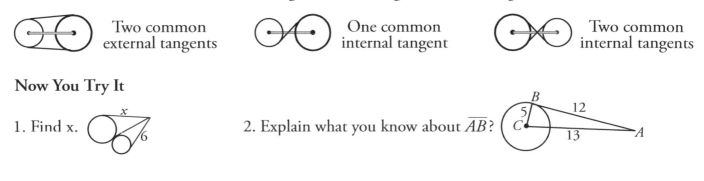

Now You Try It

1. Find x.   2. Explain what you know about $\overline{AB}$?

Inscribed and Circumscribed — Scribe means to write, inscribed means "written upon". Inscribed and circumscribed are important terms in geometry; be sure that you understand what they mean.

Inscribed Angles — The *vertex* of an *inscribed* angle *lies on* (is written upon) the circle.

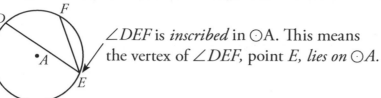

∠*DEF* is *inscribed* in ⊙A. This means the vertex of ∠*DEF*, point *E*, lies on ⊙A.

∠*GHI* is *inside* ⊙*P* but ∠*GHI* is not an inscribed angle, because its vertex does not *lie on* circle *P*.

One Picture, Two Descriptions: Inscribed Polygons — Circumscribing Circles

△*MNO* is *inscribed in* ⊙*B*, which means all three vertices of △*MNO* lie on ⊙B.

⊙*B circumscribes* △*MNO*. This means that each vertex of *MNO* lies on ⊙B. (That is, △*MNO* is *inscribed* in ⊙B.)

 A polygon is *inscribed* in a circle if *every* vertex of the polygon lies on the circle. The circle is said to *circumscribe* the polygon.

In the case above, each statement describes the illustration. Some problems will describe it one way, some problems will describe it the other way.

The quadrilateral on the left is *inside* the circle but it is not *inscribed* in the circle because not *all* of its vertices are *lying on* the circle.

One Picture, Two Descriptions: Inscribed Circles — Circumscribing Polygons

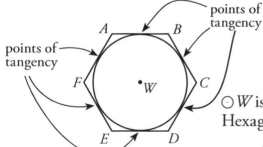

points of tangency

points of tangency

Main Idea A circle is *inscribed* in a polygon if *every* side of the polygon is *tangent* to the circle. The polygon is said to *circumscribe* the circle.

⊙*W* is *inscribed* in hexagon *ABCDEF*. Hexagon *ABCDEF* is *circumscribed about* ⊙*W*.

Each side of hexagon *ABCDEF* is *tangent* to ⊙*W*.

The circle on the right is *inside* the polygon but it is not *inscribed* in the polygon because *each* side of the polygon is not *tangent* to the circle. The polygon is not *circumscribing* the circle for the same reason.

Recall:

Tangents drawn from the same exterior point are congruent. So, if you know that a circle is *inscribed* in a polygon (or that the polygon *circumscribes* the circle), you know that the segments are tangents and therefore equal.

Given: ⊙*W* is inscribed.

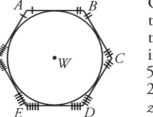

Given that the circle on the right is inscribed:
$5x = 10$, $x = 2$
$2y = 8$, $y = 4$
$z = 3$

Quick Review — We know that a circle is made up of all of the points in a plane that are equally far away from a given point called the center.

a circle & its center

Arcs

an arc

DEFINITION OF AN ARC: *An unbroken collection of points on a circle, is called an arc.*

Think of an arc as a piece of the circle's circumference.

An arc is named by its *endpoints*.

arc *AB*

A

B

The symbol for an arc is ⌒. For example, arc *AB* is written $\overset{\frown}{AB}$.

When naming an arc, it doesn't matter which endpoint appears first. For example, $\overset{\frown}{AB}$ is the same as $\overset{\frown}{BA}$ and $\overset{\frown}{AB} \cong \overset{\frown}{BA}$.

When we say the *measure* of an arc we are referring to the amount of rotation around the center of the circle. Rotation is measured in degrees. One complete revolution is 360°. This means that an entire circle measures 360°.

360°

Here are some examples of arcs and their measures:

E 90°

D

$m\overset{\frown}{ED} = 90$

30° *G*

F

$m\overset{\frown}{GF} = 30$

K

60°

J

$m\overset{\frown}{JK} = 60$

135°

M

L

$m\overset{\frown}{ML} = 135$

Notice that just as with angles, *m* stands for "measure of". The "measure of" an arc is always referring to a degree measure and when a problem says find $m\overset{\frown}{AB}$ the answer is a number.

Central Angles

When the vertex of an angle is placed at the center of a circle, the sides of the angle intercept the circle and determine, or "cut off", an arc. This kind of angle is called a *central angle*. Here are four examples:

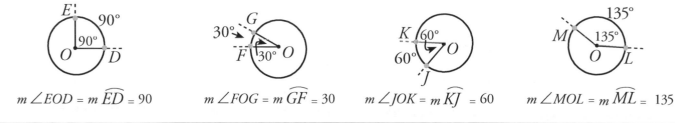

$m\angle EOD = m\overset{\frown}{ED} = 90$ $m\angle FOG = m\overset{\frown}{GF} = 30$ $m\angle JOK = m\overset{\frown}{KJ} = 60$ $m\angle MOL = m\overset{\frown}{ML} = 135$

Two Arcs Not One — Minor Arcs, Major Arcs

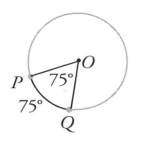

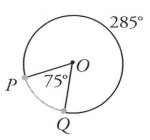

Study the circle on the left. We know that points P and Q are the endpoints of an arc which measures 75°. But points P and Q are also the endpoints of the remaining piece of the circle which must measure:

$$360° - 75° = 285°$$

This means that any two points on a circle determine *two* arcs, not one, and that the sides of a central angle intercept *two* arcs, not one.

The measure of a Minor Arc is defined to be the measure of its central angle.	**The measure of a Major Arc is equal to 360 minus the measure of its minor arc.**
Minor Arc < 180°	**Major Arc > 180°**

To name a major arc a third point is added between the endpoints of the arc and *all three points* (this means all three capital letters) are included in its name:

$\overparen{PMQ}$ is a major arc. The measure of $\overparen{PMQ}$ is 360 – 75 = 285 and is written $m\overparen{PMQ}$ = 285.

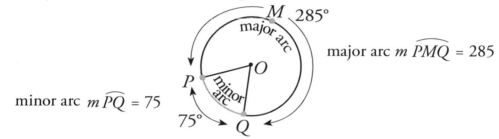

minor arc $m\overparen{PQ}$ = 75

major arc $m\overparen{PMQ}$ = 285

The endpoints of the major arc are the first and last letters in the major arc's name. It doesn't matter which point appears first and which appears last, for example $\overparen{PMQ}$ is the same as $\overparen{QMP}$.

Semicircles — If two points are at the outer endpoints of a diameter, the points divide the circle into two semicircles. A semicircle measures 180° and is named with *three* points, that is, *three* letters. Problems can indicate that a figure includes a semicircle in lots of different ways, a few of which are shown below. Beneath each figure, one of the semicircles formed (there are two in each figure) is named.

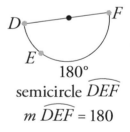

semicircle $\overparen{DEF}$

$m\overparen{DEF}$ = 180

Given diameter GH

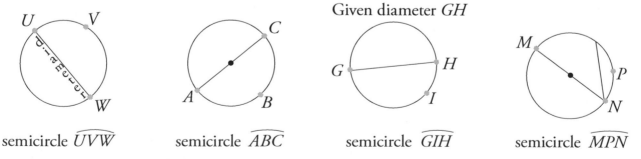

semicircle $\overparen{UVW}$ semicircle $\overparen{ABC}$ semicircle $\overparen{GIH}$ semicircle $\overparen{MPN}$

The endpoints of the semicircle are the first and last letters in the semicircle's name. It doesn't matter which point appears first and which appears last, for example $\overparen{UVW}$ is the same as $\overparen{WVU}$.

Congruent Arcs

The **2** Requirements for Arcs to be Congruent:

1. The two arcs must have the same measure.

2. The two arcs must be in the same or in congruent circles.

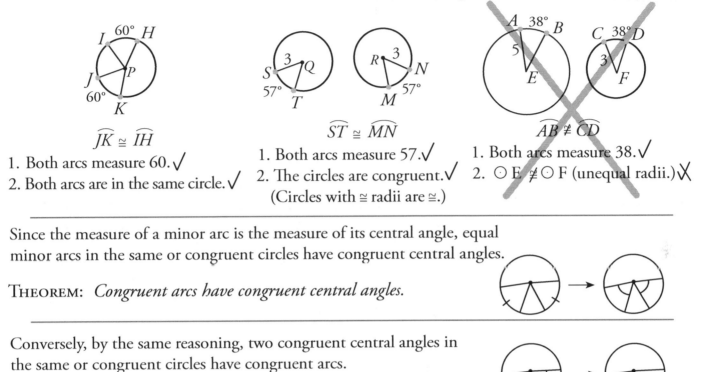

$\widehat{JK} \cong \widehat{IH}$

1. Both arcs measure 60. ✓
2. Both arcs are in the same circle. ✓

$\widehat{ST} \cong \widehat{MN}$

1. Both arcs measure 57. ✓
2. The circles are congruent. ✓
   (Circles with $\cong$ radii are $\cong$.)

$\widehat{AB} \ncong \widehat{CD}$

1. Both arcs measure 38. ✓
2. $\odot$ E $\ncong$ $\odot$ F (unequal radii.) ✗

Since the measure of a minor arc is the measure of its central angle, equal minor arcs in the same or congruent circles have congruent central angles.

THEOREM: *Congruent arcs have congruent central angles.*

Conversely, by the same reasoning, two congruent central angles in the same or congruent circles have congruent arcs.

THEOREM: *Congruent central angles have congruent arcs.*

The connection between arcs and central angles will be the key to solving many problems. Here are two examples.

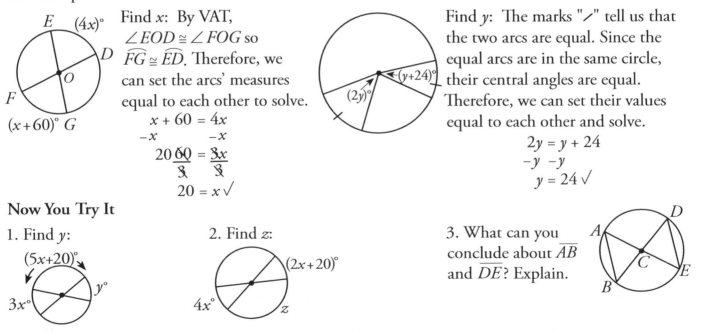

Find x: By VAT, $\angle EOD \cong \angle FOG$ so $\widehat{FG} \cong \widehat{ED}$. Therefore, we can set the arcs' measures equal to each other to solve.

$$x + 60 = 4x$$
$$-x \qquad -x$$
$$20\dfrac{60}{3} = \dfrac{3x}{3}$$
$$20 = x \checkmark$$

Find y: The marks "∕" tell us that the two arcs are equal. Since the equal arcs are in the same circle, their central angles are equal. Therefore, we can set their values equal to each other and solve.

$$2y = y + 24$$
$$-y \quad -y$$
$$y = 24 \checkmark$$

Now You Try It

1. Find y:

$(5x+20)°$

$3x°$ $y°$

2. Find z:

$(2x+20)°$

$4x°$ z

3. What can you conclude about $\overline{AB}$ and $\overline{DE}$? Explain.

Arcs and Their Chords

A chord is a line segment whose endpoints are both on a particular circle. $\overline{FG}$ is a chord of circle O. $\overline{FG}$ determines ("cuts off") two arcs, a 130° arc and a 230° arc. However, when we say the "arc of the chord" we mean the minor (smaller) arc.

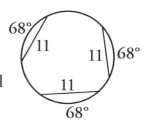

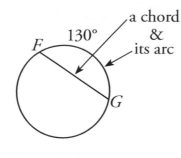

"The same or congruent circles" — You will see this phrase in many theorems about circles. What the phrase means is that the theorem applies to (works in) a single circle and that the theorem also applies to (works in) two or more circles, providing that the two circles are congruent, that is, that the radii of the two circles are congruent.

Circles are symmetrical. Because of this, a chord of a fixed length will determine ("cut off") the same size arc in a particular circle regardless of where the chord is placed. Imagine the chord being moved around the circle. As long as it remains a chord (as long as both of its endpoints lie on the circle) the arc of the chord will be the same size.

THEOREM: *In the same or in congruent circles congruent arcs have congruent chords.*

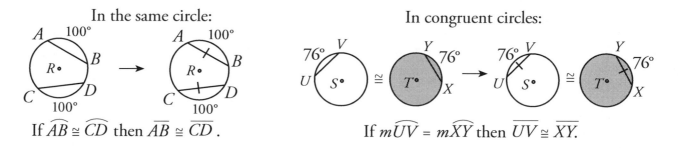

The converse of the above theorem is also true:

THEOREM: *In the same or in congruent circles congruent chords have congruent arcs.*

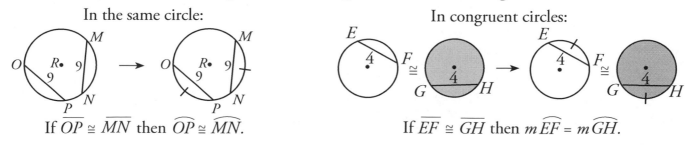

Proving Circle Theorems — The last two theorems are easy to prove. We'll use the most important fact about circles, all radii of a circle are equal, and congruent triangles:

Prove — In the same circle, congruent arcs have congruent chords.

Given: $m\widehat{AB} = m\widehat{CD}$.
Prove: $\overline{AB} \cong \overline{CD}$.

Statements	Reasons
1. $m\widehat{AB} = m\widehat{CD}$.	1. Given.
2. Draw radii from R to points A, B, C & D.	2. Two points determine a line.
3. $\overline{RA} \cong \overline{RB} \cong \overline{RC} \cong \overline{RD}$.	3. All radii of a $\odot$ are $\cong$.
4. $m\angle ARB = m\widehat{AB}$, $m\angle CRD = m\widehat{CD}$.	4. Measure of a central $\angle$ equals the arc it intercepts.
5. $m\angle ARB = m\angle CRD$.	5. Substitution. (Statements 1 & 4.)
6. $\triangle RBA \cong \triangle RCD$.	6. SAS $\cong$ Thm.
7. $\overline{AB} \cong \overline{CD}$	7. Corresp. parts of $\cong$ $\triangle$ are $\cong$ (CPCT).

To show that congruent chords have congruent arcs, the reasoning is very similar:

Mark the congruent chords as congruent ("Given"). Then by adding the congruent radii as above, we can see that the two triangles are congruent by SSS $\cong$.

Therefore, the central angles are congruent by CPCT.

Finally the arcs are equal because the measure of an arc is equal to the measure of the central angle that intercepts it.

To prove the two theorems for two or more congruent circles, we would use the definition of congruent circles, that is, congruent circles have congruent radii. This would lead to congruent triangles and then to the same reasoning as above.

$A^{\t*}_{tip}$ When doing proofs about circles, a good general strategy is:
1.) Draw in congruent radii. Add other known facts.
2.) State the resulting triangles are congruent.
3.) Use CPCT to reach or to help reach the conclusion.

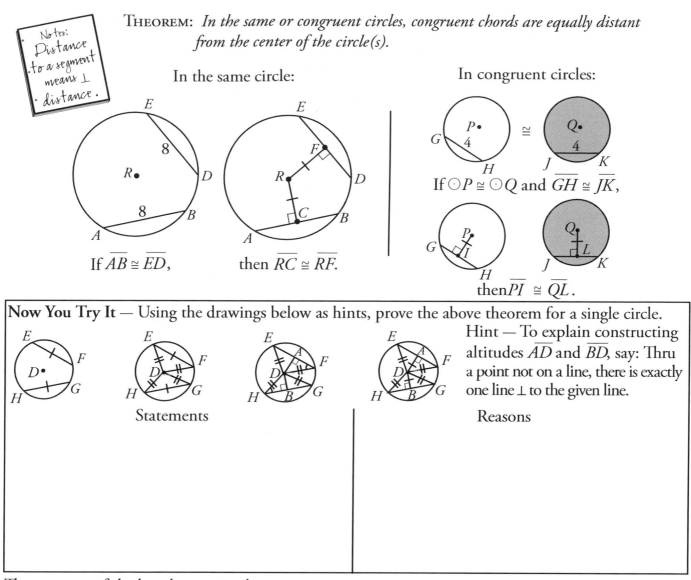

THEOREM: *In the same or congruent circles, congruent chords are equally distant from the center of the circle(s).*

In the same circle:

In congruent circles:

If $\overline{AB} \cong \overline{ED}$,

then $\overline{RC} \cong \overline{RF}$.

If $\odot P \cong \odot Q$ and $\overline{GH} \cong \overline{JK}$,

then $\overline{PI} \cong \overline{QL}$.

Notes:
Distance to a segment means ⊥ distance.

Now You Try It — Using the drawings below as hints, prove the above theorem for a single circle.

Hint — To explain constructing altitudes $\overline{AD}$ and $\overline{BD}$, say: Thru a point not on a line, there is exactly one line ⊥ to the given line.

Statements | Reasons

The converse of the last theorem is also true.

THEOREM: *In the same or congruent circles, chords which are equally distant from the center are congruent.*

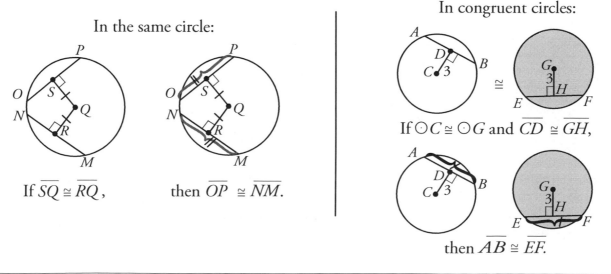

In the same circle:

If $\overline{SQ} \cong \overline{RQ}$,

then $\overline{OP} \cong \overline{NM}$.

In congruent circles:

If $\odot C \cong \odot G$ and $\overline{CD} \cong \overline{GH}$,

then $\overline{AB} \cong \overline{EF}$.

Putting it all Together — Problems involving circles and chords follow certain patterns. Here are a few of them:

Find x:

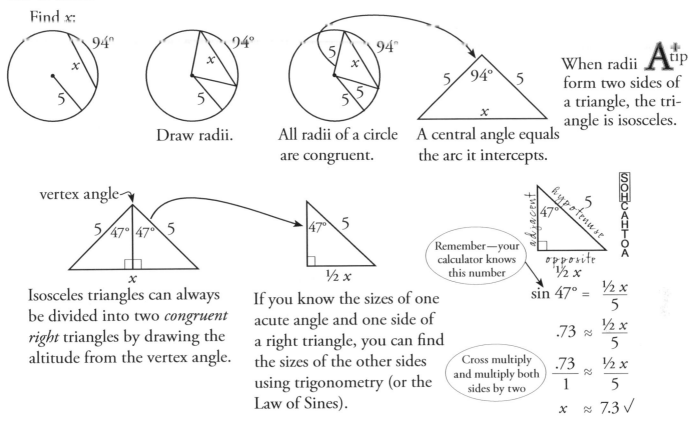

Draw radii.

All radii of a circle are congruent.

A central angle equals the arc it intercepts.

When radii **A**tip form two sides of a triangle, the triangle is isosceles.

Isosceles triangles can always be divided into two *congruent right* triangles by drawing the altitude from the vertex angle.

If you know the sizes of one acute angle and one side of a right triangle, you can find the sizes of the other sides using trigonometry (or the Law of Sines).

Remember—your calculator knows this number

$$\sin 47° = \frac{\frac{1}{2}x}{5}$$

$$.73 \approx \frac{\frac{1}{2}x}{5}$$

Cross multiply and multiply both sides by two

$$\frac{.73}{1} \approx \frac{\frac{1}{2}x}{5}$$

$$x \approx 7.3 \checkmark$$

In the above problem, if we had been given the size of the chord we would have used the same technique to find the length of the radius.

___

In the following problem, we are given the radius and the perpendicular distance from the center to a chord. The problem is asking for the measure of the arc of the chord.

Find y:

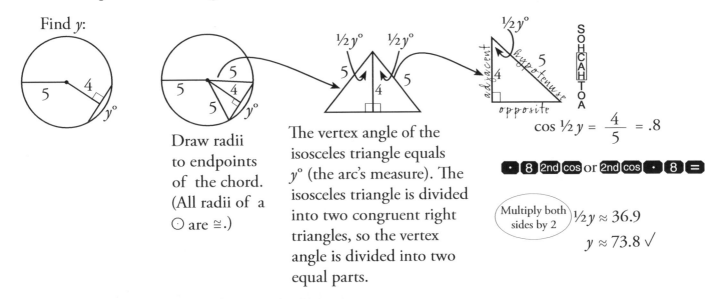

Draw radii to endpoints of the chord. (All radii of a ○ are ≅.)

The vertex angle of the isosceles triangle equals $y°$ (the arc's measure). The isosceles triangle is divided into two congruent right triangles, so the vertex angle is divided into two equal parts.

$$\cos \frac{1}{2}y = \frac{4}{5} = .8$$

[·] [8] [2nd] [cos] or [2nd] [cos] [·] [8] [=]

Multiply both sides by 2

$$\frac{1}{2}y \approx 36.9$$

$$y \approx 73.8 \checkmark$$

In the previous problem we divided the isosceles triangle into two right triangles which were congruent (by HL).

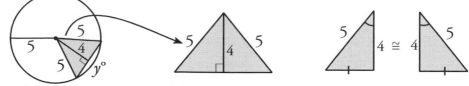

CPCT tells us that the altitude divided the chord and the vertex angle in half. This leads to the following theorem:

THEOREM: *A diameter (or radius) that is perpendicular to a chord bisects the chord and its arc.*

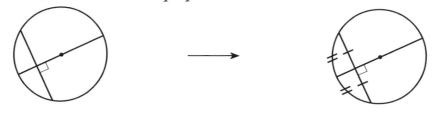

This theorem is easy to prove. First draw radii to the endpoints of the chord. Since all radii of a circle are equal, we have again formed an isosceles triangle which we can divide into two congruent right triangles. Triangle $\triangle ADC \cong \triangle BDC$ (by HL). CPCT tells us that $\overline{AD} \cong \overline{DB}$, that is, that the chord is bisected. (Definition of a segment bisector. Definition of a midpoint.)

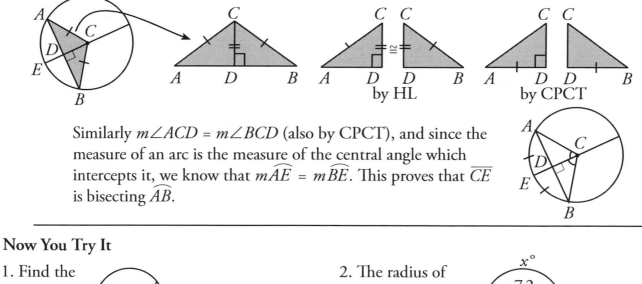

Similarly $m\angle ACD = m\angle BCD$ (also by CPCT), and since the measure of an arc is the measure of the central angle which intercepts it, we know that $m\overset{\frown}{AE} = m\overset{\frown}{BE}$. This proves that $\overline{CE}$ is bisecting $\overset{\frown}{AB}$.

Now You Try It

1. Find the length of a diameter of circle C.

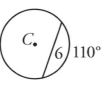

2. The radius of circle W is 4, find x.

Angles and The Arcs They Intercept

When the vertex of an angle is placed at the center of a circle, the sides of the angle intercept the circle and determine (cut off) an arc. This kind of angle is called a central angle. A central angle and the minor arc it intercepts, or cuts off, are equal.

DEFINITION: *The measure of a minor arc equals the measure of its central angle.*

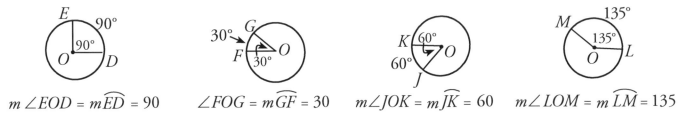

$m\angle EOD = m\widehat{ED} = 90$ $\angle FOG = m\widehat{GF} = 30$ $m\angle JOK = m\widehat{JK} = 60$ $m\angle LOM = m\widehat{LM} = 135$

When the vertex of an angle lies on a circle and the sides of the angle are chords of the circle, the angle is called an inscribed angle.

INSCRIBED ANGLE THEOREM: 1. *An arc equals two times its inscribing angle.* $\angle = \frac{1}{2}\widehat{\textbf{Arc}}$

Here are some examples: 2. *An inscribed angle = ½ the arc it intercepts.*

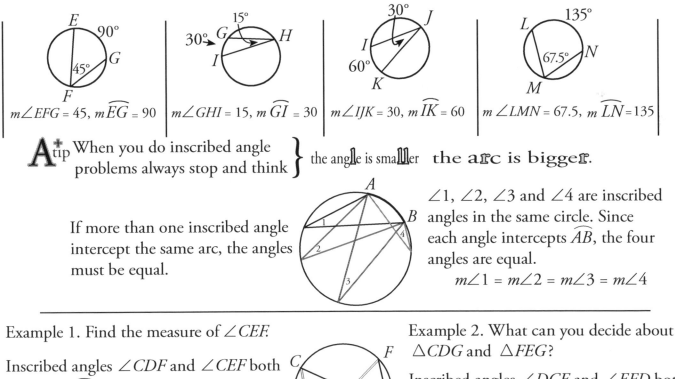

$m\angle EFG = 45, m\widehat{EG} = 90$ | $m\angle GHI = 15, m\widehat{GI} = 30$ | $m\angle IJK = 30, m\widehat{IK} = 60$ | $m\angle LMN = 67.5, m\widehat{LN} = 135$

A tip When you do inscribed angle problems always stop and think } the angle is smaller **the arc is bigger.**

If more than one inscribed angle intercept the same arc, the angles must be equal.

$\angle 1, \angle 2, \angle 3$ and $\angle 4$ are inscribed angles in the same circle. Since each angle intercepts $\widehat{AB}$, the four angles are equal.
$$m\angle 1 = m\angle 2 = m\angle 3 = m\angle 4$$

Example 1. Find the measure of $\angle CEF$.

Inscribed angles $\angle CDF$ and $\angle CEF$ both intercept $\widehat{CF}$ so the angles must be equal. Setting their measures equal to each other:
$$x + 30 = 3x, 2x = 30,$$
$$x = 15$$
$$m\angle CEF = 3x = 45 \checkmark$$

Example 2. What can you decide about $\triangle CDG$ and $\triangle FEG$?

Inscribed angles $\angle DCE$ and $\angle EFD$ both intercept $\widehat{DE}$ so the two angles are equal. Since $\angle CDF$ and $\angle CEF$ both intercept $\widehat{CF}$, they are also equal. Therefore, by the AA ~ Postulate: $\triangle CDG \sim \triangle FEG$. $\checkmark$

The sides of an inscribed angle are lines which include chords of the circle. As a chord gets smaller and smaller its limit is a tangent.

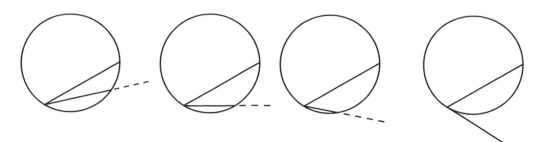

The chord and the tangent in the fourth drawing above form an angle, and the measure of the angle formed is equal to ½ the measure of the arc of the chord. This is an example of drawing a conclusion by taking a limit.

$$\angle = ½ \overset{\frown}{Arc}$$

THEOREM: *The angle formed by a chord and a tangent equals one half the measure of the arc of the chord.*

Here are two examples:

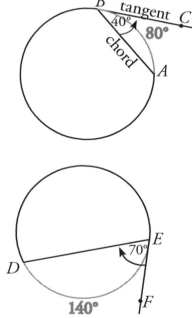

If we know that $\overline{BC}$ is tangent to the circle and that $m\overset{\frown}{AB} = 80$ then we also know that $m\angle ABC = 40$.

Notes:
• "arc of the chord" means
• the arc cut off by the chord.

If we know that $\overline{EF}$ is tangent to the circle and that $m\angle DEF = 70$ then we also know that $m\overset{\frown}{DE} = 140$.

A⁺tip It's important to notice that in order to use the last theorem, a problem must tell you or you must be able to figure out, that a line segment is tangent. For example, in the figure on the right, we know nothing about the measure of $\angle GHI$ because we don't know (and can never assume) that segment $\overline{HI}$ is tangent to the circle.

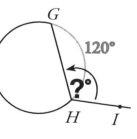

Quick Review			
Angle	Location of Vertex	Sides	∠Relationship to Arc(s)
CENTRAL ANGLE	CENTER OF THE CIRCLE	2 RADII	=
INSCRIBED ANGLE	ON THE CIRCLE	2 CHORDS	½
CHORD/TANGENT	ON THE CIRCLE	1 CHORD, 1 TANGENT	½

Now You Try It

1. Given the angle and arc measures shown in the figure and that $\overrightarrow{BA}$ and $\overline{DC}$ are tangents, find all of the numbered measurements. (The point at 4 is the center of the circle in the figure.)

Note: If you can, always answer this kind of problem in numerical order. Usually, the numbers are ordered to lead you through the problem in the easiest, (most straight forward) way.

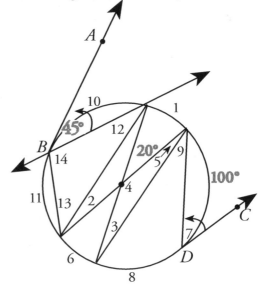

2. Given $\overrightarrow{BC}$ is tangent to the circle at right, $m\,\widehat{AB} = (4x + 32)°$ and $m\angle ABC = (x + 30)°$, find x.

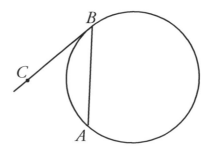

155

When solving angle and arc problems, always ask yourself, *"where is the vertex of the angle?"* $\mathbf{A}^{+}_{tip}$

More Angles and the Arcs They Intercept

So far we have studied angles with the vertex in the center of the circle and on the circle itself. Now we will learn about angles with vertices somewhere inside the circle, but not necessarily at the center. Here's an example:

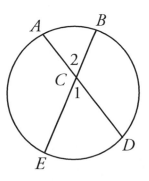

$\angle 1$ and $\angle 2$ are formed by the intersection of two chords of the circle.

Although 4 arcs are formed, for this problem we will focus on $\overarc{AB}$ and $\overarc{ED}$.

Notice that $m\angle 1 = m\angle 2$ by the Vertical Angle Theorem. Therefore, if we find the measure of one angle, we are finding the measure of the other.

Here's the connection that $\angle 1$ and $\angle 2$ have to the intercepted arcs, $\overarc{AB}$ and $\overarc{ED}$:

$$m\angle 1 = m\angle 2 = \frac{m\overarc{AB} + m\overarc{ED}}{2}$$

This means that each angle is equal to the average of the two intercepted arcs.

THEOREM: *If two chords intersect in a circle, the angles formed equal the average of the two intercepted arcs.* $\angle = \dfrac{\overarc{Arc} + \overarc{Arc}}{2}$

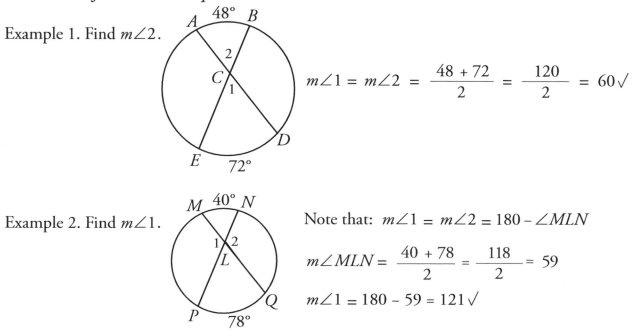

Example 1. Find $m\angle 2$.

$m\angle 1 = m\angle 2 = \dfrac{48 + 72}{2} = \dfrac{120}{2} = 60 \checkmark$

Example 2. Find $m\angle 1$.

Note that: $m\angle 1 = m\angle 2 = 180 - \angle MLN$

$m\angle MLN = \dfrac{40 + 78}{2} = \dfrac{118}{2} = 59$

$m\angle 1 = 180 - 59 = 121 \checkmark$

It's important to know the connection between the measurements of the angles and arcs. Some problems will put all three in different variable expressions. An example of a variable expression is $(3x + 5)$. You need to memorize the formulas and know where each value goes:

Example 3. Solve for x.

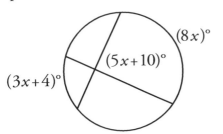

Use the relationship: $\angle = \dfrac{\overset{\frown}{Arc} + \overset{\frown}{Arc}}{2}$

Substitute in the given expressions: $5x + 10 = \dfrac{(3x+4) + (8x)}{2}$

Form a proportion: $\dfrac{(5x+10)}{1} = \dfrac{(3x+4) + (8x)}{2}$

Cross multiply: $2(5x+10) = 1((3x+4) + (8x))$

Distribute: $10x + 20 = 3x + 4 + 8x$

Combine like terms: $10x + 20 = 11x + 4$

Solve for x: $x = 16 \checkmark$

Some problems will give the angle and one of the arcs and ask you to find the missing arc. Here's an example:

Example 4. Solve for x.

Use the relationship: $\angle = \dfrac{\overset{\frown}{Arc} + \overset{\frown}{Arc}}{2}$

Substitute in the given expressions: $52 = \dfrac{38 + x}{2}$

Form a proportion: $\dfrac{(52)}{1} = \dfrac{38 + x}{2}$

Cross multiply: $(52)2 = 1(38 + x)$

Distribute: $104 = 38 + x$

Solve for x: $x = 66 \checkmark$

Now You Try It

1. Find x.

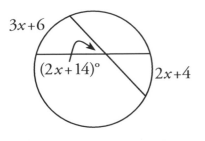

2. Find $m\overset{\frown}{DC}$.

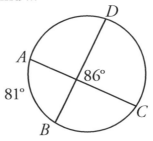

Quick Review

ANGLE	LOCATION OF VERTEX	SIDES	$\angle$ RELATIONSHIP TO ARC(S)
CENTRAL ANGLE	CENTER OF THE CIRCLE	2 RADII	=
INSCRIBED ANGLE	ON THE CIRCLE	2 CHORDS	½
CHORD/TANGENT	ON THE CIRCLE	1 CHORD, 1 TANGENT	½
INTERSECTION OF TWO CHORDS	INTERIOR OF THE CIRCLE	PARTS OF CHORDS	$\dfrac{\overset{\frown}{Arc} + \overset{\frown}{Arc}}{2}$

Angles With Vertices Outside the Circle — So far, we have studied angles with the vertex at the center of the circle, on the circle itself and with vertices somewhere in the interior of the circle. Now we will learn about angles with vertices *outside* the circle. Here's an example:

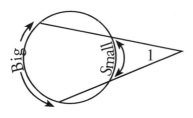

The vertex of $\angle 1$ is *outside* the circle.

The sides of $\angle 1$ are two sccants, however, they could also be a secant and a tangent, or two tangents.

Although $\angle 1$ divides the circle into four arcs, we are interested in the arcs farthest from and closest to $\angle 1$.

The arc farthest from the angle will always be bigger than the arc closest to the angle. The arcs are labeled accordingly.

Here's the connection that the size of the angle has to the intercepted arcs.

$$m\angle 1 = \frac{\overset{\frown}{Big} - \overset{\frown}{Sml}}{2}$$

THEOREM: *The angle formed by two secants, a secant and a tangent, or two tangents, equals one half the difference of the larger intercepted arc minus the smaller.* $\quad \angle = \dfrac{\overset{\frown}{Big} - \overset{\frown}{Sml}}{2}$

For example:

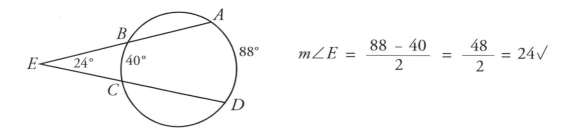

$$m\angle E = \frac{88 - 40}{2} = \frac{48}{2} = 24 \checkmark$$

If the sides of the angle are a tangent and a secant, the problem works in the same way:

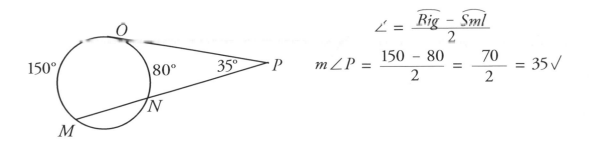

$$\angle = \frac{\overset{\frown}{Big} - \overset{\frown}{Sml}}{2}$$

$$m\angle P = \frac{150 - 80}{2} = \frac{70}{2} = 35\checkmark$$

Now with variables, solve for x.

Use the relationship:	$\angle = \dfrac{\overset{\frown}{Big} - \overset{\frown}{Sml}}{2}$
Substitute in the given expressions:	$22 = \dfrac{124 - x}{2}$
Form a proportion:	$\dfrac{22}{1} = \dfrac{124 - x}{2}$
Cross multiply:	$2(22) = 1(124 - x)$
Distribute:	$44 = 124 - x$
Solve for x:	$x = 80\checkmark$

The Two Tangent Case — If the sides of the angle are two tangents, the connection between the angle and the arcs is the same, but the arcs have a special connection with each other. The two arcs make one complete circle which means that their measures will always add up to 360°.

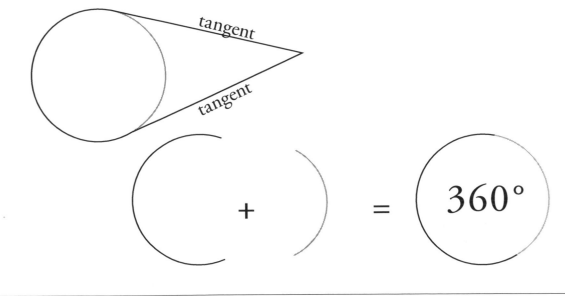

Here's an example of a two tangent problem:

Given tangents $\overline{AB}$ and $\overline{AD}$, and $m\angle A = 56$, find $m\widehat{BD}$, and $m\widehat{BCD}$.

Let $m\widehat{BD} = x$, then $m\widehat{BCD} = 360 - x$.

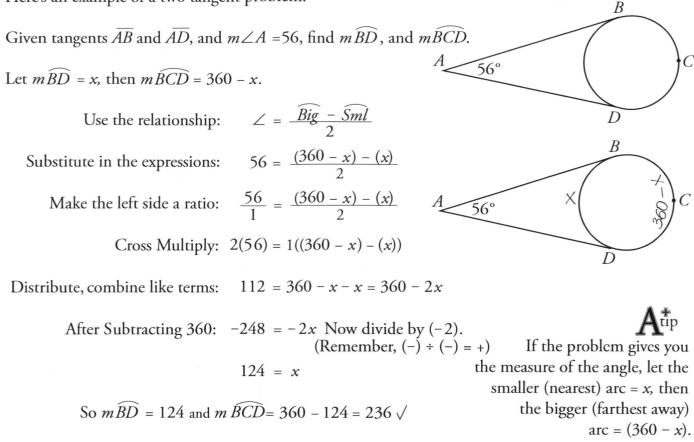

Use the relationship: $\angle = \dfrac{\widehat{Big} - \widehat{Sml}}{2}$

Substitute in the expressions: $56 = \dfrac{(360 - x) - (x)}{2}$

Make the left side a ratio: $\dfrac{56}{1} = \dfrac{(360 - x) - (x)}{2}$

Cross Multiply: $2(56) = 1((360 - x) - (x))$

Distribute, combine like terms: $112 = 360 - x - x = 360 - 2x$

After Subtracting 360: $-248 = -2x$ Now divide by (-2).
(Remember, $(-) \div (-) = +$)

$124 = x$

So $m\widehat{BD} = 124$ and $m\widehat{BCD} = 360 - 124 = 236 \checkmark$

A⁺tip

If the problem gives you the measure of the angle, let the smaller (nearest) arc = x, then the bigger (farthest away) arc = $(360 - x)$.

Now You Try It

1. Find the value of x in the figure on the right:

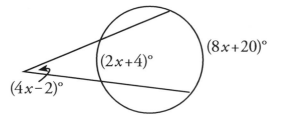

$(8x+20)°$
$(2x+4)°$
$(4x-2)°$

2. Given tangents $\overline{AD}$ and $\overline{CD}$, and $m\angle D = 70$, find $m\widehat{AC}$, and $m\widehat{ABC}$.

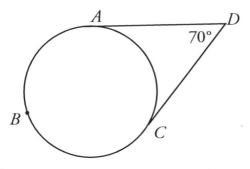

Be sure to check in the back of the book to see how you did!

A^+_{tip} Remember! When solving angle and arc problems, ask yourself,
"where is the vertex?"

The location of the vertex:

The angle measure's relationship to the arc measure:

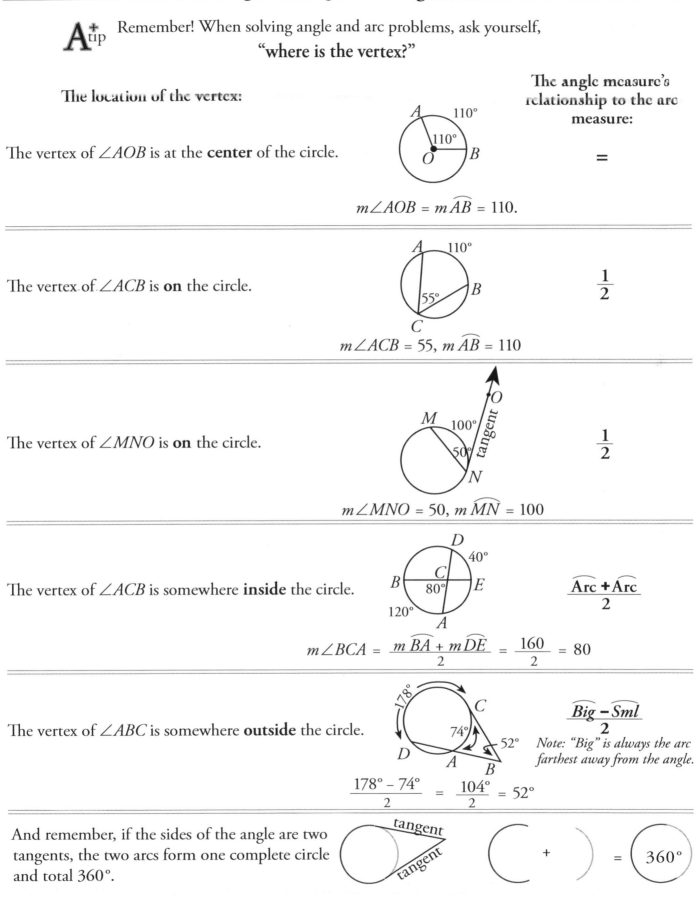

The vertex of $\angle AOB$ is at the **center** of the circle.

=

$$m\angle AOB = m\widehat{AB} = 110.$$

The vertex of $\angle ACB$ is **on** the circle.

$\dfrac{1}{2}$

$$m\angle ACB = 55, \; m\widehat{AB} = 110$$

The vertex of $\angle MNO$ is **on** the circle.

$\dfrac{1}{2}$

$$m\angle MNO = 50, \; m\widehat{MN} = 100$$

The vertex of $\angle ACB$ is somewhere **inside** the circle.

$\dfrac{\widehat{Arc} + \widehat{Arc}}{2}$

$$m\angle BCA = \dfrac{m\widehat{BA} + m\widehat{DE}}{2} = \dfrac{160}{2} = 80$$

The vertex of $\angle ABC$ is somewhere **outside** the circle.

$\dfrac{\widehat{Big} - \widehat{Sml}}{2}$

Note: "Big" is always the arc farthest away from the angle.

$$\dfrac{178° - 74°}{2} = \dfrac{104°}{2} = 52°$$

And remember, if the sides of the angle are two tangents, the two arcs form one complete circle and total 360°.

$+$ $=$ $360°$

Segments and Circles

Chord Segments — When two chords of a circle intersect, each chord is divided into two parts. The lengths of the parts of one chord have a connection with the lengths of the parts of the other chord. The connection is due to the symmetry of circles. Study the figures; can you discover the pattern?

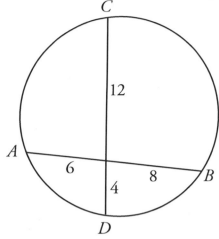

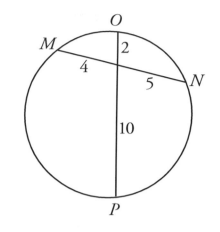

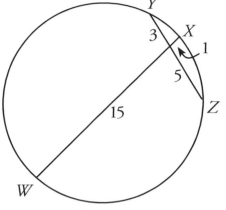

$AB = 14$ and $DC = 16$

The *product* of the *parts* of each chord are equal.

$(6)(8) = 48$
$(4)(12) = 48 \checkmark$

$MN = 9$ and $OP = 12$

The *product* of the *parts* of each chord are equal.

$(4)(5) = 20$
$(10)(2) = 20 \checkmark$

$WX = 16$ and $YZ = 8$

The *product* of the *parts* of each chord are equal.

$(3)(5) = 15$
$(15)(1) = 15 \checkmark$

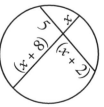

THEOREM: *If two chords intersect in a circle, the products of each chord's segments are equal.*

$$(a)(b) = (c)(d)$$

Here's an example that uses the above theorem:

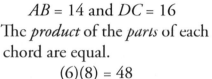

Find x.

Since the products of each chord's parts are equal, set them equal to each other:

$(x - 3)(x + 3) = (4)(4)$ — Now, foil and distribute.

$x^2 - 3x + 3x - 9 = 16$ — Now, collect like terms.

$x^2 - 9 = 16$ — Add 9 to each side.

$x^2 = 25$ — Take the square root of each side.

$x = 5 \checkmark$ — Done!

Now You Try It

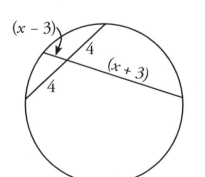

1. Find y.

2. Find the length of the shorter chord.

162

Secant and Tangent Segments

Because of the symmetry of circles, there is a connection between the lengths of secant parts and/or tangents when both objects start from the same point outside of the circle. First, here's a review of some circle terms:

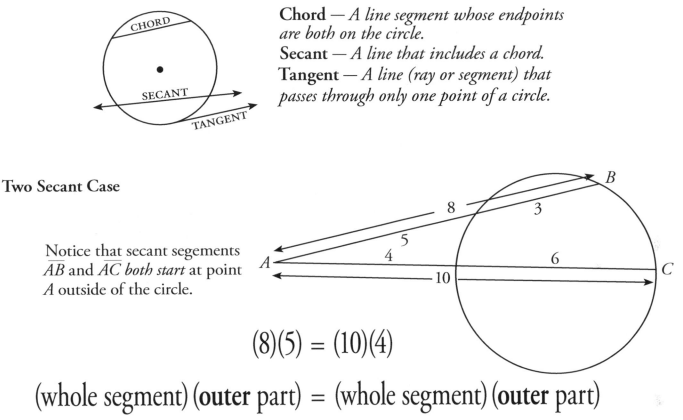

Chord — *A line segment whose endpoints are both on the circle.*
Secant — *A line that includes a chord.*
Tangent — *A line (ray or segment) that passes through only one point of a circle.*

Two Secant Case

Notice that secant segements $\overline{AB}$ and $\overline{AC}$ *both start* at point *A* outside of the circle.

$$(8)(5) = (10)(4)$$

$$(\text{whole segment})\,(\textbf{outer}\ \text{part}) = (\text{whole segment})\,(\textbf{outer}\ \text{part})$$

These problems are easy as long as you pay attention to the *details* of the rule:

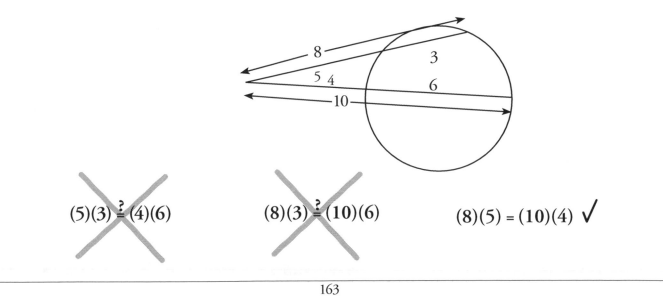

$$(5)(3) \overset{?}{=} (4)(6)$$

$$(8)(3) \overset{?}{=} (10)(6)$$

$$(8)(5) = (10)(4) \ \checkmark$$

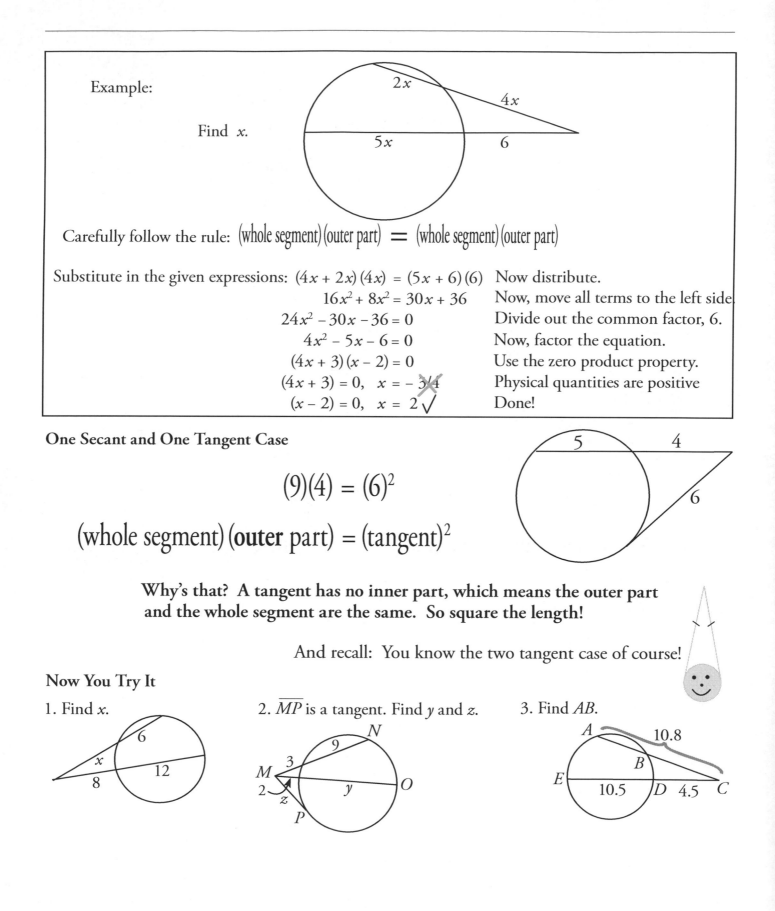

Example:

Find *x*.

Carefully follow the rule: (whole segment)(outer part) = (whole segment)(outer part)

Substitute in the given expressions: $(4x + 2x)(4x) = (5x + 6)(6)$ Now distribute.

$$16x^2 + 8x^2 = 30x + 36$$ Now, move all terms to the left side.

$$24x^2 - 30x - 36 = 0$$ Divide out the common factor, 6.

$$4x^2 - 5x - 6 = 0$$ Now, factor the equation.

$$(4x + 3)(x - 2) = 0$$ Use the zero product property.

$(4x + 3) = 0, \quad x = -\cancel{3/4}$ Physical quantities are positive

$(x - 2) = 0, \quad x = 2 \checkmark$ Done!

One Secant and One Tangent Case

$$(9)(4) = (6)^2$$

$$\text{(whole segment)}\,\textbf{(outer part)} = \text{(tangent)}^2$$

Why's that? A tangent has no inner part, which means the outer part and the whole segment are the same. So square the length!

And recall: You know the two tangent case of course!

Now You Try It

1. Find *x*.

2. $\overline{MP}$ is a tangent. Find *y* and *z*.

3. Find *AB*.

12. AREA

Squares

A square is a polygon with 4 equal sides and 4 right angles.

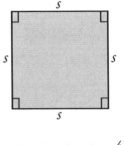

If the side of a square is *s* units long, its perimeter is 4*s*.

What's a perimeter? Perimeter is a "linear" measurement. The perimeter is the total of the lengths of the sides of a figure. Think of starting at one corner of the figure and walking all the way around until you get back to your starting point. The distance you've walked is the perimeter of the figure.

Think of a **P**erson walking when you are working with **P**erimeters.

$$p = 8 + 8 + 8 + 8 = 32$$

$$p = s + s + s + s = 4s$$

If the side of a square is *s* units long, its **area** is $s \cdot s = s^2$

When we say area, we mean surface area. Area answers questions such as, how much paint do you need to cover a surface, or how much grass seed do you need to seed a lawn. Area is a square measurement. The answer should be in square units, for example, square inches or square centimeters. However, if no unit is given in the problem, sometimes the unit is left out in the answer.

$$A = 8 \cdot 8 = 64 \, un^2$$

$$A = s^2$$

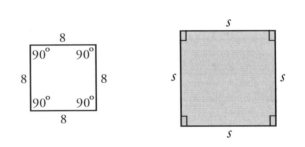

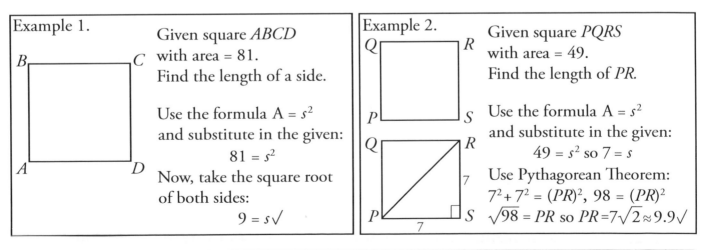

Example 1. Given square *ABCD* with area = 81. Find the length of a side.

Use the formula $A = s^2$ and substitute in the given:
$$81 = s^2$$
Now, take the square root of both sides:
$$9 = s \checkmark$$

Example 2. Given square *PQRS* with area = 49. Find the length of *PR*.

Use the formula $A = s^2$ and substitute in the given:
$$49 = s^2 \text{ so } 7 = s$$
Use Pythagorean Theorem:
$$7^2 + 7^2 = (PR)^2, \ 98 = (PR)^2$$
$$\sqrt{98} = PR \text{ so } PR = 7\sqrt{2} \approx 9.9 \checkmark$$

Rectangles

A rectangle is a quadrilateral with 4 right angles. A rectangle has two pairs of equal sides. Squares are rectangles with four equal sides. This means that a square is a "special case" of a rectangle.

The perimeter of a rectangle with base b and height h is 2b + 2h.

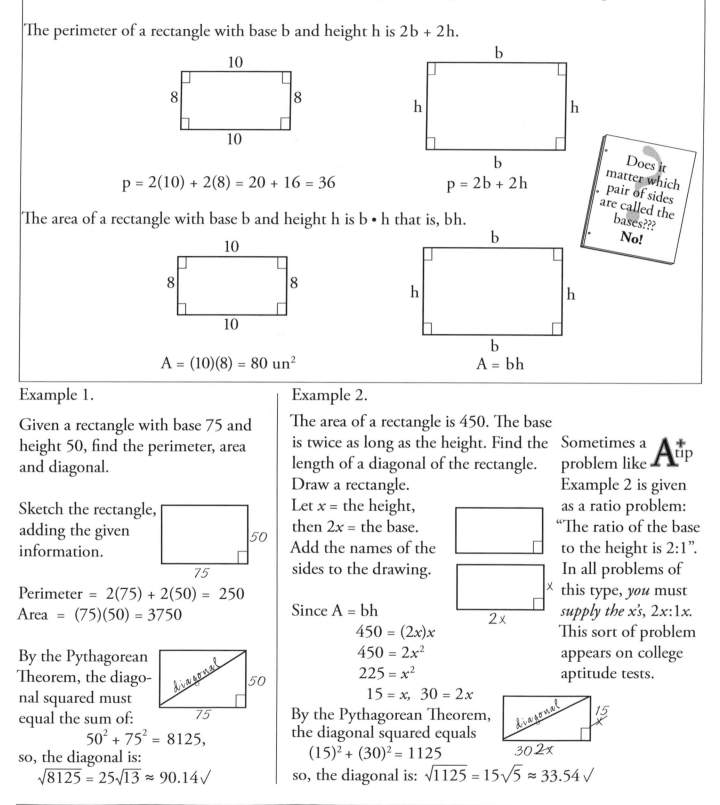

$p = 2(10) + 2(8) = 20 + 16 = 36$ $p = 2b + 2h$

The area of a rectangle with base b and height h is b • h that is, bh.

Does it matter which pair of sides are called the bases??? **No!**

$A = (10)(8) = 80 \text{ un}^2$ $A = bh$

Example 1.

Given a rectangle with base 75 and height 50, find the perimeter, area and diagonal.

Sketch the rectangle, adding the given information.

Perimeter $= 2(75) + 2(50) = 250$
Area $= (75)(50) = 3750$

By the Pythagorean Theorem, the diagonal squared must equal the sum of:
$$50^2 + 75^2 = 8125,$$
so, the diagonal is:
$$\sqrt{8125} = 25\sqrt{13} \approx 90.14 \checkmark$$

Example 2.

The area of a rectangle is 450. The base is twice as long as the height. Find the length of a diagonal of the rectangle. Draw a rectangle.
Let x = the height, then $2x$ = the base.
Add the names of the sides to the drawing.

Since $A = bh$
$$450 = (2x)x$$
$$450 = 2x^2$$
$$225 = x^2$$
$$15 = x, \quad 30 = 2x$$
By the Pythagorean Theorem, the diagonal squared equals
$$(15)^2 + (30)^2 = 1125$$
so, the diagonal is: $\sqrt{1125} = 15\sqrt{5} \approx 33.54 \checkmark$

Sometimes a problem like A⁺ tip Example 2 is given as a ratio problem: "The ratio of the base to the height is 2:1". In all problems of this type, *you* must *supply the x's, 2x:1x.* This sort of problem appears on college aptitude tests.

Example 3.

The area of a rectangle is 96. The ratio of the base to the height is 2:3. Find the base, the height, the perimeter and the diagonal of the rectangle.

1. Draw a rectangle that roughly matches the given information.

2. The ratio is 2:3, you supply the x's, $2x:3x$. Now, mark up the drawing, showing all the information that you know.

3. Area = (base)(height)
$$96 = (2x)(3x) = 6x^2$$
$$16 = x^2, \quad x = 4.$$
base = 8 √   height = 12 √
perimeter = 8 + 8 + 12 + 12 = 40 √

By the Pythagorean Theorem,
$$(\text{diagonal})^2 = (8)^2 + (12)^2$$
$$(\text{diagonal})^2 = 64 + 144$$
$$(\text{diagonal})^2 = 208$$
$$\text{diagonal} = 4\sqrt{13} \approx 14.4 \,\checkmark$$

In problems like Example 3 that ask for several pieces of information, always be sure to go back and check that you've given all of the requested answers:

base √   height √   perimeter √   diagonal √

Example 4.

Find the area of the figure at right.

A good strategy for solving this kind of problem is to break the figure into rectangular pieces:
Mark up the drawing, dividing the figure into rectangles.
Label each rectangle.
Find the missing dimensions:
Find x: $40 = x + x + 11 + 5$, $40 = 2x + 16$, $24 = 2x$, $x = 12$
Find y: $y = 4 + 10 + 8 + 24 = 46$, $y = 46$
Now find the area of each rectangle:

$\boxed{1}$12 (4)(12) = 48

$\boxed{2}$12 (14)(12) = 168

$\boxed{3}$11 (22)(11) = 242

$\boxed{4}$5 (46)(5) = 230

Whole Figure = Rectangles $1 + 2 + 3 + 4 = 48 + 168 + 242 + 230 = 688$ √

Do these problems step by step.

Another way to divide the figure.

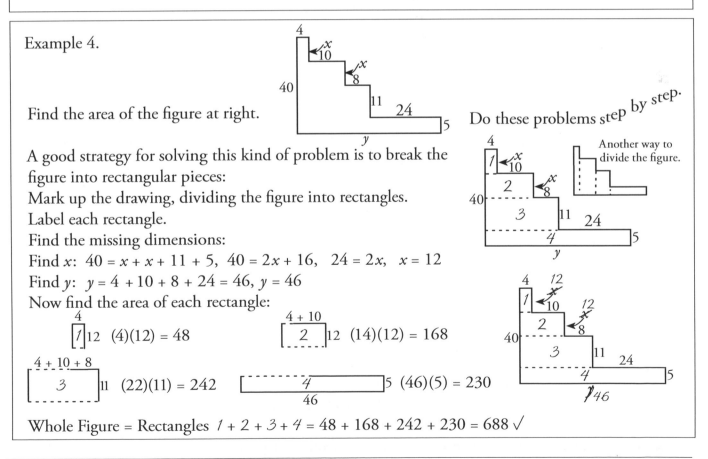

Parallelograms

Quick Review — When you see the word or the symbol for a parallelogram, here's what you're supposed to know about the figure:

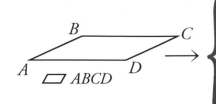

▱ *ABCD*

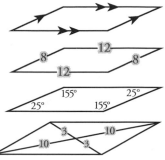

1. Both pairs of opposite sides are parallel.

2. Both pairs of opposite sides are congruent.

3. Both pairs of opposite angles are equal.

4. The diagonals break each other in half.

Remember, whenever you see the ▱ symbol you know all of the above facts about the figure. An extra fact about parallelograms: same side interior (consecutive) angles always add up to 180°, that is, they are supplementary.

Altitudes of Parallelograms

An **altitude** of a parallelogram is the (⊥) distance between a side and the *line that includes the opposite side*. The length of the altitude is the **height**. Since the altitude is ⊥, we have to say "the line that includes" the opposite side. Here's why:

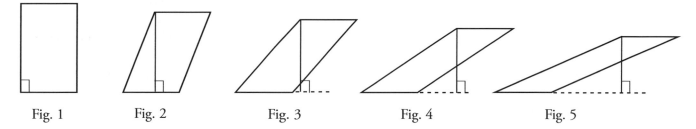

Fig. 1   Fig. 2   Fig. 3   Fig. 4   Fig. 5

The important thing is the perpendicularity. In figures 3, 4, and 5, a (⊥) altitude cannot be drawn from the top side to the opposite side *within* the figure. So extend the opposite side to show the line of which it is a part, and then draw in the altitude. Remember, each side of a parallelogram is a segment and every segment is just a small part of an infinite line.

Since opposite sides are parallel, the distance between them is always the same. This means that it doesn't matter where the altitude is drawn. the length of every altitude drawn from a side (or the line which includes the side) to the opposite side (or the line which includes the opposite side) is always the same.

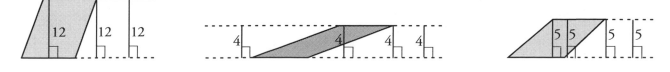

Since each parallelogram has two pairs of parallel sides, each parallelogram has two altitudes. Here are some examples:

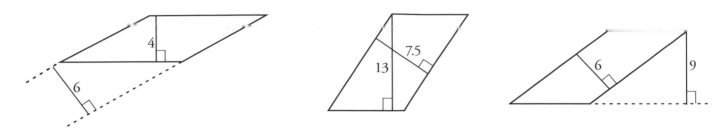

Finding the Area of a Parallelogram

We know that the area of a *rectangle* is A = (base) ✕ (height).
We can use this fact to find the area of a parallelogram.

Given ▱   Draw an altitude.   Note the right triangle.   Slide the right triangle over and you have a rectangle!

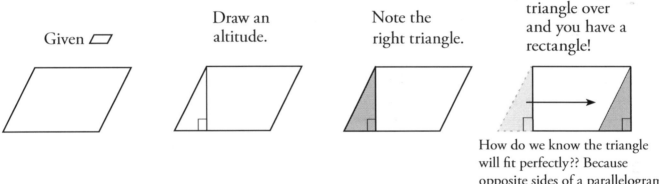

How do we know the triangle will fit perfectly?? Because opposite sides of a parallelogram are congruent and parallel. The triangles are congruent by HL.

Since we know that the area of a rectangle is

A = (base) ✕ (height)

The area of a parallelogram must be

A = (base) ✕ (height to *that* base)

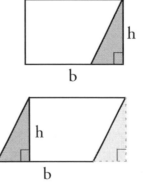

Example 1. Find the area of the parallelogram below.

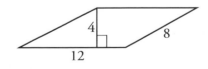

A = (12)(8)   A = (8)(4)

A = (base) ✕ (height to *that* base) = (12)(4) = 48 √

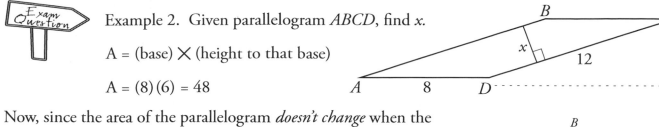

Example 2. Given parallelogram *ABCD*, find *x*.

A = (base) ✕ (height to that base)

A = (8)(6) = 48

Now, since the area of the parallelogram *doesn't change* when the other base and height are used:

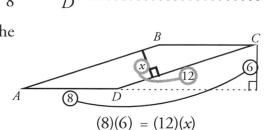

A = 48 = (12)(*x*)   The (other) base times the height to *that* base.

$$\frac{\cancel{48}^{\,4}}{\cancel{12}} = \frac{(\cancel{12})(x)}{\cancel{12}} \qquad x = 4$$

$$(8)(6) = (12)(x)$$

Example 3. Find the area of parallelogram *MNOP*.

The problem is giving us both bases but neither height. Since finding the area requires a height, there must be a way to find one with the given information. Because the problem gives an angle measure, we should think about using trigonometry (or special triangles or the Law of Sines) to find the height.

Draw an altitude from *M* and label the point of intersection *R*. Constructing *MR* creates right triangle *MRP* and because opposite sides of a parallelogram are equal, we know that *MP* = 28. Since the problem gave us *m∠P* = 60°, we can solve the triangle for *MR*.

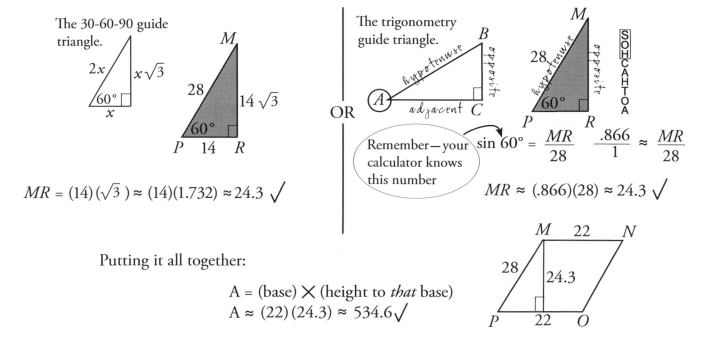

Finding *MR* using special triangles:

The 30-60-90 guide triangle.

$$MR = (14)(\sqrt{3}) \approx (14)(1.732) \approx 24.3 \ \checkmark$$

Finding *MR* using trigonometry:

The trigonometry guide triangle.

OR

Remember—your calculator knows this number

$$\sin 60° = \frac{MR}{28} \qquad \frac{.866}{1} \approx \frac{MR}{28}$$

$$MR \approx (.866)(28) \approx 24.3 \ \checkmark$$

Putting it all together:

A = (base) ✕ (height to *that* base)
A ≈ (22)(24.3) ≈ 534.6 ✓

Now You Try It

1. Find the perimeter of square *UVWX*.

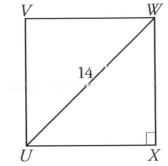

2. The area of a rectangle is 180. The ratio of the base to the height is 4:5. Find the perimeter and the diagonal of the rectangle.

3. Find the area of the figure on the right.

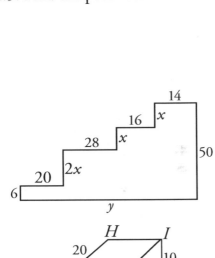

4. Find the area of parallelogram *GHIJ*.

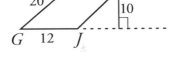

5. Given parallelogram *ABCD*, find *x*.

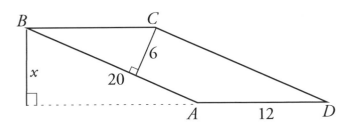

6. Find the area of parallelogram *PQRS*.

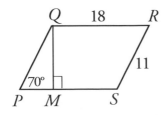

Triangles

A triangle is ½ of a parallelogram. Here are some examples:

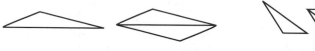

Because a triangle is ½ of a parallelogram, it makes sense that the area of a triangle is:

$$\frac{1}{2}\left[(\text{base})\times(\text{height to }that\text{ base})\right] = \frac{(\text{base})\times(\text{height to }that\text{ base})}{2}$$

Notes:
½ x is the same as $\frac{x}{2}$

The height is the length of the altitude. The altitude to a base is the (⊥) distance from the opposite vertex to the line that includes the base.

All triangles have 3 altitudes. Here are the altitudes of an acute, right and obtuse triangle.

Example 1. Find the area of the triangle on the right.

$$A = \frac{1}{2}(15)(4) = 30 \checkmark$$

Example 2. Find the area of triangle *ABC*.

$$A = \frac{1}{2}(20)(12) \quad \text{(Wrong way.)}$$

$$A = \frac{1}{2}(20)(5) = 50 \checkmark \text{ (Right way!)}$$

A Special Formula for the Area of <u>Right Triangles Only</u>

The two legs of a right triangle are perpendicular. This means that each leg is the altitude (height) to the other leg.

$$A_\triangle = \frac{(\text{Leg})\times(\text{Leg})}{2}$$

In your geometry class you will need to find the area of many right triangles. Just think "leg times leg divided by two". Notice, there is no hypotenuse in this formula.

Example 1. Find the area of triangle *MNO*.

$$A_\triangle = \frac{(\text{Leg})\times(\text{Leg})}{2}$$

$$A_\triangle = \frac{(5)(12)}{2} = 30\checkmark$$

Example 2. Find the area of triangle *FED*. There is no perpendicular symbol (⌐) and the problem does not say that triangle *FED* is a right triangle. We can use the converse of the Pythagorean Theorem to test the triangle to see if it is a right triangle:

$$6^2 + 8^2 \stackrel{?}{=} 10^2$$
$$36 + 64 = 100 \checkmark$$

Since *FED* is a right triangle, we can use the special formula:

$$A_\triangle = \frac{(6)(8)}{2} = 24\checkmark$$

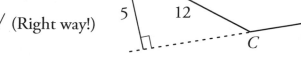

A⁺tip Become an expert with Isosceles Triangles!

Finding the Area of Isosceles and Equilateral Triangles
A triangle with at least two equal sides is an isosceles triangle.
A triangle with three equal sides is an equilateral triangle.

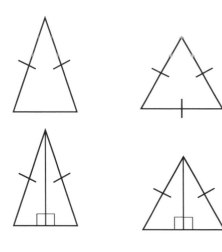

Draw an altitude between two equal sides of each triangle.

Each triangle is divided into **2** congruent right triangles.

To find the area of a right triangle only we can use the special formula: $A_{\triangle} = \dfrac{(\text{Leg}) \times (\text{Leg})}{2}$

Example 1. Find the area of the isosceles triangle below.

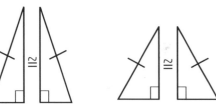

The given triangle.	Draw the altitude between the equal sides.	The altitude divides the original triangle into two ≅ right triangles.	By Pythagorean Theorem: $x^2 + 7^2 = 10^2$ $x^2 = 100 - 49 = 51$ $x = \sqrt{51}$

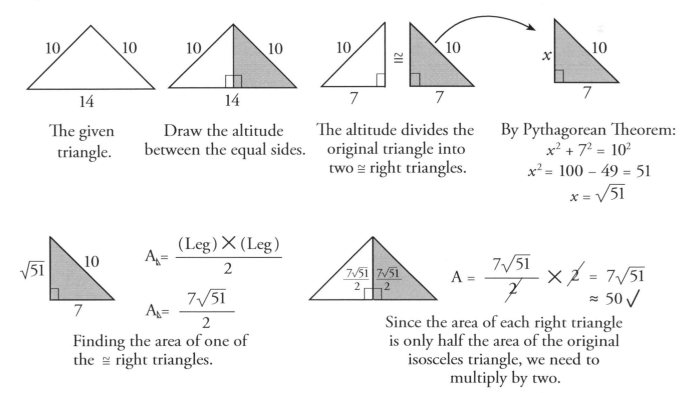

$A_{\triangle} = \dfrac{(\text{Leg}) \times (\text{Leg})}{2}$

$A_{\triangle} = \dfrac{7\sqrt{51}}{2}$

Finding the area of one of the ≅ right triangles.

$A = \dfrac{7\sqrt{51}}{2} \times 2 = 7\sqrt{51}$
$\approx 50 \checkmark$

Since the area of each right triangle is only half the area of the original isosceles triangle, we need to multiply by two.

Example 2. Find the area of the given triangle.

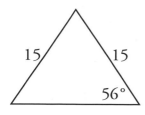

The given.

Step 1.

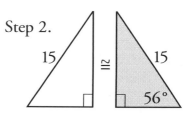

Draw the altitude between the equal sides.

Step 2.

The altitude divides the original triangle into two congruent right triangles.

Step 3.

If you know the size of one acute angle and one side of a right triangle, you can find the lengths of the other sides using trigonometry. (You can also use the Law of Sines. Be sure to see Example 3 on the next page.)

Step 4.

We need to find the lengths of the legs of the triangle, that is, the opposite and adjacent sides.

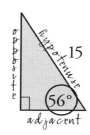

Step 5.

Find "opposite".

$$\sin 56° = \frac{opp}{15}$$

$$.829 \approx \frac{opp}{15}$$

$$\frac{.829}{1} \approx \frac{opp}{15}$$

$$opp \approx 12.4 \checkmark$$

Find "adjacent".

$$\cos 56° = \frac{adj}{15}$$

$$.559 \approx \frac{adj}{15}$$

$$\frac{.559}{1} \approx \frac{adj}{15}$$

$$adj \approx 8.4 \checkmark$$

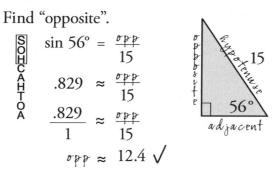

Step 6.

$$A_\triangle = \frac{(\text{Leg}) \times (\text{Leg})}{2}$$

$$A_\triangle \approx \frac{(12.4)(8.4)}{2} \approx 52.1 \checkmark$$

Step 7.

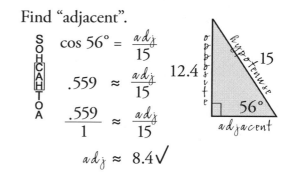

$$A \approx 52.1 \times 2 \approx 104.2$$

Since one right triangle is only half the area of the original triangle, multiply by two.

Done !

Example 3. Find the area of the given triangle using the Law of Sines.

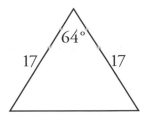

The given.

Step 1.

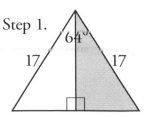

Draw the altitude beween the equal sides.

Step 2.

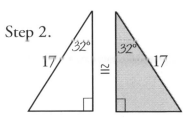

The altitude divides the original triangle into two congruent right triangles.

Step 3.

If you know the size of one acute angle and one side of a right triangle, you can always find the lengths of the other sides using the Law of Sines. (You can also use trigonometry. See Example 2 on the previous page.)

Step 4.

To find its area, we need to find the lengths of the legs of the triangle. Label the sides and angles of the triangle.

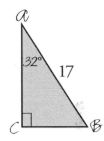

Step 5.

Find the length of leg $\overline{CB}$:

$$\frac{\sin 32°}{CB} = \frac{\sin 90°}{17}$$

$$\frac{.53}{\cancel{\sin 32°}} \approx \frac{1.0}{\cancel{\sin 90°}}$$
$$\frac{.53}{CB} \approx \frac{1.0}{17}$$

$$(.53)\,17 \approx CB(1.0)$$

$$9.01 \approx CB \checkmark$$

Find the length of leg $\overline{AC}$:

$$\frac{\sin 58°}{AC} = \frac{\sin \angle 90°}{17}$$

$$\frac{.85}{\cancel{\sin 58°}} \approx \frac{1.0}{\cancel{\sin 90°}}$$
$$\frac{.85}{AC} \approx \frac{1.0}{17}$$

$$(.85)\,17 \approx AC(1.0)$$

$$14.5 \approx AC \checkmark$$

Step 6.

$$A_{\triangle} = \frac{(\text{Leg}) \times (\text{Leg})}{2}$$

$$A_{\triangle} \approx \frac{(14.5)(9.01)}{2} \approx 65.3 \checkmark$$

Step 7.

Since one right triangle is only half the area of the original triangle, multiply by two.

$$A \approx 65.3 \times 2 \approx 130.6$$

130.6 ✓

Done !

Question: What if the problem had given the length of the base? Answer: You're half done. Use trig or another method to find the length of the altitude from the vertex angle and then you'll have the two legs that you need to find the area.

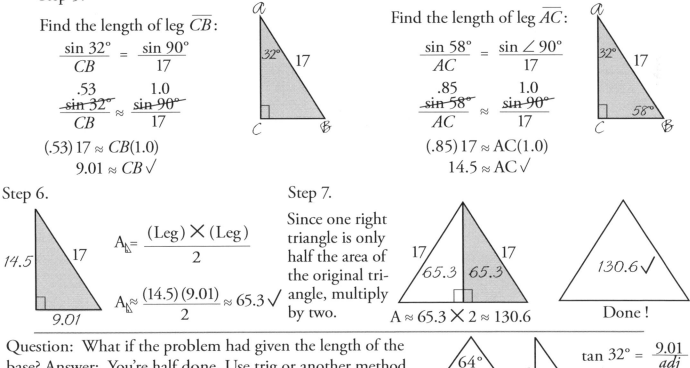

$$\tan 32° = \frac{9.01}{adj}$$

$$\frac{.62}{\cancel{\tan 32°}} \approx \frac{9.01}{adj}$$
$$\frac{.62}{1} \approx \frac{9.01}{adj}$$

$$adj \approx 14.5$$

Heron's Formula for Finding the Area of a Triangle

> If your teacher doesn't require you to know Heron's Formula, you may want to skip this page.

If you know the lengths of all three sides of any triangle, you can find its area. First we need to learn a new term:

SEMIPERIMETER: *The semiperimeter of a triangle with sides a, b, and c is:*

$$S = \frac{a + b + c}{2}$$

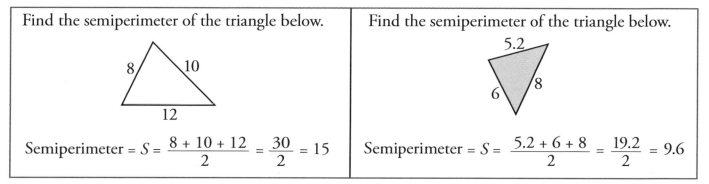

That is, one half of the perimeter of a figure. S is the symbol for semiperimeter.

Here are two examples:

Find the semiperimeter of the triangle below.	Find the semiperimeter of the triangle below.
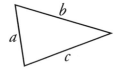	
Semiperimeter = $S = \dfrac{8 + 10 + 12}{2} = \dfrac{30}{2} = 15$	Semiperimeter = $S = \dfrac{5.2 + 6 + 8}{2} = \dfrac{19.2}{2} = 9.6$

HERON'S FORMULA: *The area of a triangle with sides a, b and c is*

$$A = \sqrt{S(S-a)(S-b)(S-c)}$$

Heron's Formula A^{+}_{tip} is not as scary as it looks and is very easy to use once you get used to it!

Here are two examples:

Find the area of the triangle below

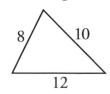

Semiperimeter = $S = \dfrac{8 + 10 + 12}{2} = \dfrac{30}{2} = 15$

$A = \sqrt{S(S-a)(S-b)(S-c)}$

$A = \sqrt{15(15-8)(15-10)(15-12)}$

$A = \sqrt{15(7)(5)(3)} = \sqrt{1575} \approx 39.7 \checkmark$

Find the area of the triangle below

Semiperimeter = $S = \dfrac{5.2 + 6 + 8}{2} = \dfrac{19.2}{2} = 9.6$

$A = \sqrt{S(S-a)(S-b)(S-c)}$

$A = \sqrt{9.6(9.6-5.2)(9.6-6)(9.6-8)}$

$A = \sqrt{9.6(4.4)(3.6)(1.6)} = \sqrt{243.3} \approx 15.6 \checkmark$

Putting it All Together — The Area of Triangles

Finding the Area of Any Triangle

$$A = \frac{1}{2}[(\text{base}) \times (\text{height to that base})] = \frac{(\text{base}) \times (\text{height to that base})}{2}$$

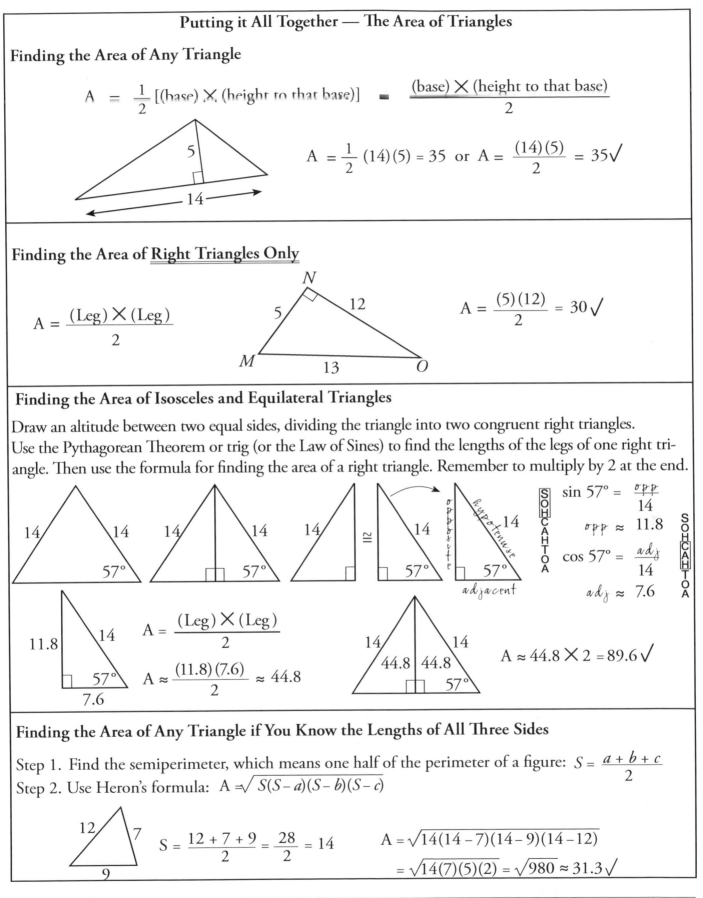

$$A = \frac{1}{2}(14)(5) = 35 \quad \text{or} \quad A = \frac{(14)(5)}{2} = 35 \checkmark$$

Finding the Area of <u>Right Triangles Only</u>

$$A = \frac{(\text{Leg}) \times (\text{Leg})}{2}$$

$$A = \frac{(5)(12)}{2} = 30 \checkmark$$

Finding the Area of Isosceles and Equilateral Triangles

Draw an altitude between two equal sides, dividing the triangle into two congruent right triangles.
Use the Pythagorean Theorem or trig (or the Law of Sines) to find the lengths of the legs of one right triangle. Then use the formula for finding the area of a right triangle. Remember to multiply by 2 at the end.

$$\sin 57° = \frac{opp}{14}$$
$$opp \approx 11.8$$
$$\cos 57° = \frac{adj}{14}$$
$$adj \approx 7.6$$

$$A = \frac{(\text{Leg}) \times (\text{Leg})}{2}$$

$$A \approx \frac{(11.8)(7.6)}{2} \approx 44.8$$

$$A \approx 44.8 \times 2 = 89.6 \checkmark$$

Finding the Area of Any Triangle if You Know the Lengths of All Three Sides

Step 1. Find the semiperimeter, which means one half of the perimeter of a figure: $S = \frac{a + b + c}{2}$

Step 2. Use Heron's formula: $A = \sqrt{S(S-a)(S-b)(S-c)}$

$$S = \frac{12 + 7 + 9}{2} = \frac{28}{2} = 14$$

$$A = \sqrt{14(14-7)(14-9)(14-12)}$$
$$= \sqrt{14(7)(5)(2)} = \sqrt{980} \approx 31.3 \checkmark$$

The Area of a Rhombus

Rhombus Review

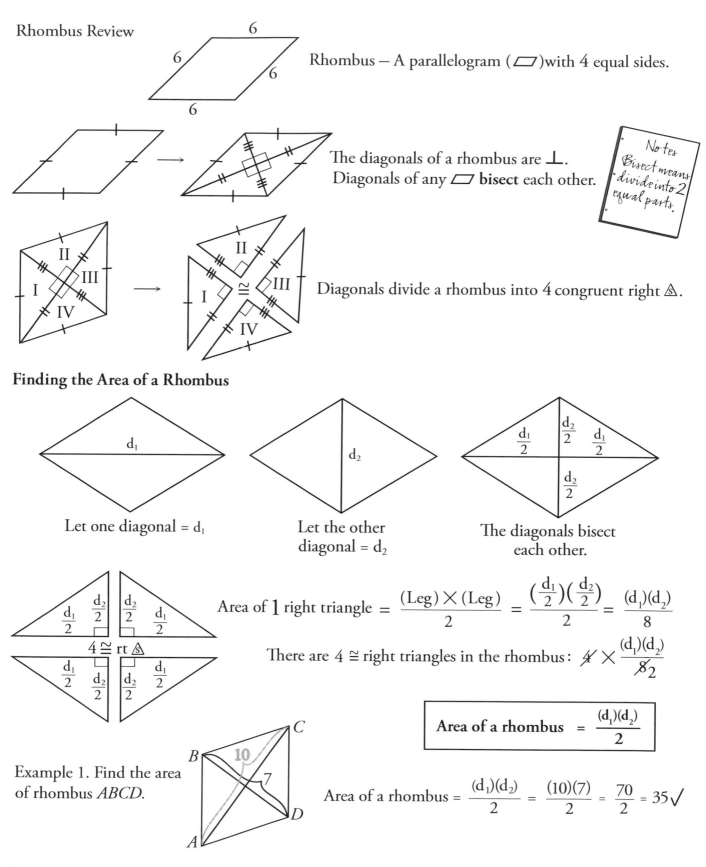

Rhombus – A parallelogram ($\square$) with 4 equal sides.

The diagonals of a rhombus are $\perp$.
Diagonals of any $\square$ **bisect** each other.

Notes
Bisect means
divide into 2
equal parts.

Diagonals divide a rhombus into 4 congruent right $\triangle$.

Finding the Area of a Rhombus

Let one diagonal = d_1

Let the other diagonal = d_2

The diagonals bisect each other.

$$\text{Area of 1 right triangle} = \frac{(\text{Leg}) \times (\text{Leg})}{2} = \frac{\left(\frac{d_1}{2}\right)\left(\frac{d_2}{2}\right)}{2} = \frac{(d_1)(d_2)}{8}$$

There are 4 $\cong$ right triangles in the rhombus: $\cancel{4} \times \dfrac{(d_1)(d_2)}{\cancel{8}2}$

$$\boxed{\text{Area of a rhombus} = \frac{(d_1)(d_2)}{2}}$$

Example 1. Find the area of rhombus *ABCD*.

$$\text{Area of a rhombus} = \frac{(d_1)(d_2)}{2} = \frac{(10)(7)}{2} = \frac{70}{2} = 35 \checkmark$$

178

Example 2. Find the area of rhombus *QRST*.

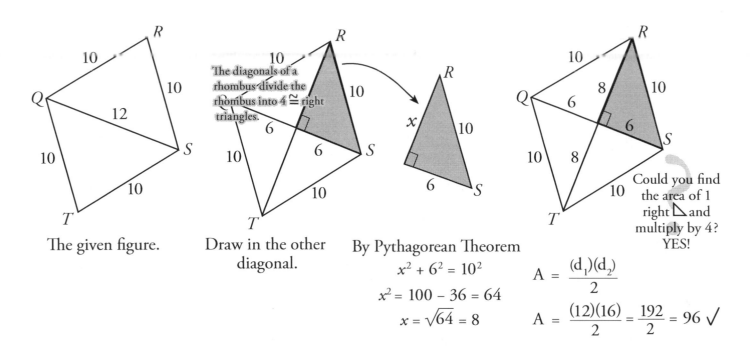

The given figure.	Draw in the other diagonal.

The diagonals of a rhombus divide the rhombus into 4 ≅ right triangles.

By Pythagorean Theorem

$$x^2 + 6^2 = 10^2$$
$$x^2 = 100 - 36 = 64$$
$$x = \sqrt{64} = 8$$

Could you find the area of 1 right △ and multiply by 4? YES!

$$A = \frac{(d_1)(d_2)}{2}$$

$$A = \frac{(12)(16)}{2} = \frac{192}{2} = 96 \checkmark$$

Now You Try It

1. The measure of each side of the equilateral triangle on the right is 5. Find the area of the triangle.

2. Find the area of the rhombus below. (Hint, draw the diagonals, which divide the rhombus into four congruent right triangles, then use trigonometry.)

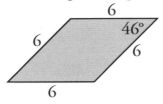

3. Find the area of triangle *ABC*.

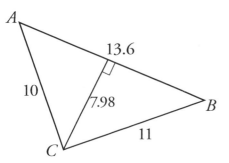

Be sure to check in the back of the book to see how you did!

The Area of a Trapezoid

Trapezoid Review

A trapezoid is a quadrilateral with *exactly* 2 parallel sides, called the bases.
The two non-parallel sides of a trapezoid are called the legs.

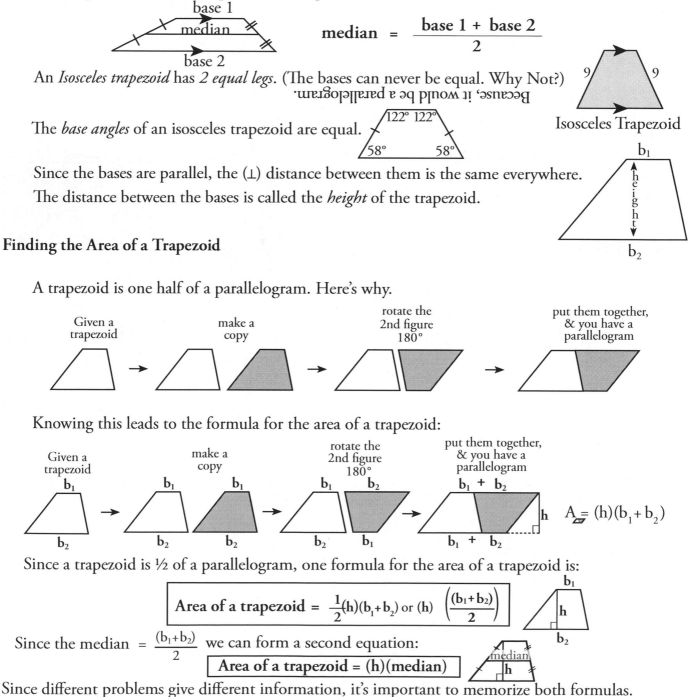

The *median* of a trapezoid connects the midpoints of the legs and is parallel to the two bases.
The length of the median equals the average of the two bases:

$$\text{median} = \frac{\text{base 1} + \text{base 2}}{2}$$

An *Isosceles trapezoid* has *2 equal legs*. (The bases can never be equal. Why Not?)
Because, it would be a parallelogram.

The *base angles* of an isosceles trapezoid are equal.

Isosceles Trapezoid

Since the bases are parallel, the (⊥) distance between them is the same everywhere.
The distance between the bases is called the *height* of the trapezoid.

Finding the Area of a Trapezoid

A trapezoid is one half of a parallelogram. Here's why.

Given a trapezoid → make a copy → rotate the 2nd figure 180° → put them together, & you have a parallelogram

Knowing this leads to the formula for the area of a trapezoid:

Given a trapezoid → make a copy → rotate the 2nd figure 180° → put them together, & you have a parallelogram

$$A_{\square} = (h)(b_1 + b_2)$$

Since a trapezoid is ½ of a parallelogram, one formula for the area of a trapezoid is:

$$\text{Area of a trapezoid} = \frac{1}{2}(h)(b_1+b_2) \text{ or } (h)\left(\frac{(b_1+b_2)}{2}\right)$$

Since the median $= \frac{(b_1+b_2)}{2}$ we can form a second equation:

$$\text{Area of a trapezoid} = (h)(\text{median})$$

Since different problems give different information, it's important to memorize both formulas.

Here are some examples:

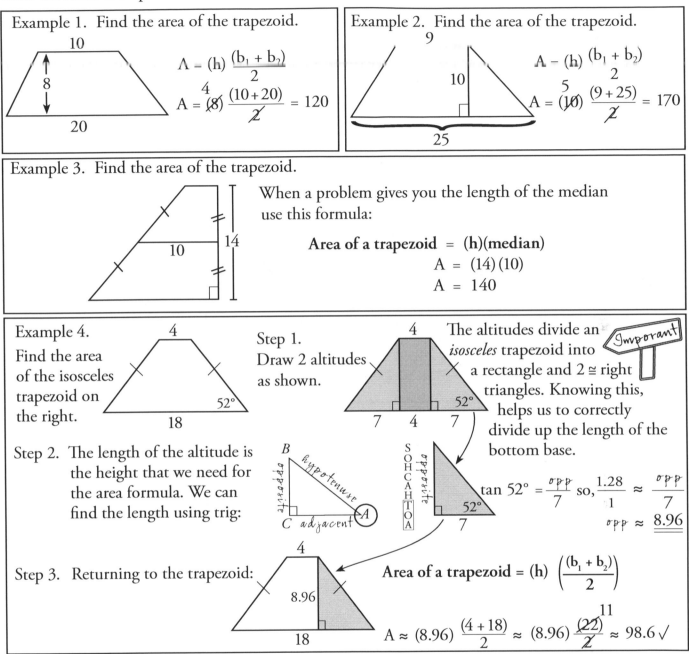

Example 1. Find the area of the trapezoid.

10

8

20

$A = (h) \dfrac{(b_1 + b_2)}{2}$

$A = (\cancel{8})^4 \dfrac{(10 + 20)}{\cancel{2}} = 120$

Example 2. Find the area of the trapezoid.

9

10

25

$A = (h) \dfrac{(b_1 + b_2)}{2}$

$A = (\cancel{10})^5 \dfrac{(9 + 25)}{\cancel{2}} = 170$

Example 3. Find the area of the trapezoid.

10   14

When a problem gives you the length of the median use this formula:

Area of a trapezoid = (h)(median)

A = (14)(10)

A = 140

Example 4.

Find the area of the isosceles trapezoid on the right.

4

52°

18

Step 1.
Draw 2 altitudes as shown.

4

52°

7   4   7

The altitudes divide an *isosceles* trapezoid into a rectangle and 2 ≅ right triangles. Knowing this, helps us to correctly divide up the length of the bottom base.

Important

Step 2. The length of the altitude is the height that we need for the area formula. We can find the length using trig:

B
hypotenuse
opposite
C adjacent
A

S
O
H
C
A
H
T
O
A
opposite
52°
7

$\tan 52° = \dfrac{opp}{7}$ so, $\dfrac{1.28}{1} \approx \dfrac{opp}{7}$

$opp \approx \underline{8.96}$

Step 3. Returning to the trapezoid:

4

8.96

18

Area of a trapezoid = (h) $\left(\dfrac{(b_1 + b_2)}{2} \right)$

$A \approx (8.96) \dfrac{(4 + 18)}{2} \approx (8.96) \dfrac{(\cancel{22})^{11}}{\cancel{2}} \approx 98.6 \checkmark$

Now You Try It

1. Find the area of the trapezoid below.

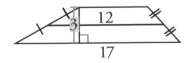

12
3
17

2. Find the area of the trapezoid below.

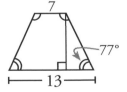

7
77°
13

★ Regular Means Equal Sides and Equal Angles ★

A polygon with equal sides and equal angles is called a **regular** polygon. This means, a regular polygon is both equilateral and equiangular. Here are some examples:

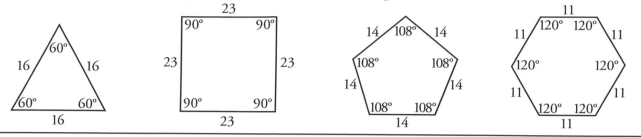

Any *regular* polygon can be *circumscribed* by a circle. Circumscribed means that each vertex lies on the circle.

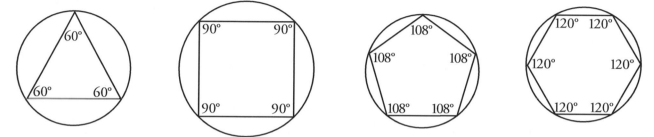

When you see "circumscribed" think about the vertices *lying on* the circle.

NewTerms The *center* of the circumscribing circle is the **center** of the polygon. A **r**adius of the circumscribing circle which goes to a ve**r**tex of the polygon, is a **radius** of the polygon.

If a regular polygon has *n* sides, the radiuses divide the polygon into *n* congruent isosceles triangles.

NewTerm An **apothem** (rhymes with rap'-ah-them) is a perpendicular segment from the center of a regular polygon to a side.

The radiuses and the apothems divide a regular polygon into *2n* congruent right triangles.

Working with Regular (equal sides and equal angles) Polygons

1. For a given *regular* polygon, all radiuses are equal.

Each Radius goes to a veRtex.

2. The radiuses divide a polygon with n sides into n congruent isosceles triangles:

n = 5

3. The radiuses meet in the center of the polygon.
   One entire rotation is 360°.

360°

4. The radiuses divide the 360° angle into n equal angles.
   Each angle at the center is 360 ÷ n degrees.

$$\frac{360°}{5} = 72°$$

72° 72°
72° 72°
72°

5. Each isosceles triangle 72° can be divided into two congruent triangles by drawing in the altitude from the vertex angle (the angle between the two equal sides):

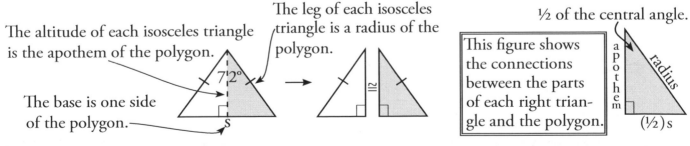

The altitude of each isosceles triangle is the apothem of the polygon.

The leg of each isosceles triangle is a radius of the polygon.

½ of the central angle.

The base is one side of the polygon.

72°

s

This figure shows the connections between the parts of each right triangle and the polygon.

apothem

radius

(½)s

Finding the Area of a Regular Polygon

Example. Find the area of the regular pentagon below.	Mark all sides equal and draw in the (also equal) radiuses.	One complete rotation is 360°.	The radiuses divide the 360° into five 72° angles (360÷5).

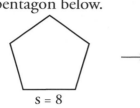 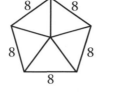  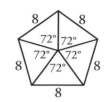

The radiuses divide the pentagon into 5 ≅ isosceles △.	Study one isosceles △.	Draw the altitude (the apothem) between the equal sides.	This altitude divides the isosceles triangle into two ≅ right triangles.

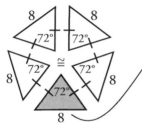

To do this problem, we need the lengths of both legs of the triangle.

We have one acute angle and one side, so we can use trig (or the Law of Sines) to find the other leg.

Find the "adjacent" side.

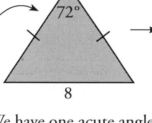

$$\tan 36° \approx \frac{4}{adj}$$

$$\frac{.727}{1} \approx \frac{4}{adj}$$

$$.727\,(adj) \approx 4$$

$$\frac{(.727)\,adj}{.727} \approx \frac{4}{.727}$$

$$adj \approx 5.5 \checkmark$$

SOHCAHTOA

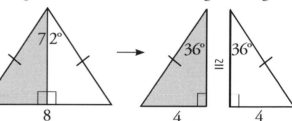

2 Ways to Finish the Problem:

1st Way —

Area of a Regular Pentagon = (Area of one right triangle) × (Number of right triangles in the figure)

n = 5

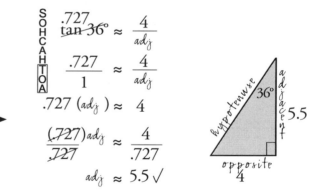

The area of one right triangle = $\frac{(\text{Leg}) \times (\text{Leg})}{2}$ $A_{\triangle} \approx \frac{(4) \times (5.5)}{2} \approx 11$

The radiuses and apothems divide the polygon into 2n = 10 congruent right triangles.

Area of the pentagon ≈ 10 × 11 ≈ 110 √

2nd Way — THEOREM: *The area of a **regular** polygon = ½ (apothem)(perimeter)*   **A = ½ ap**

The approximate length of the apothem is 5.5 and the perimeter = (10)(4) = 40.

Area of the pentagon ≈ (½)(5.5)(40) ≈ 110 √

Example. Find the area of a regular triangle with radius equal to 8.

A regular triangle is another name for an equilateral (equiangular) triangle. "n" is the number of sides, so n = 3. In order to find the area, break the figure down into parts, each of which is also a triangle.

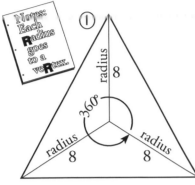

One complete rotation equals 360°.

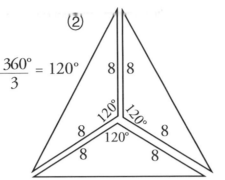

The radiuses divide the triangle into three congruent isosceles triangles.

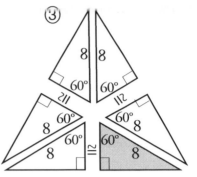

The apothems divide the isosceles triangles into 2n = 6 congruent right triangles.

④ To do this problem, we need to find the lengths of both legs of one of the congruent right triangles. The problem involves a 30°-60°-90° triangle, so we can use special right triangles (or trig or the Law of Sines) to solve.

⑤ Comparing the problem to the 30°-60°-90° model allows us to find the lengths of the legs.

2 Ways to Finish the Problem

1st way — The total area of the figure equals:

(Area of one right triangle) × (Number of right triangles in the figure)

The area of one right triangle = $\frac{(\text{Leg}) \times (\text{Leg})}{2}$ $A_{\triangle} = \frac{(4) \times (4\sqrt{3})}{2} = 8\sqrt{3}$

The radiuses and apothems divide the figure into 2n = 6 congruent right triangles.

Area of the regular triangle = $(8\sqrt{3}) \times (6) = 48\sqrt{3} \approx 83.14$ √

2nd way — The area of a regular polygon = ½ (apothem) x (perimeter):

The apothem is the distance from the center to the side, which is the leg from the center and is equal to 4

The perimeter is made up of the other leg ×6

$(4\sqrt{3}) \times (6) = 24\sqrt{3}$

Area of the regular triangle = ½ (4)(24√3) = 48√3 ≈ 83.14 √

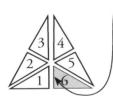

Example. Find the area of a
regular hexagon with apothem = 5.

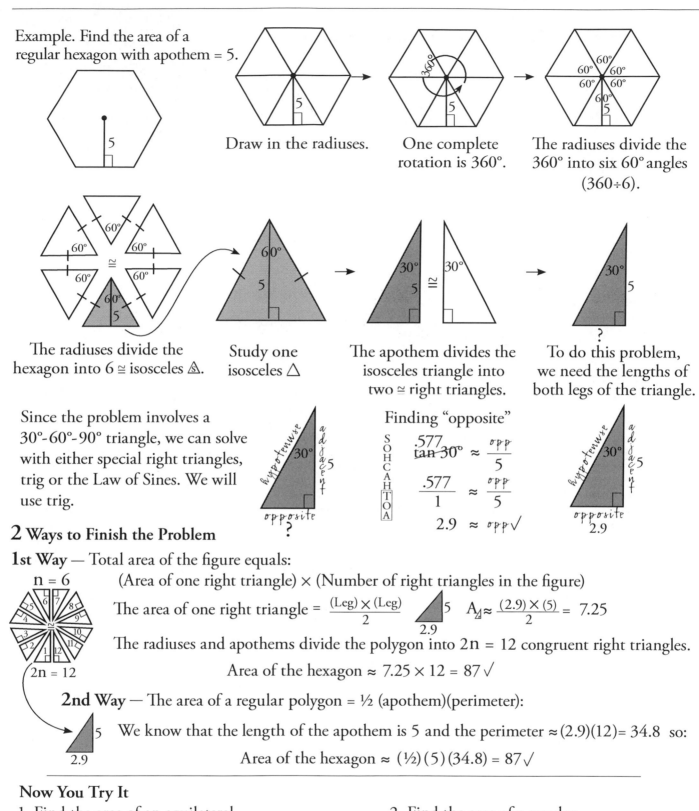

Draw in the radiuses.

One complete
rotation is 360°.

The radiuses divide the
360° into six 60° angles
(360÷6).

The radiuses divide the
hexagon into 6 ≅ isosceles △.

Study one
isosceles △

The apothem divides the
isosceles triangle into
two ≅ right triangles.

To do this problem,
we need the lengths of
both legs of the triangle.

Since the problem involves a
30°-60°-90° triangle, we can solve
with either special right triangles,
trig or the Law of Sines. We will
use trig.

Finding "opposite"

S O H
C A H
T O A

$\tan 30° \approx \dfrac{opp}{5}$

$\dfrac{.577}{1} \approx \dfrac{opp}{5}$

$2.9 \approx opp \checkmark$

2 Ways to Finish the Problem

1st Way — Total area of the figure equals:

n = 6

(Area of one right triangle) × (Number of right triangles in the figure)

The area of one right triangle = $\dfrac{(Leg) \times (Leg)}{2}$ $A_{△} \approx \dfrac{(2.9) \times (5)}{2} = 7.25$

The radiuses and apothems divide the polygon into 2n = 12 congruent right triangles.

2n = 12

Area of the hexagon ≈ 7.25 × 12 = 87 √

2nd Way — The area of a regular polygon = ½ (apothem)(perimeter):

We know that the length of the apothem is 5 and the perimeter ≈ (2.9)(12)= 34.8 so:

Area of the hexagon ≈ (½)(5)(34.8) = 87 √

Now You Try It

1. Find the area of an equilateral
triangle with apothem = 10.

2. Find the area of a regular
pentagon with radius = 6

Pi (pronounced pie) is a number that occurs naturally in mathematics. The symbol for Pi is π. Since Pi is an irrational number, we have to approximate its value in calculations.

One good approximate value for Pi is 3.14, another is $\frac{22}{7}$. When you see π, remember, it's just a symbol that stands for a number.

Circumference — The *circumference* of a circle is a special word for the perimeter, that is, the distance from one point all the way around the circle and back to the starting point. The formula for the circumference is C = 2πr. Since two radiuses make a diameter, another form of the formula is C = πd. It's important to memorize these formulas.

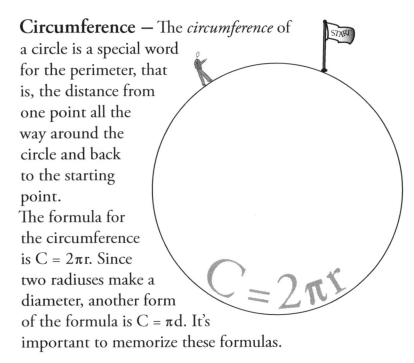

Area — The *area* of a circle is the size of its surface. The formula for the area is A = πr^2. Since all areas are found by multiplying a length times a length, it makes sense that there is a *square* in the formula. It's important to memorize this formula and to use it correctly.

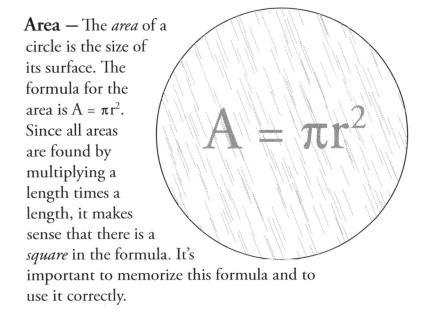

When you use one of the circle formulas, stop and think, which formula do you want? Remember, the area formula is the one with the squared term.

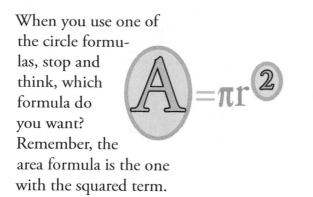
$A = \pi r^2$

Example 1. Find the circumference and area of a circle with radius equal to 3.

$$C = 2\pi r \implies C = 2\pi(3) \implies C = 6\pi \approx 18.8 \checkmark$$
$$A = \pi r^2 \implies A = \pi(3)^2 \implies A = 9\pi \approx 28.3 \checkmark$$

Example 2. The diameter of a circle is 7. Find its circumference and area.

For this kind of problem, it's a good idea to find the radius first. Since the diameter is equal to two times the radius, we know that the radius equals 3.5. (There are two formulas for the circumference, but it's best to find one correct way to do a problem and to stay with that method.)

$$C = 2\pi r \implies C = 2\pi(3.5) \implies C = 7\pi \approx 22 \checkmark$$
$$A = \pi r^2 \implies A = \pi(3.5)^2 \implies A = 12.25\pi \approx 38.5 \checkmark$$

Example 3. The area of a circle is 20, find its circumference.

The answer to the area formula is given, so set the answer equal to the formula and work backwards to solve for the radius:

$$A = \pi r^2 \implies 20 = \pi r^2 \implies 20 \div \pi = r^2 \implies 6.4 \approx r^2 \implies 2.5 \approx r \implies C = 2\pi r \implies C \approx 5\pi \approx 15.7 \checkmark$$

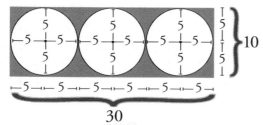

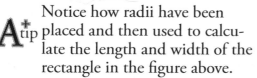

30

Example 4. The three circles in the figure on the left are congruent. Find the:
a. length, width and area of the rectangle, and
b. the area of the shaded region only.

Solution: All radii of a circle are equal, so always add radii anywhere and everywhere where they may be helpful:
a. The length of the rectangle is 30 and the width is 10, so the area is 300. $\checkmark$
b. The area of each circle is $\pi r^2 = \pi(5)^2 = 25\pi \approx 78.5$. There are 3 circles, so the total area of the 3 circles is approximately: $3 \times 78.5 = 235.5$

A⁺tip Notice how radii have been placed and then used to calculate the length and width of the rectangle in the figure above.

300   –   235.5   =   64.5 $\checkmark$

188

Arc Length

The length of an arc is the linear measure of a part of the circumference.

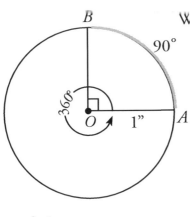

We know the $m\overset{\frown}{AB} = 90$ and that the total degree measure of a circle is 360. How does this help us find the *length* of $\overset{\frown}{AB}$?

Since the measure of $\overset{\frown}{AB}$ is $\frac{90}{360} = \frac{1}{4}$ of the total measure of the circle, it makes sense that we have the same part, $\frac{1}{4}$, of the circle's circumference:

$$C = 2\pi r = 2\pi (1 \text{ in.}) = 2\pi \text{ in.} \approx 6.28 \text{ in.}$$

$$\left(\frac{1}{4}\right)C \approx \left(\frac{1}{4}\right)(6.28) \text{ in.} \approx 1.6 \text{ in.} \checkmark$$

$B \bullet$ 1.6" A

A⁺tip

The 2 Measurements of an Arc:

1. The "measure of" or "m" $\overset{\frown}{AB}$ means the *degree* measure of $\overset{\frown}{AB}$. "m"$\overset{\frown}{AB}$ is giving or asking for the number of degrees (by definition, the same as the central angle intercepting or "cutting off" $\overset{\frown}{AB}$). In the example above, the measure of $\overset{\frown}{AB}$ or $m\overset{\frown}{AB} = 90$.

2. The "length of" or the "arc length" of $\overset{\frown}{AB}$ means the *linear measure* of $\overset{\frown}{AB}$. It's as if you were to take a string, lay it on the arc, and then measure the string. The length of $\overset{\frown}{AB}$ is giving or asking for a length. In the example above, the length or arc length of $\overset{\frown}{AB} \approx 1.6 \text{ in.}$

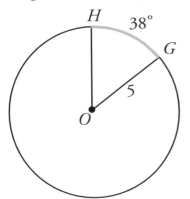

What is the *length* of $\overset{\frown}{GH}$? We are given that the $m\overset{\frown}{GH} = 38$.

Since $m\overset{\frown}{GH}$ is $\frac{38}{360} = \frac{19}{180} \approx .11$ of the total degree measure of the circle, , it makes sense that we have the same part, .11, of the circumference:

$$C = 2\pi r = 2\pi (5) = 10\pi$$

$$(.11)C = (.11)(10)\pi = 1.1\pi \approx 3.5 \checkmark$$

{ Where did the inch go? Unless your teacher wants you to add one, if the problem doesn't have a unit of measure, neither does the answer.

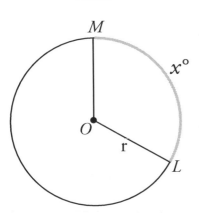

What is the *length* of $\overset{\frown}{LM}$? We are given that the $m\overset{\frown}{LM} = x$.

Since the $m\overset{\frown}{LM}$ is $\frac{x}{360}$ of the total number of degrees in a circle, it makes sense that we have the same part, $\frac{x}{360}$, of the circumference.

We know the formula for the circumference is $C = 2\pi r$ so,

$$\boxed{\text{ARC LENGTH} = \frac{x}{360}(2\pi r)}$$

Memorize

Example 1. Find the length of $\overset{\frown}{UV}$.

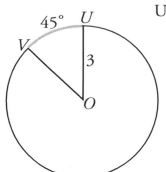

Using the formula for arc length, $\dfrac{x^\circ}{360}(2\pi r)$, and given $x = 45$ and $r = 3$:

Calculate the fraction: $\dfrac{45}{360} = \dfrac{1}{8} = .125$

Calculate the circumference: $C = 2\pi r = 2\pi(3) = 6\pi$

Put the parts together: $(.125)(6\pi) = .75\pi \approx 2.4\ \checkmark$

Example 2. The length of $\overset{\frown}{GH}$ is 10. The area of circle O is 64π. Find the measure of $\angle GOH$.

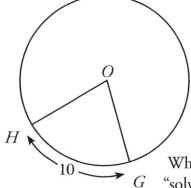

In the problem, we are given the answers to the formula for the arc length and the formula for the area. To get started, we form the two equations:

① arc length:
$$10 = \left(\frac{m\angle GOH}{360}\right)(2\pi r)$$

② area:
$$64\pi = \pi r^2$$

Which of the two equations can be solved? Most of the time, when we say "solve", we mean that we need to get a number for our answer, (not one variable "in terms of" another). And for an equation to be solvable, it can only have *one* unknown (one unknown variable). The first equation has two unknowns, $m\angle GOH$ and r, which means that it can't be solved until we find at least one of the 2 unknowns. Because the second equation has only one unknown (r), it can be solved (for r) and we can then use *that* solution to solve the first equation.

Starting with the second equation, find r:

$$\frac{64\pi}{\pi} = \frac{\pi r^2}{\pi} \rightarrow 64 = r^2 \rightarrow 8 = r$$

Now substitute r = 8 into the first equation to finish the problem:

$$10 = \left(\frac{m\angle GOH}{360}\right)(2\pi 8)$$

$$360\left[10 = (16\pi)\left(\frac{m\angle GOH}{360}\right)\right]360$$

$$\frac{3600}{16\pi} = \frac{16\pi\,(m\angle GOH)}{16\pi}$$

$$71.6 \approx m\angle GOH\ \checkmark$$

Sectors and Their Areas — A *sector* of a circle is formed by two radiuses and an arc.

Think Pie!

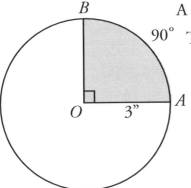

A sector is named by 3 letters.

90° The shaded area in the circle on the left is sector *AOB*.

The *area* of a sector is a part of the circle's total area.

What is the area of sector *AOB*?

Since $m\angle AOB$ is $\dfrac{90}{360} = \dfrac{1}{4}$ of the circle's total degrees, it makes

sense that the area of the sector equals $\dfrac{1}{4}$ of the circle's total area:

$$A = \pi r^2 = \pi(3 \text{ in})^2 = 9\pi \text{ in}^2$$

$$\frac{1}{4} A = \frac{1}{4}(9)\pi \text{ in}^2 = 2.25\pi \text{ in}^2 \approx 7 \text{ in}^2$$

7 in²√

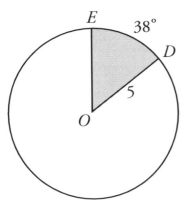

What is the area of sector *DOE*?

Since $m\angle DOE$ is $\dfrac{38}{360} = \dfrac{19}{180} \approx .11$ of the total degrees in the circle,

it makes sense that the area of the sector is .11 of the circle's total area:

$$A = \pi r^2 = \pi(5)^2 = 25\pi$$

$$(.11)A = (.11)(25)\pi = 2.75\pi \approx 8.6 \text{ un}^2$$

8.6un²√

What is the area of sector *LOM*?

Since $m\angle LOM$ is $\dfrac{x}{360}$ of the total degrees, it makes sense that

the area of the sector is $\dfrac{x}{360}$ of the circle's total area.

The formula for the area of circle is $A = \pi r^2$ so:

$\left(\dfrac{x}{360}\right)\pi r^2$√

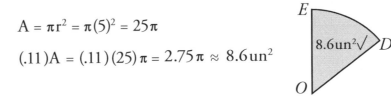

Memorize

$$\boxed{\text{AREA OF SECTOR} = \frac{x}{360}(\pi r^2)}$$

Example 1. Find the area of sector *KOL* in the circle on the right.

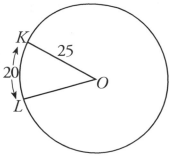

To find the area of sector *KOL*, we need the $m\angle KOL$. Since the problem is giving us two of the three variables from the arc length formula, we can start with the arc length formula and work backwards to solve for the $m\angle KOL$:

$$\text{Arc length } = \frac{x°}{360}(2\pi r), \text{ so: } 20 = \left(\frac{m\angle KOL}{360}\right)(2\pi)(25)$$

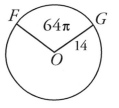

A tip Whenever approximations are being made at each step, we accumulate rounding errors. Don't be concerned if you get a slightly different answer. Remember, the main idea is to understand the method.

$$360\left[20 = (50\pi)\left(\frac{m\angle KOL}{360}\right)\right]360$$

$$7200 = (50\pi)m\angle KOL$$

$$\frac{7200}{50\pi} = \frac{50\pi\,(m\angle KOL)}{50\pi}$$

$$45.9 \approx m\angle KOL$$

Now we have enough information to use the formula for the area of a sector:

$$\text{Area of a sector } = \frac{x°}{360}(\pi r^2), \text{ so: } \text{Area of } KOL \approx \left(\frac{45.9}{360}\right)(\pi)(25)^2 \approx (.128)(\pi)(625) \approx 251.2 \checkmark$$

Example 2. The radius of circle *O* is 14 and the area of sector *FOG* is 64π. Find the length of $\overarc{FG}$.

If we substitute the value of the radius into the formula for arc length we have:

$$\overarc{FG} = \left(\frac{m\angle FOG}{360}\right)2\pi(14)$$

So, in order to find the length of $\overarc{FG}$, we need to find $m\angle FOG$. Since we were given the area of the sector, we can set that value equal to the sector area formula and solve for $m\angle FOG$:

$$64\pi = \left(\frac{m\angle FOG}{360}\right)(\pi)(14)^2$$

$$64\pi = (196\pi)\left(\frac{m\angle FOG}{360}\right)$$

$$360\left[64\pi = (196\pi)\left(\frac{m\angle FOG}{360}\right)\right]360$$

$$\frac{23040\pi}{196\pi} = \frac{196\pi\,(m\angle FOG)}{196\pi}$$

$$117.6 \approx m\angle FOG$$

Now we can substitute 117.6 into the equation for arc length and solve for $\overarc{FG}$:

$$\overarc{FG} \approx \left(\frac{117.6}{360}\right)2\pi(14) \approx (.33)(28)(\pi) \approx 29 \checkmark$$

Circle Segments

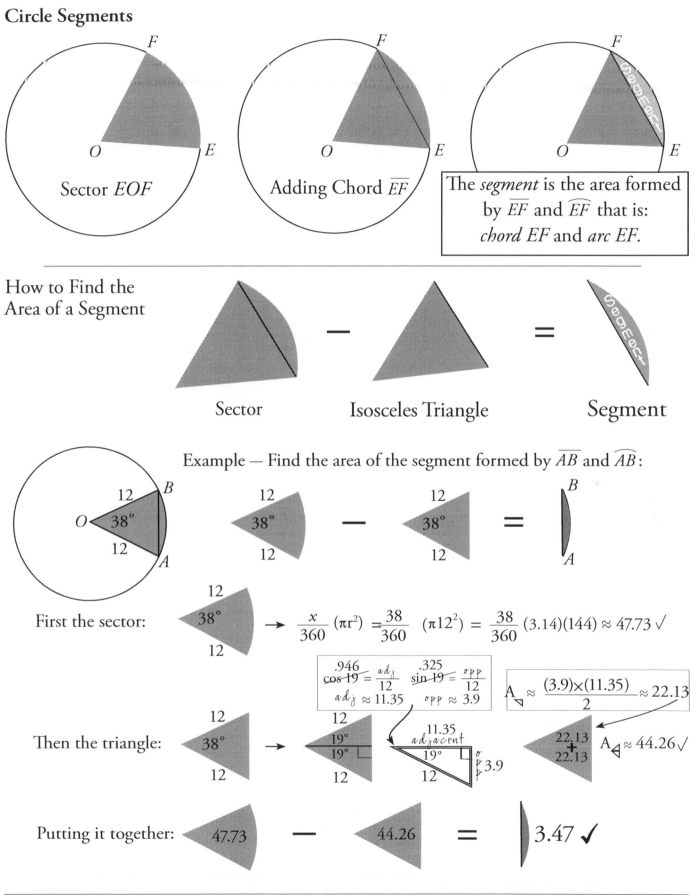

Sector *EOF*

Adding Chord $\overline{EF}$

The *segment* is the area formed by $\overline{EF}$ and $\overparen{EF}$ that is: *chord EF* and *arc EF*.

How to Find the Area of a Segment

Sector — Isosceles Triangle = Segment

Example — Find the area of the segment formed by $\overline{AB}$ and $\overparen{AB}$:

First the sector: → $\dfrac{x}{360}(\pi r^2) = \dfrac{38}{360}(\pi 12^2) = \dfrac{38}{360}(3.14)(144) \approx 47.73\ \checkmark$

Then the triangle: →

$\cos 19 = \dfrac{.946}{12} = \dfrac{adj}{12}$ $\sin 19 = \dfrac{.325}{12} = \dfrac{opp}{12}$
$adj \approx 11.35$ $opp \approx 3.9$

$A_\triangleleft \approx \dfrac{(3.9)\times(11.35)}{2} \approx 22.13$

$A_\triangleleft \approx 44.26\ \checkmark$

Putting it together: 47.73 — 44.26 = 3.47 $\checkmark$

Now You Try It — The figures on this page are made up of circles and parts of circles. Figures that appear congruent are congruent. Students will need extra scratch paper to do these problems.

1. Find the total area of the black parts of the figure on the right.

2. Find the total area of the gray parts of the figure.

3. Find the total length of the perimeters of the black parts of the figure.

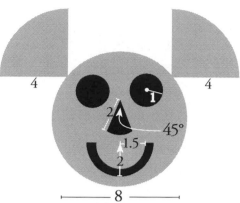

4. Find the total length of the perimeters (including all inner and outer edges) of the gray areas.

Problems 1-4

5. The figure is placed on a 14 × 10 rectangular black board. Find the total of the perimeters (including all inner and outer edges) of the black areas of the figure including the board.

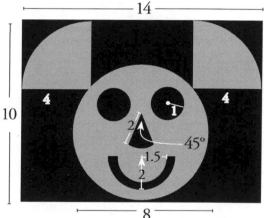

6. Find the total area of the black portions of the figure including those on the black board.

Problems 5-6

Problem 7

7. a. What is the area of the black part of the figure on the left? b. What is the perimeter of the black part of the figure including inside and outside edges?

8. Find the area of the circle segment shown in the figure on the right.

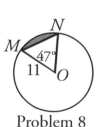

Problem 8

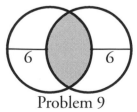

Problem 9

9. Find the area of the gray portion of the figure on the left.

194

Probability Problems Using A Simple Counting Argument

If you have five tops, three pairs of shorts and two jackets, how many different outfits can you make?

$$5 \times 3 \times 2 = 30 \checkmark$$

This simple method works for many problems. The key is to count the paths. Here are some examples:

How many outfits can you make that have a short sleeved (not sleeveless) top and your lighter jacket? You can count the paths that satisfy the requirements or you can do the math:

$$2 \times 3 \times 1 = 6 \checkmark$$

What is the probability that you are wearing a short sleeved (not sleeveless) top and your lighter jacket? We know that six paths satisfy that requirement and that there are a total of 30 paths. Therefore, the probability is:

$$\frac{\text{Number Of Winning Paths}}{\text{Total number of Paths}}$$

$$\frac{6}{30} = \frac{1}{5} \checkmark$$

What is the probability that you are wearing a long sleeved top and your black and grey sweatshirt? Six paths satisfy that requirement and there are a total of 30 paths. Therefore, the probability is:

$$\frac{\text{Number Of Winning Paths}}{\text{Total number of Paths}}$$

$$\frac{6}{30} = \frac{1}{5} \checkmark$$

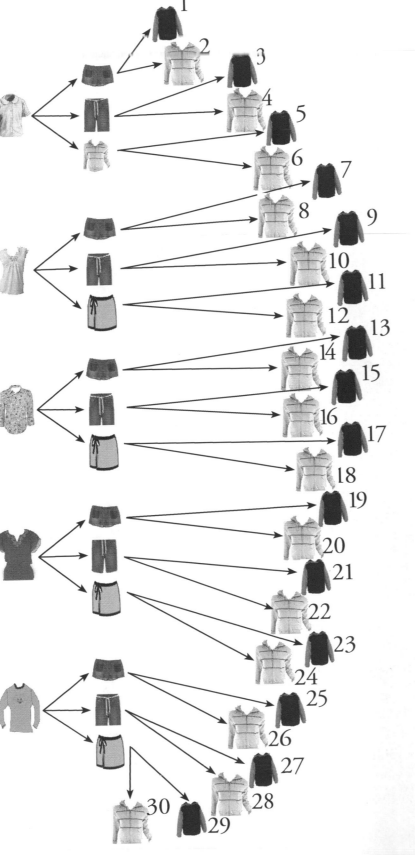

Geometric Probability

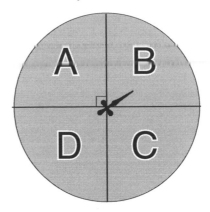

What is the probability of the spinner landing in sector B? We know it's ¼ or 25%.

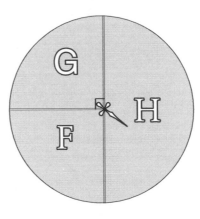

What is the probability of the spinner landing in sector H? We know it's ½ or 50%.

How do we "know" the probability in the two examples above? We naturally understand and use the principles of geometric probability. In these examples:

$$\text{Probability} = \frac{\text{Area of a win}}{\text{Area of whole figure}}$$

The rules in probability problems are very simple. Here is how they would apply in the above examples:

1. The spinner is fair and is equally likely to land in any position.

2. The spinner can only land somewhere on the circle. For example, the probability in either example above of landing on sector R equals 0. (There is no sector R.)

3. The probability of each spinner landing somewhere in its circle is equal to 1 (100%). In fact, the *sum* of the probabilities of all possible outcomes in any problem equals 1.

A⁺ₜᵢₚ Probability is locked into the number interval [0,1] for example, ¼, .007, ½, 3%, .20, 0, .01, 100%, ¾, and 1. If you get an answer smaller than 0 or larger than 1, you need to go back and do the problem again.

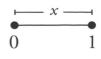

Example 1. On the math teacher's dart board on the right, all squares are congruent, all darts must land somewhere on the board and all tosses are random.

△	⬠	2	5
◉	△	⬠	2
1	◉	△	⬠
3	1	★	△

Find the probability of getting:
1. An odd number, 2. A circle, 3. A triangle, 4. A convex polygon.

There are 16 squares on the board so "everything" (the denominator) will always be 16. To find the numerator, count the number of "win" boxes.

Probability (odd number)

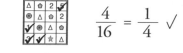 $\dfrac{4}{16} = \dfrac{1}{4}$ ✓

Probability (circle)

$\dfrac{2}{16} = \dfrac{1}{8}$ ✓

Probability (triangle)

$\dfrac{4}{16} = \dfrac{1}{4}$ ✓

Probability (convex polygon)

$\dfrac{7}{16}$ ✓

(Pentagons, triangles and stars are all polygons, but stars are non-convex.)

Example 2. What is the probability of breaking a balloon on the regular hexagonal dartboard on the right? Each side equals 26", the radius of each balloon is 3", each dart hits somewhere on the board, and if the round part of a balloon is hit, it will break.

$$\text{Probability} = \frac{\text{Area of a win}}{\text{Area of whole figure}}$$

Break the problem down into 2 parts:

1. The numerator — Each balloon is a circle with radius equal to 3". There are 11 balloons and the formula for the area of one circle is $A = \pi r^2$:

Area of 11 balloons $= 11 \times \pi(3)^2 = 99\pi \approx 311$ ✓

2. The denominator — The dartboard is a regular (equal sides, equal angles) hexagon:

 Finding "adjacent"

$\dfrac{.577}{\tan 30} = \dfrac{13}{adj}$ $adj \approx 22.5$

$A_\triangle \approx \dfrac{(13) \times (22.5)}{2} \approx 146.3$

12 △'s in a hexagon: $12 \times 146.3 \approx 1755.6$ ✓

$$\text{Probability} \approx \frac{311}{1755.6} \approx .18 \checkmark$$

Question — An amusement park charges $1 for one dart. If you break a balloon you win a $3 stuffed animal. Is it a good bet? What is the probability of not breaking a balloon? Hint: The total of all probabilities add up to 1. Explain your answers. (Check in the Answer Section to see how you did!)

Example — Assuming you have no skill in archery and only arrows that hit somewhere on the target count, what is the probability of your arrow hitting the white area in the target?

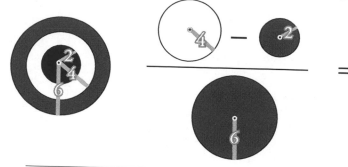

$$\frac{}{} = \frac{\pi(4)^2 - \pi(2)^2}{\pi(6)^2} = \frac{1}{3} \checkmark$$

Geometric Probability also works for lengths. The idea is the same:

$$\text{Probability} = \frac{\text{Total length of a win}}{\text{Entire length}}$$

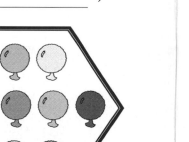 Here's an example. Your kitty has tangled up your shoelace into a ball. The shoelace is 18" long but you need at least a 12" piece to lace up your running shoes. You are going to have to cut the lace to untangle it. If you make one cut, what is the probability that at least one of the two pieces will be 12" long?

Below is a sketch of an untangled shoelace. In which segments do cuts result in a "win"? A cut in the 0-1" interval results in a 1" piece and a 17" piece (a "win"). Continue with this logic marking off those 1" segments in which a cut would result in one of the pieces being at least 12" long:

There are twelve 1" segments in which a cut would result in a "win".

$$\text{Probability} = \frac{\text{Total length of a win}}{\text{Entire length}} = \frac{12}{18} = \frac{2}{3} \checkmark$$

Time problems work in the same way. Just think of elapsed time as lengths. Here's an example:

A 24 hour movie theatre continuously shows: a 2 hour movie, then a 15 minute preview, then hour movie and then a 5 minute preview. If you go to the theatre at a random time, what is t ability that you'll get there during a preview? A "win" in this problem is arriving during eith minute or 5 minute preview for a total of 20 minutes. One total time cycle is 3 hours and 5(which equals 230 minutes:

$$\text{Probability} = \frac{20}{230} = \frac{2}{23} \checkmark$$

Geometric Probability

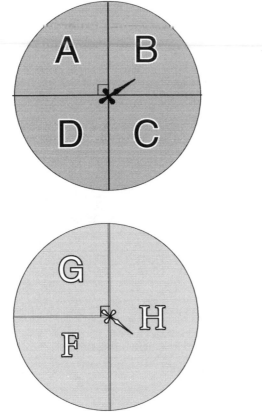

What is the probability of the spinner landing in sector B? We know it's ¼ or 25%.

What is the probability of the spinner landing in sector H? We know it's ½ or 50%.

How do we "know" the probability in the two examples above? We naturally understand and use the principles of geometric probability. In these examples:

$$\text{Probability} \quad = \quad \frac{\text{Area of a win}}{\text{Area of whole figure}}$$

The rules in probability problems are very simple. Here is how they would apply in the above examples:

1. The spinner is fair and is equally likely to land in any position.

2. The spinner can only land somewhere on the circle. For example, the probability in either example above of landing on sector R equals 0. (There is no sector R.)

3. The probability of each spinner landing somewhere in its circle is equal to 1 (100%). In fact, the *sum* of the probabilities of all possible outcomes in any problem equals 1.

A⁺tip Probability is locked into the number interval [0,1] for example, ¼, .007, ½, 3%, .20, 0, .01, 100%, ¾, and 1. If you get an answer smaller than 0 or larger than 1, you need to go back and do the problem again.

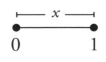

Example 1. On the math teacher's dart board on the right, all squares are congruent, all darts must land somewhere on the board and all tosses are random.

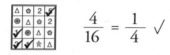

Find the probability of getting:
1. An odd number, 2. A circle, 3. A triangle, 4. A convex polygon.

There are 16 squares on the board so "everything" (the denominator) will always be 16. To find the numerator, count the number of "win" boxes.

Probability (odd number)

$$\frac{4}{16} = \frac{1}{4} \checkmark$$

Probability (circle)

$$\frac{2}{16} = \frac{1}{8} \checkmark$$

Probability (triangle)

$$\frac{4}{16} = \frac{1}{4} \checkmark$$

Probability (convex polygon)

$$\frac{7}{16} \checkmark$$

(Pentagons, triangles and stars are all polygons, but stars are non-convex.)

Example 2. What is the probability of breaking a balloon on the regular hexagonal dartboard on the right? Each side equals 26", the radius of each balloon is 3", each dart hits somewhere on the board, and if the round part of a balloon is hit, it will break.

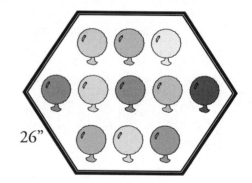

26"

$$\text{Probability} = \frac{\text{Area of a win}}{\text{Area of whole figure}}$$

Break the problem down into 2 parts:

1. The numerator — Each balloon is a circle with radius equal to 3". There are 11 balloons and the formula for the area of one circle is $A = \pi r^2$:

Area of 11 balloons $= 11 \times \pi(3)^2 = 99\pi \approx 311 \checkmark$

2. The denominator — The dartboard is a regular (equal sides, equal angles) hexagon:

Finding "adjacent"

$$\tan 30 = \frac{13}{adj} \qquad adj \approx 22.5$$

$$A_{\triangle} \approx \frac{(13) \times (22.5)}{2} \approx 146.3$$

12 $\triangle$'s in a hexagon: $12 \times 146.3 \approx 1755.6 \checkmark$

$$\text{Probability} \approx \frac{311}{1755.6} \approx .18 \checkmark$$

Question — An amusement park charges $1 for one dart. If you break a balloon you win a $3 stuffed animal. Is it a good bet? What is the probability of not breaking a balloon? Hint: The total of all probabilities add up to 1. Explain your answers. (Check in the Answer Section to see how you did!)

Example — Assuming you have no skill in archery and only arrows that hit somewhere on the target count, what is the probability of your arrow hitting the white area in the target?

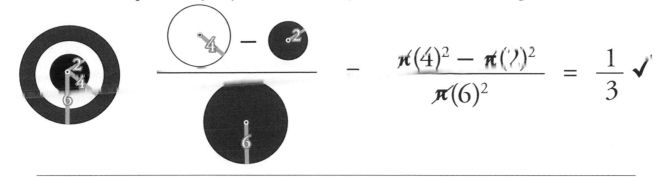

$$\frac{\pi(4)^2 - \pi(2)^2}{\pi(6)^2} = \frac{1}{3} \checkmark$$

Geometric Probability also works for lengths. The idea is the same:

$$\text{Probability} = \frac{\text{Total length of a win}}{\text{Entire length}}$$

Here's an example. Your kitty has tangled up your shoelace into a ball. The shoelace is 18" long but you need at least a 12" piece to lace up your running shoes. You are going to have to cut the lace to untangle it. If you make one cut, what is the probability that at least one of the two pieces will be 12" long?

Below is a sketch of an untangled shoelace. In which segments do cuts result in a "win"? A cut in the 0-1" interval results in a 1" piece and a 17" piece (a "win"). Continue with this logic marking off those 1" segments in which a cut would result in one of the pieces being at least 12" long:

w w w w w w   w w w w w w

```
|++|++|++|++|++|++|++|++|++|++|++|++|++|++|++|++|++|++|
0   1   2   3   4   5   6   7   8   9  10  11  12  13  14  15  16  17  18
```

There are twelve 1" segments in which a cut would result in a "win".

$$\text{Probability} = \frac{\text{Total length of a win}}{\text{Entire length}} = \frac{12}{18} = \frac{2}{3} \checkmark$$

Time problems work in the same way. Just think of elapsed time as lengths. Here's an example:

A 24 hour movie theatre continuously shows: a 2 hour movie, then a 15 minute preview, then a 1½ hour movie and then a 5 minute preview. If you go to the theatre at a random time, what is the probability that you'll get there during a preview? A "win" in this problem is arriving during either the 15 minute or 5 minute preview for a total of 20 minutes. One total time cycle is 3 hours and 50 minutes which equals 230 minutes:

$$\text{Probability} = \frac{20}{230} = \frac{2}{23} \checkmark$$

Probability Problems Using A Simple Counting Argument

If you have five tops, three pairs of shorts and two jackets, how many different outfits can you make?

$$5 \times 3 \times 2 = 30 \checkmark$$

This simple method works for many problems. The key is to count the paths. Here are some examples:

How many outfits can you make that have a short sleeved (not sleeveless) top and your lighter jacket? You can count the paths that satisfy the requirements or you can do the math:

$$2 \times 3 \times 1 = 6 \checkmark$$

What is the probability that you are wearing a short sleeved (not sleeveless) top and your lighter jacket? We know that six paths satisfy that requirement and that there are a total of 30 paths. Therefore, the probability is:

$$\frac{\text{Number Of Winning Paths}}{\text{Total number of Paths}}$$

$$\frac{6}{30} = \frac{1}{5} \checkmark$$

What is the probability that you are wearing a long sleeved top and your black and grey sweatshirt? Six paths satisfy that requirement and there are a total of 30 paths. Therefore, the probability is:

$$\frac{\text{Number Of Winning Paths}}{\text{Total number of Paths}}$$

$$\frac{6}{30} = \frac{1}{5} \checkmark$$

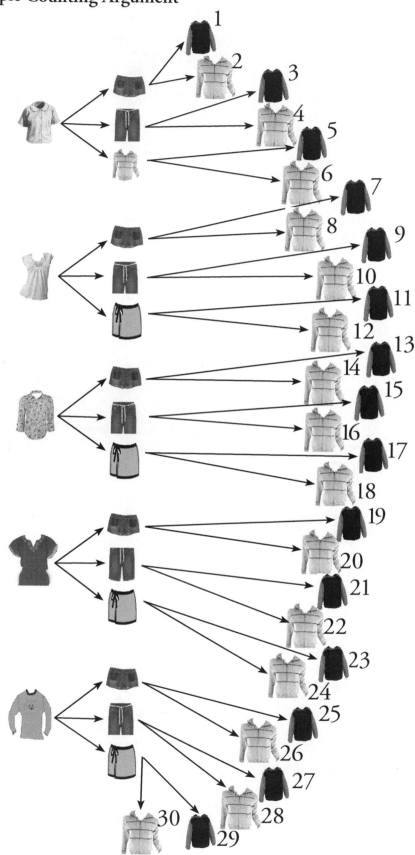

Now You Try It

1. What is the probability of hitting a diamond on the dart board on the right? The diamonds are congruent, each side is equal to 5" and the shorter diagonal is equal to 6". Assume that only darts that hit the board count and that all tosses are random.

2. Study the archery target on the right, with measurements as indicated. Assuming you are a beginning archer and only arrows that hit the target count, find the probability of:

    a. Hitting the bulls eye.

    b. Hitting either white area.

3. Your hero, Tiger, has promised to stop by and help dedicate the new community center sometime this afternoon. He will stay for 15 minutes during the 3 hour opening ceremony but can't be sure what time he can get there. During the same 3 hours, you can only stop by for a minute on your way home from school because you have to get home and do your darned (but awesome) geometry project. What is the probability that you will get to see Tiger?

4. The local pizza place has a special which includes a large pizza with or without extra cheese and the choice of one meat topping and one vegetable topping. There are 3 kinds of meat, pepperoni, sausage and ham; and there are 4 different kinds of vegetables, green peppers, onions, mushrooms and pineapple. You want to buy the special and need to decide what to choose.

    a. How many different kinds of special pizzas are there to choose from?

    b. If you always choose pepperoni for the meat choice, how many types of pizza can you choose from?

    c. If the pizza place decides to offer thin and thick crusts, how would the answers to the above questions change?

13. Volume and Area of Solid Objects

Solid objects are 3-dimensional. A prism is a simple solid object:

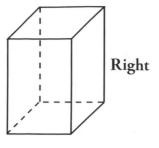

2 Kinds of Prisms **Right**

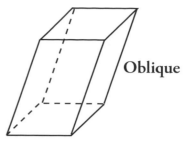

Oblique

Right Prisms stand up straight, they are perpendicular to the ground.

Oblique Prisms *lean* and are not perpendicular to the ground.

Right Prisms — In high school geometry classes most prism problems are about right prisms.

A right prism has **2 bases.** Bases come in pairs.

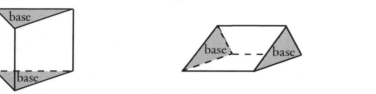

It doesn't matter which way a prism is turned, the bases might not be on the top and bottom.

The 2 bases are: 1. Congruent 2. Polygons 3. In parallel planes.

Know these facts!

Some prisms have more than one pair of sides that could be considered bases:

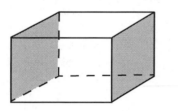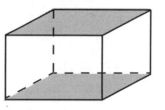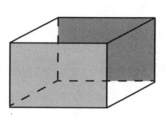

Sometimes you can decide which pair to use as the bases, sometimes the problem will say which pair to use as the bases.

Prisms are named by the *shape* of their *bases*:

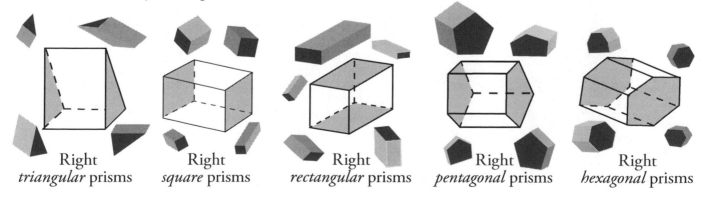

Right *triangular* prisms   Right *square* prisms   Right *rectangular* prisms   Right *pentagonal* prisms   Right *hexagonal* prisms

The sides of prisms that are *not* bases are *lateral faces* or more simply, *faces.*

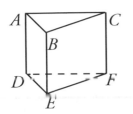

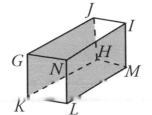

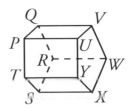

The **2** bases are *ABC* and *DEF.* The faces are *ABED, BCFE* and *CADF.*

Given: The **2** bases are *GJHK* and *NIML.* The faces are *GNLK, JING, HMIJ* and *KLMH.*

The **2** bases are *PQRST* and *UVWXY.* The faces are *PUYT, TYXS, SXWR, RWVQ* and *PUVQ.*

 For right prisms, the **faces are always rectangles.**

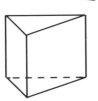

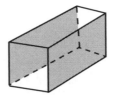

The bases are triangles.
The *faces are rectangles.*

The bases are rectangles.
The *faces are rectangles.*
(Squares *are* rectangles!)

The bases are pentagons.
The *faces are rectangles.*

A *lateral edge* is a line segment where two lateral faces meet. Here are some examples:

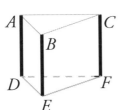

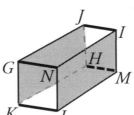

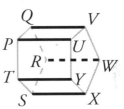

Lateral edges
AD, BE and *CF.*

Lateral edges
GN, KL, HM and *JI.*

Lateral edges
PU, TY, SX, RW, QV.

A lateral edge of a right prism is an *altitude* of the prism. The length of an altitude is the *height* of the prism. Notice that the height is the (⊥) distance between the bases. Here are some examples:

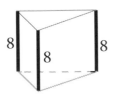

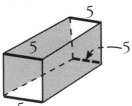

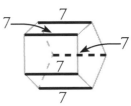

The height of the triangular prism is 8.

The height of the rectangular prism is 5.

The height of the pentagonal prism is 7.

Remember for right prisms, think:

$$\text{Altitude} \rightleftharpoons \text{Lateral Edge} \rightleftharpoons \text{Height}$$

With prisms there are **3** measurements you will be asked to find:
      I.   Lateral Area (the area of the faces).
      II. Total Area (the area of the faces and the 2 bases).
      III. Volume.

I. Lateral Area: *The area of the (lateral) faces.*   L.A. = ph

There are two ways to find the lateral area:
    1. Since the faces are rectangles, find the area of each rectangle and add them up, or
    2. Find the perimeter of the base and multiply it by the length of a lateral edge (the height).

Example:
Find the lateral area of the
right triangular prism.

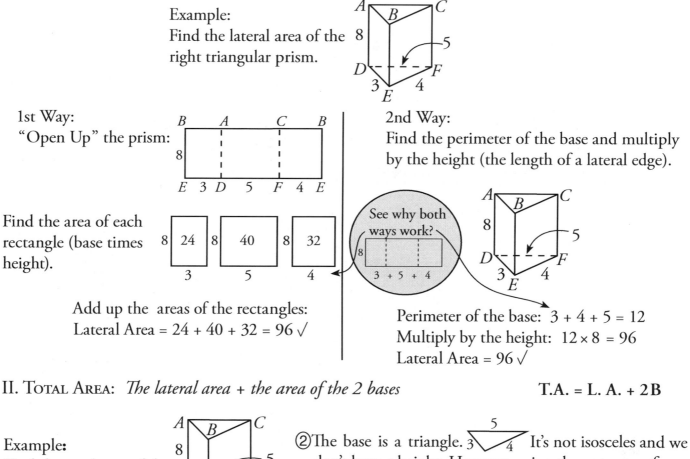

1st Way:
"Open Up" the prism:

Find the area of each
rectangle (base times
height).

Add up the areas of the rectangles:
Lateral Area = 24 + 40 + 32 = 96 √

See why both ways work?

2nd Way:
Find the perimeter of the base and multiply
by the height (the length of a lateral edge).

Perimeter of the base: 3 + 4 + 5 = 12
Multiply by the height: 12 × 8 = 96
Lateral Area = 96 √

II. Total Area: *The lateral area + the area of the 2 bases*   T.A. = L. A. + 2B

Example:
Find the total area of the
right triangular prism.

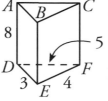

① The lateral area is the area of the faces.
From above, the lateral area = 96.
Since all prisms have two congruent
bases, we must find the area of a base.

② The base is a triangle. It's not isosceles and we
don't have a height. However, using the converse of
the Pythagorean Theorem, we can show that it is a right
triangle. So:
$$\text{Area} = \frac{(\text{Leg}) \times (\text{Leg})}{2} = \frac{(3) \times (4)}{2} = 6$$

Total Area = "the faces + the 2 bases"

Total Area =   96   +   6 + 6 = 108 √

III. Volume: *The volume of a prism = Area of the base × height* $\quad$ **V = Bh**

In the formula, the B is capitalized. Capital B means the *area* of *one* of the two congruent bases. The h stands for height, the distance between the bases.

Units — Volume is measured in *cubic units*. If the units in a problem are centimeters, the answer is in cubic centimeters which is written as cm^3. If the units are in inches, the answer is in cubic inches which is written as in^3. Even when a physical unit is not given in volume problems, your teacher might want you to give the answer in units cubed which is written as un^3.

Here are some examples of volume problems:

Example 1. Find the volume of the rectangular prism below.

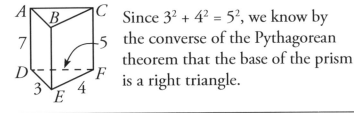

Since each pair of opposite sides are congruent rectangles, we can choose which pair of sides are the bases. We will use the top and bottom sides as the bases.

$$B = (3)(2) = 6$$
$$V = Bh = (6)(2) = 12 \ cm^3 \ \checkmark$$

A⁺ tip Notice that the area of **only 1** of the bases is used in the volume formula.

Example 2. Find the volume of the triangular prism below.

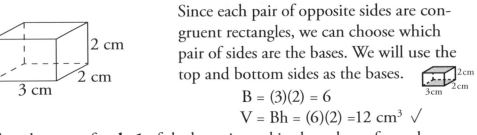

Since $3^2 + 4^2 = 5^2$, we know by the converse of the Pythagorean theorem that the base of the prism is a right triangle.

$$A_{\triangle} = \frac{(Leg) \times (Leg)}{2} = \frac{(3) \times (4)}{2} = 6$$

Since B is the area of the base, $B = A_{\triangle} = 6$ and $V = Bh = (6)(7) = 42 \ \checkmark$

Example 3. Find the volume of a regular pentagonal prism with base edge = 6 and height = 8.

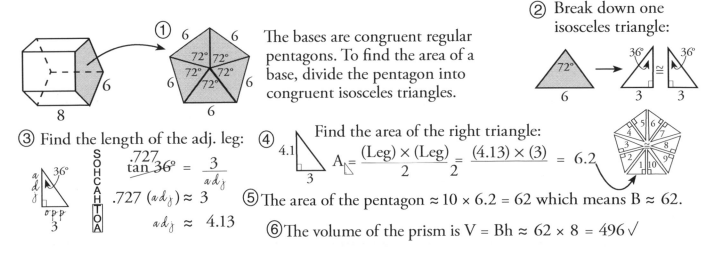

① The bases are congruent regular pentagons. To find the area of a base, divide the pentagon into congruent isosceles triangles.

② Break down one isosceles triangle:

③ Find the length of the adj. leg:
$$\tan 36° = \frac{3}{adj}$$
$$.727$$
$$.727 \ (adj) \approx 3$$
$$adj \approx 4.13$$

SOHCAHTOA

④ Find the area of the right triangle:
$$A_{\triangle} = \frac{(Leg) \times (Leg)}{2} = \frac{(4.13) \times (3)}{2} = 6.2$$

⑤ The area of the pentagon ≈ 10 × 6.2 = 62 which means B ≈ 62.

⑥ The volume of the prism is $V = Bh \approx 62 \times 8 = 496 \ \checkmark$

Example 4.   The volume of the trapezoidal prism on the right, is 3960 un³.
Find the total area of the prism.

In order to find the total area of the prism we need to know the height of the prism which is not given. The problem *is* giving us the volume and a lot of information about the trapezoidal base of the prism. If we set the given value for the volume equal to the volume formula we have:

$$3960 = Bh$$

B is the area of one of the congruent trapezoidal bases of the prism and h is the ($\perp$) distance between the bases. In order to solve, we can only have a single variable, and since the problem gives us enough information ($13\underset{10}{\overset{20}{\diagdown\diagup}}13$) to find B, that is where we should start.

 It's important to understand that there are 2 different heights in this problem and we need to keep them straight. The first height h, belongs to the prism itself. The second height which we've labeled h_B, belongs to each of the congruent bases (top and bottom).

In order to find B, the area of one of the trapezoidal bases, we need to find h_B:

Using the Pythagorean theorem we can find h_B:

$$(h_B)^2 + 5^2 = 13^2$$
$$(h_B)^2 = 144, \quad h_B = 12 \ \checkmark$$

Knowing h_B we can now find B, the area of the base:

Area of the trapezoid	$= (h_B)\dfrac{(b_1+b_2)}{2}$

$$A_{\triangledown} = B = (12)\frac{(10+20)}{2} = 180 \ \checkmark$$

Now we can find h, the height of the prism itself:

$$3960 = Bh = (180)h$$
$$22 = h \ \checkmark$$

Now that we have found the value of h, we can find the lateral and therefore, the total area of the prism:

Notes
L.A. = lateral area (area of the faces)

$$T.A. = L.A. + 2B$$

$$L.A. = ph \quad \text{(p is the perimeter of the base, h is the height of the prism)}$$

$$p = 13 + 20 + 13 + 10 = 56$$

$$L.A. = (56)(22)$$

Putting it all together: $\boxed{T.A. = (56)(22) + 2(180) = 1592 \ un^2 \ \checkmark}$

The kinds of solids shown on this page are often used in homework and test problems. Here are some keys to working with them:

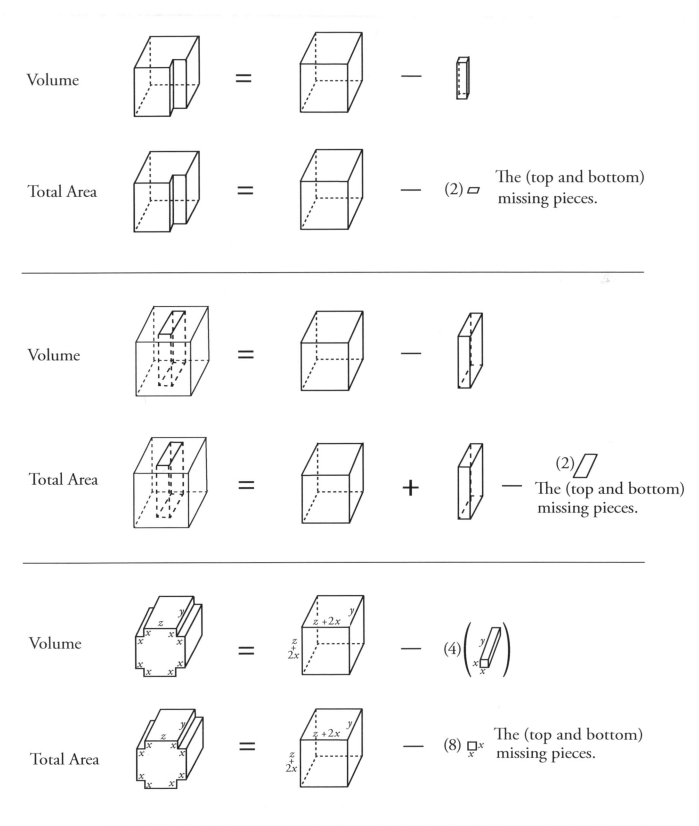

CUBES

All the edges of a cube are equal.
Given that a side measures s,
$B = s^2$, $h = s$ and $V = s^3$
Here are some examples of cubes:

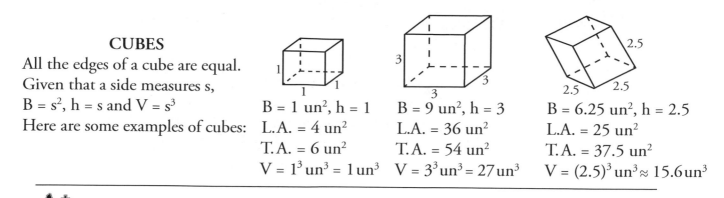

$B = 1$ un^2, $h = 1$
L.A. = 4 un^2
T.A. = 6 un^2
$V = 1^3$ un$^3 = 1$ un^3

$B = 9$ un^2, $h = 3$
L.A. = 36 un^2
T.A. = 54 un^2
$V = 3^3$ un$^3 = 27$ un^3

$B = 6.25$ un^2, $h = 2.5$
L.A. = 25 un^2
T.A. = 37.5 un^2
$V = (2.5)^3$ un$^3 \approx 15.6$ un^3

A^+_{tips}
For area and volume problems:

1. Carefully *memorize* the formulas (even if your teacher allows notes on exams).
2. Draw a sketch (doesn't have to be perfect) of each problem.
3. When you find the value of a variable, add it to your sketch.
4. Use different colors in your sketch for the bases and the lateral edges.

Now You Try It

1. Find the lateral area, total area and volume of the square prism on the right.

2. The volume of the regular hexagonal prism on the right = 300 un^3. Find the prism's lateral area and total area.

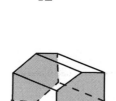

3. The total area of the triangular prism on the right is 768 cm^2. Find the volume of the prism.

4. The volume of a cube is 125 un^3. Find the total area of the cube.

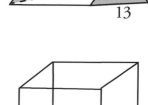

Pyramids

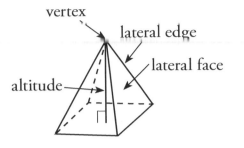

1. Pyramids come to a point called a vertex.
2. The sides of the pyramid are triangles called lateral faces.
3. Each edge where two lateral faces meet, is called a lateral edge.
4. The altitude is the ⊥ segment from the vertex to the base.
5. The length of the altitude is the height of the pyramid.

Pyramid or Prism? Stop and think and be sure you have the right figure in mind. A^+_{tip}

Pyramids are like the tombs in Egypt where the pharaohs are buried.

Pyramids come to a *Point!*

Regular Pyramids — In high school Geometry classes, most pyramid problems are about regular pyramids. Here are the properties of a regular pyramid:

1. The base is a regular (equal sides, equal angles) polygon.

2. The faces are (1)congruent, (2)isosceles, (3)triangles.

3. The vertex is directly above the center of the (regular polygonal) base and therefore the altitude meets the base at its center.

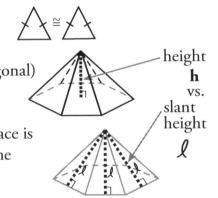

New Term

4. The *slant height* of a pyramid is the height of a face (each face is one of the congruent isosceles triangles). The symbol for the slant height is a lower case cursive L (ℓ).

Right triangles and the Pythagorean Theorem are the key to solving many pyramid problems. Be sure that you know the theorem and that you are using it correctly:

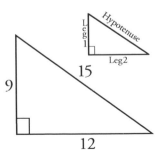

Quick Review — The Pythagorean Theorem: $(\text{Leg1})^2 + (\text{Leg2})^2 = (\text{Hypotenuse})^2$

1. The hypotenuse is opposite the right angle.

2. The hypotenuse is always the longest side. $15 > 9$ and $15 > 12$ ✓

3. The length of the hypotenuse sits all by itself on one side of the equation.

$$9^2 + 12^2 = \boxed{15}^2$$

$$9^2 + 12^2 = 15^2$$

4. You must square each of the lengths.   ~~9 + 12 = 15~~ $81 + 144 = 225$ ✓

The different parts of pyramids combine to form triangles. Notice how each part of the pyramid appears in more than one triangle.

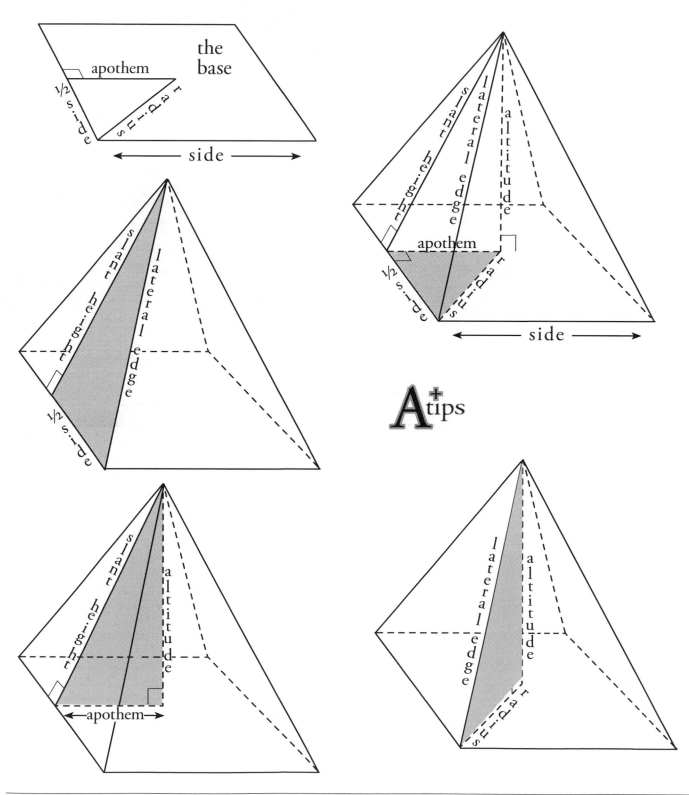

Pyramids are *named by their bases.*

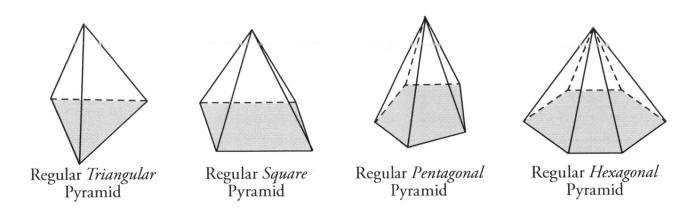

Regular *Triangular* Pyramid   Regular *Square* Pyramid   Regular *Pentagonal* Pyramid   Regular *Hexagonal* Pyramid

There are **3** measurements of regular pyramids that you will be asked to find:

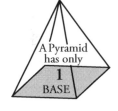

     I.   The Lateral Area (the area of the faces).

     II.   The Total Area (the area of the faces and the *single* base).

     III.   The Volume.

I. LATERAL AREA — *The area of the (lateral) faces.*

There are two ways to find the lateral area:

1st way — Since the faces are congruent isosceles triangles, find the area of one triangle and multiply by the number of faces (which is the same as the number of sides of the base).

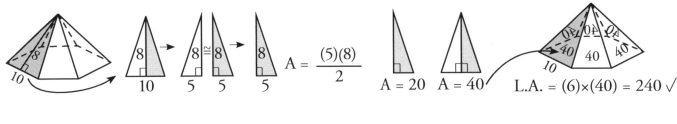

or, **2nd way** — Find the *slant height* of one triangle and the perimeter of the *base.*

LATERAL AREA OF A PYRAMID = ½ pℓ

Notice that the p stands for the perimeter of the base.

$p = (6) \times (10) = 60,$ $\ell = 8$ Lateral Area $= (½) \times (60) \times (8) = 240 \checkmark$

II. TOTAL AREA — *The area of the (lateral) faces and the single base.*   **T = L.A. + 1B**

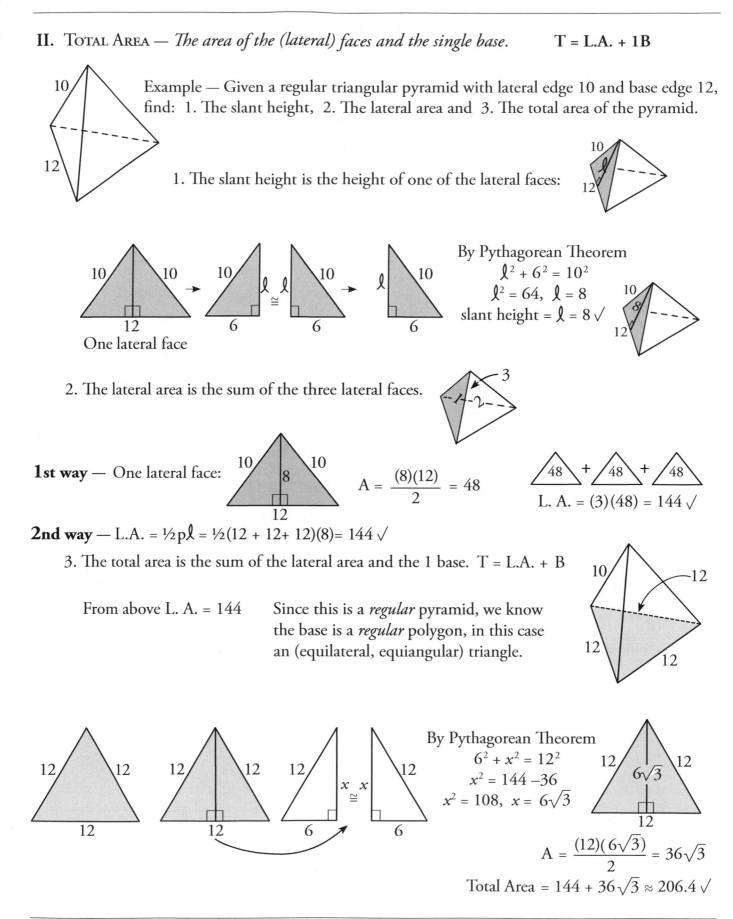

Example — Given a regular triangular pyramid with lateral edge 10 and base edge 12, find: 1. The slant height, 2. The lateral area and 3. The total area of the pyramid.

1. The slant height is the height of one of the lateral faces:

One lateral face

By Pythagorean Theorem
$\ell^2 + 6^2 = 10^2$
$\ell^2 = 64, \ \ell = 8$
slant height = $\ell = 8 \checkmark$

2. The lateral area is the sum of the three lateral faces.

1st way — One lateral face: $A = \dfrac{(8)(12)}{2} = 48$

L. A. = (3)(48) = 144 $\checkmark$

2nd way — L.A. = $\frac{1}{2}p\ell = \frac{1}{2}(12 + 12 + 12)(8) = 144 \checkmark$

3. The total area is the sum of the lateral area and the 1 base. T = L.A. + B

From above L. A. = 144

Since this is a *regular* pyramid, we know the base is a *regular* polygon, in this case an (equilateral, equiangular) triangle.

By Pythagorean Theorem
$6^2 + x^2 = 12^2$
$x^2 = 144 - 36$
$x^2 = 108, \ x = 6\sqrt{3}$

$A = \dfrac{(12)(6\sqrt{3})}{2} = 36\sqrt{3}$

Total Area = $144 + 36\sqrt{3} \approx 206.4 \checkmark$

III. Volume

$V_{\triangle} = \frac{1}{3} Bh$

The volume of a pyramid = 1/3 area of the Base × height of the pyramid

One of my students said, "You mean pyramids lose ⅔ of their stuff on the way up?"

Yes, they do! Remember, if a solid comes to a **point**, its volume is $\frac{1}{3}$!

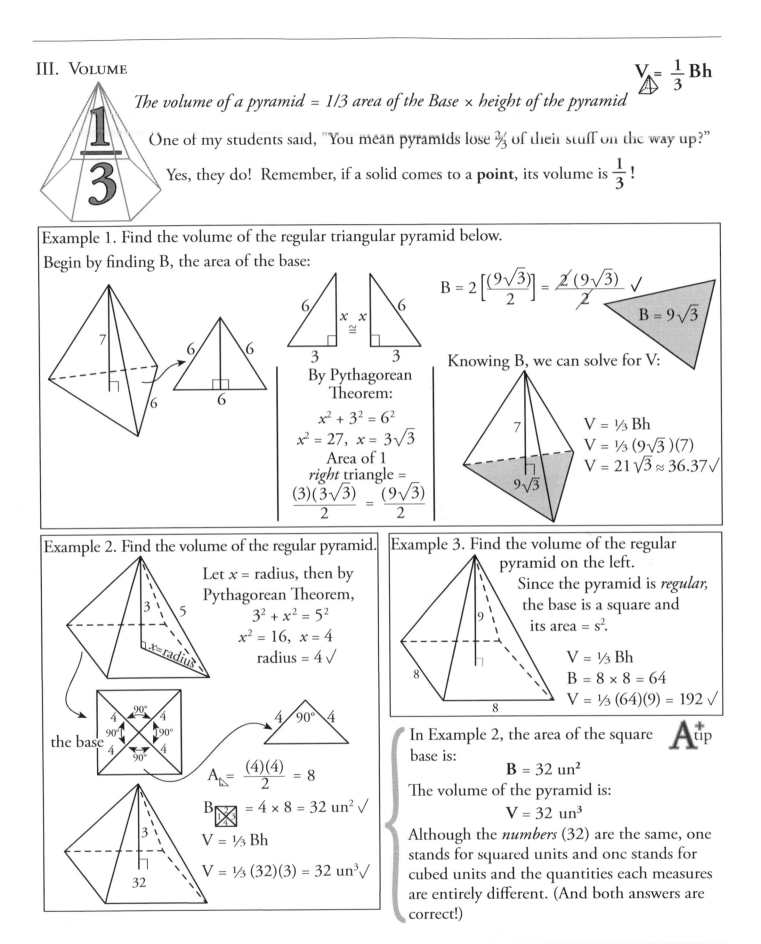

Example 1. Find the volume of the regular triangular pyramid below.

Begin by finding B, the area of the base:

By Pythagorean Theorem:

$$x^2 + 3^2 = 6^2$$
$$x^2 = 27, \quad x = 3\sqrt{3}$$

Area of 1 *right* triangle =
$$\frac{(3)(3\sqrt{3})}{2} = \frac{(9\sqrt{3})}{2}$$

$$B = 2\left[\frac{(9\sqrt{3})}{2}\right] = \frac{2(9\sqrt{3})}{2} \checkmark$$

$$B = 9\sqrt{3}$$

Knowing B, we can solve for V:

$$V = \tfrac{1}{3} Bh$$
$$V = \tfrac{1}{3}(9\sqrt{3})(7)$$
$$V = 21\sqrt{3} \approx 36.37 \checkmark$$

Example 2. Find the volume of the regular pyramid.

Let x = radius, then by Pythagorean Theorem,
$$3^2 + x^2 = 5^2$$
$$x^2 = 16, \quad x = 4$$
$$\text{radius} = 4 \checkmark$$

the base

$$A_{\triangle} = \frac{(4)(4)}{2} = 8$$

$$B_{\square} = 4 \times 8 = 32 \text{ un}^2 \checkmark$$

$$V = \tfrac{1}{3} Bh$$

$$V = \tfrac{1}{3}(32)(3) = 32 \text{ un}^3 \checkmark$$

Example 3. Find the volume of the regular pyramid on the left.

Since the pyramid is *regular,* the base is a square and its area = s^2.

$$V = \tfrac{1}{3} Bh$$
$$B = 8 \times 8 = 64$$
$$V = \tfrac{1}{3}(64)(9) = 192 \checkmark$$

A⁺ tip

In Example 2, the area of the square base is:

$$B = 32 \text{ un}^2$$

The volume of the pyramid is:

$$V = 32 \text{ un}^3$$

Although the *numbers* (32) are the same, one stands for squared units and one stands for cubed units and the quantities each measures are entirely different. (And both answers are correct!)

211

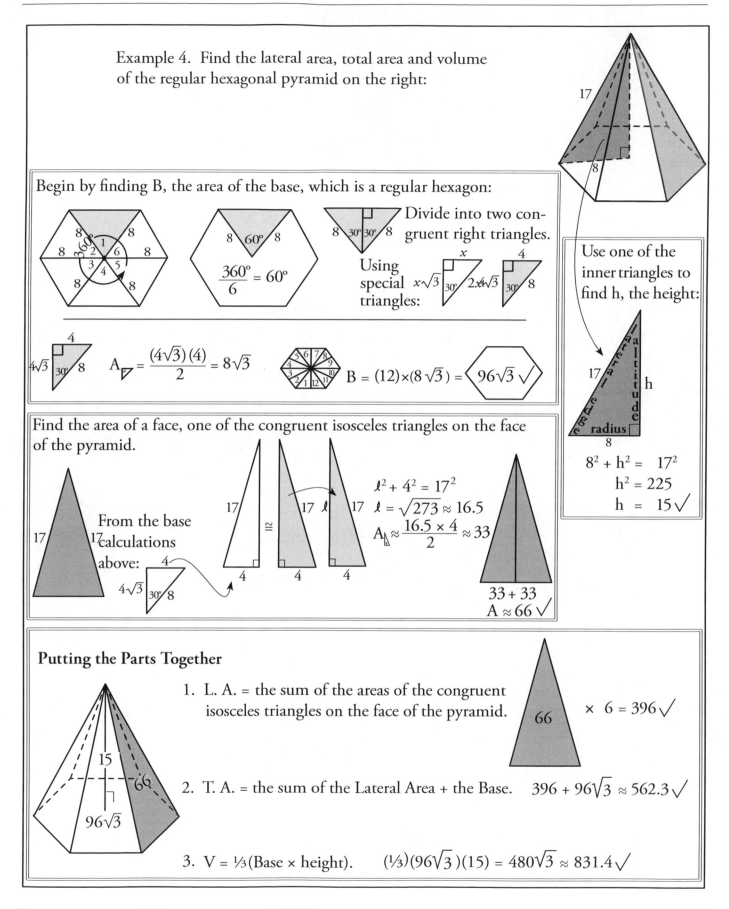

Example 4. Find the lateral area, total area and volume of the regular hexagonal pyramid on the right:

Begin by finding B, the area of the base, which is a regular hexagon:

Divide into two congruent right triangles.

$$\frac{360°}{6} = 60°$$

Using special triangles:

$$A_{\triangledown} = \frac{(4\sqrt{3})(4)}{2} = 8\sqrt{3}$$

$$B = (12) \times (8\sqrt{3}) = \boxed{96\sqrt{3}} \checkmark$$

Use one of the inner triangles to find h, the height:

lateral edge / altitude / radius

$$8^2 + h^2 = 17^2$$
$$h^2 = 225$$
$$h = 15 \checkmark$$

Find the area of a face, one of the congruent isosceles triangles on the face of the pyramid.

From the base calculations above:

$$\ell^2 + 4^2 = 17^2$$
$$\ell = \sqrt{273} \approx 16.5$$
$$A_{\triangle} \approx \frac{16.5 \times 4}{2} \approx 33$$

$$33 + 33$$
$$A \approx 66 \checkmark$$

Putting the Parts Together

1. L. A. = the sum of the areas of the congruent isosceles triangles on the face of the pyramid.

$$66 \times 6 = 396 \checkmark$$

2. T. A. = the sum of the Lateral Area + the Base. $396 + 96\sqrt{3} \approx 562.3 \checkmark$

3. V = ⅓(Base × height). $(⅓)(96\sqrt{3})(15) = 480\sqrt{3} \approx 831.4 \checkmark$

Quick Pyramid Review

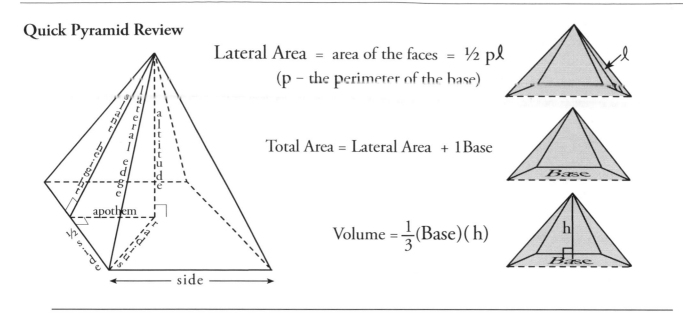

Lateral Area = area of the faces = ½ pℓ
(p – the perimeter of the base)

Total Area = Lateral Area + 1 Base

Volume = $\frac{1}{3}$(Base)(h)

Now You Try It

1. A regular square pyramid fits perfectly in a cube with edges equal to 4. What is the volume of the unused space in the solid?

2. A regular triangular pyramid has height = 10. If the apothem of the base equals 5, find the volume of the pyramid.

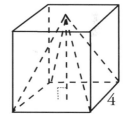

3. The storage tank pictured on the right is a square pyramid. The tank has an upper reserve tank as shown in the figure. Find the volume of the upper reserve tank.

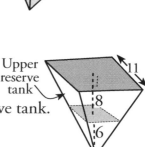

Cylinders — Cylinders are like prisms except, their bases are circles:

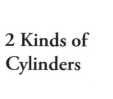

2 Kinds of Cylinders

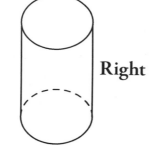 **Right**

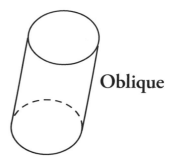 **Oblique**

Right Cylinders stand up straight, they are perpendicular to the ground.

Oblique Cylinders *lean* and are not perpendicular to the ground.

RIGHT CYLINDERS: In high school Geometry classes, most cylinder problems are about right cylinders. The bases are congruent circles. The bases are in parallel planes.

There are **3** measurements of cylinders that you will be asked to find:

I. The Lateral Area   II. The Total Area   III. The Volume

I. LATERAL AREA OF A CYLINDER = $2\pi rh$

A can is a good example of a cylinder.

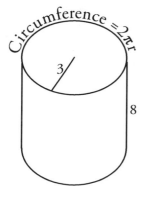

If you were to cut off the can's label and lay it out flat, it would be a *rectangle*.

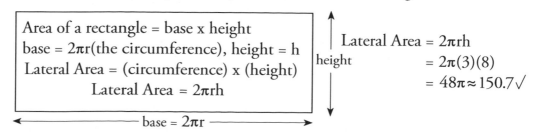

The area of the can's label is the *lateral area* of the cylinder.

Example: Find the lateral area of a cylinder with radius = 3 and height = 8.

If the cylinder had a label, it would be a *rectangle*.
The lateral area of a cylinder equals the area of the rectangle.

Area of a rectangle = base x height
base = $2\pi r$(the circumference), height = h
Lateral Area = (circumference) x (height)
Lateral Area = $2\pi rh$

Lateral Area = $2\pi rh$
  = $2\pi(3)(8)$
  = $48\pi \approx 150.7\sqrt{}$

base = $2\pi r$

II. Total Area of a Cylinder = *The lateral area + the area of the bases.*

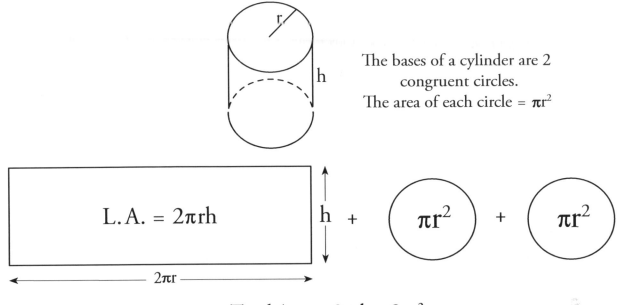

The bases of a cylinder are 2 congruent circles.
The area of each circle = πr^2

L.A. = $2\pi rh$

h + πr^2 + πr^2

2πr

Total Area = $2\pi rh + 2\pi r^2$

Example:

Find the total area of a cylinder with radius = 3 and height = 8.

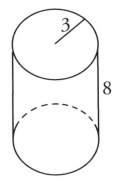

From the previous page, we know that the lateral area of the given cylinder = $48\pi \approx 150.8$.

The area of each circle is πr^2, that is, $\pi 3^2 = 9\pi$.

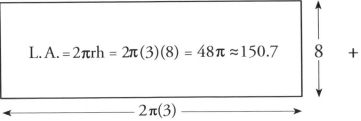

L.A. = $2\pi rh = 2\pi(3)(8) = 48\pi \approx 150.7$

8

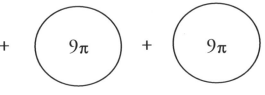

+ 9π + 9π

$2\pi(3)$

Total Area = $48\pi + (2)9\pi = 48\pi + 18\pi = 66\pi \approx 207.24$ √

III. Volume of a Cylinder = *Area of the base × height*

The basic formula for volumes is V = Bh, where B stands for the **area** of the base which in the case of a cylinder, is a circle and the area of each circle = πr².

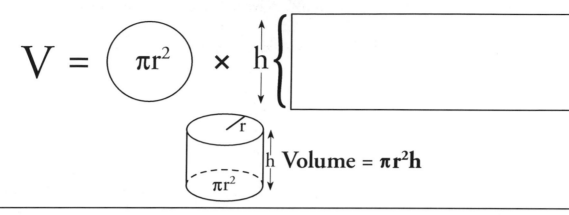

$$V = \left(\pi r^2 \right) \times h \Bigg\{$$

Volume = **πr²h**

Example:

Find the volume of the cylinder shown at right.

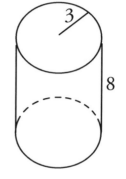

The area of each base is **πr²**,
that is, **π3² = 9π.**

$$V = \left(9\pi \right) \times 8 \Bigg\{$$

Volume $= \boldsymbol{\pi r^2 h} = \boldsymbol{\pi 3^2} \times 8 = 9\boldsymbol{\pi} \times 8 = 72\boldsymbol{\pi} \approx 226.1 \ \checkmark$

Putting It All Together

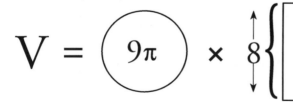

$\begin{cases} \text{I. Lateral Area of a Cylinder} = 2\pi rh \\ \text{II. Total Area of a Cylinder} = 2\pi rh + 2\pi r^2 \\ \text{III. Volume of a Cylinder} = \pi r^2 h \end{cases}$

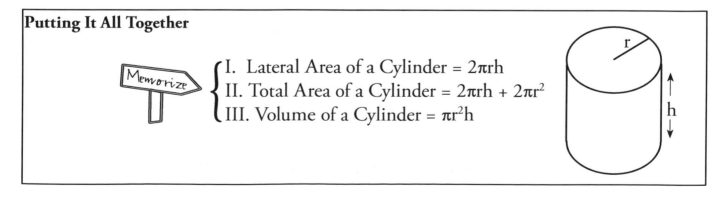

Example 1.

The total area of a cylinder - 240π and the height is 7. Find the lateral area of the cylinder.

In the problem, the answer to the total area formula is given, so set the answer equal to the formula:

$$\text{Total area} = 2\pi rh + 2\pi r^2$$
$$240\pi = 2\pi rh + 2\pi r^2$$

Since h is given, r is the unknown; substitute $h = 7$
into the equation and solve for r:

$$240\pi = 2\pi r(7) + 2\pi r^2$$
$$240\pi = 14\pi r + 2\pi r^2$$

$$\frac{\overset{120}{\cancel{240\pi}}}{\cancel{2\pi}} = \frac{\overset{7r}{\cancel{14\pi r}}}{\cancel{2\pi}} + \frac{\overset{r^2}{\cancel{2\pi r^2}}}{\cancel{2\pi}}$$

There is a quadratic term (r^2), a linear term (7r) and a constant (120), so move all the terms to one side, factor and solve.

$$r^2 + 7r - 120 = 0$$
$$(r + 15)(r - 8) = 0$$

$$r + 15 = 0, r = \cancel{-15} \qquad r - 8 = 0, r = 8 \;\checkmark$$

r can't be negative

The problem asks for the lateral area and the formula for the lateral area is $2\pi rh$:

$$\text{L.A.} = 2\pi(8)(7) = 112\pi \approx 351.68 \;\checkmark$$

Example 2.

A square with area = 16 is rotated around one edge. a.) What solid is formed? b.) Find its volume.

a.) Imagine one edge of the square fixed in place:

Now imagine the square being spun around the fixed side:

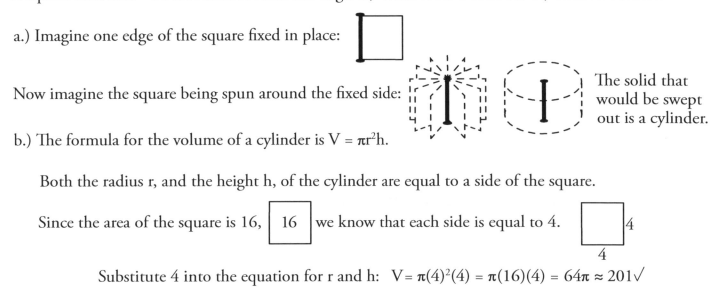

The solid that would be swept out is a cylinder.

b.) The formula for the volume of a cylinder is $V = \pi r^2 h$.

Both the radius r, and the height h, of the cylinder are equal to a side of the square.

Since the area of the square is 16, | 16 | we know that each side is equal to 4.

Substitute 4 into the equation for r and h: $V = \pi(4)^2(4) = \pi(16)(4) = 64\pi \approx 201\checkmark$

Example 3.

If the radius of a cylinder is doubled, what is the volume of the cylinder multiplied by?

To do this type of problem:

      1. Start with the regular formula, in this case $V = \pi r^2 h$
      2. Substitute in the modified variable, in this case, replace r with 2r,
$$V = \pi(2r)^2 h = \pi 2^2 r^2 h = \pi 4 r^2 h = 4\pi r^2 h$$
      3. Compare the solution, $4\pi r^2 h$ to the original formula, $V = \pi r^2 h$,
         $4\pi r^2 h = 4$ times the original volume.

> Notice that the entire quantity, 2r, is squared:
> $(2r)^2 = 2^2 r^2 = 4r^2$

Therefore, when the radius of a cylinder is doubled, the volume is multiplied by 4. ✓

A⁺tip With this type of problem, always go through each of the steps shown above. Estimating and/or guessing is *not* recommended.

Now You Try It

1. If the radius of a cylinder is halved, what is the effect on the size of the volume of the cylinder?

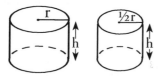

2. The lateral area of a cylinder is 100π and the height is 5. Find the circumference of the base of the cylinder.

3. Find the total area and the volume of the cylinder in problem 2.

4. What is the amount (volume) of metal needed to make a 10 inch length of pipe that is ½ inch thick and has inner diameter equal to 5 in?

Be sure to check in the back of the book to see how you did!

Cones

A cone comes to a ***Point***. The base of a cone is a (circle.) Think of an ice cream **cone**.

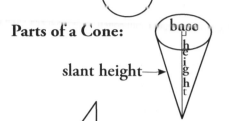

Parts of a Cone:

slant height→

Although cones can be right or oblique, high school geometry problems are usually about right cones, and when textbooks say "cone" the term usually means a right cone.

With cones, there are **3** measurements you will be asked to find:

      I.   The Lateral Area.
     II.   The Total Area (the lateral area plus the single base, a circle).
    III. The Volume.

It's best to carefully memorize each of the following formulas:

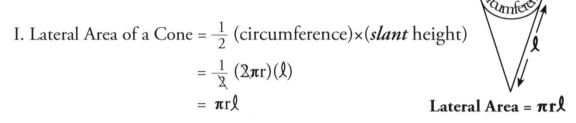

I. Lateral Area of a Cone $= \frac{1}{2}$ (circumference)×(***slant*** height)

$$= \frac{1}{2}(2\pi r)(\ell)$$

$$= \pi r \ell$$

Lateral Area = πrℓ

II. Total Area of a Cone = lateral area + area of the **single** base (a circle)

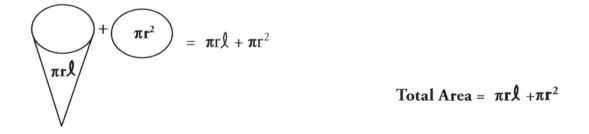

$\pi r \ell$ + πr^2 $= \pi r \ell + \pi r^2$

Total Area = πrℓ +πr²

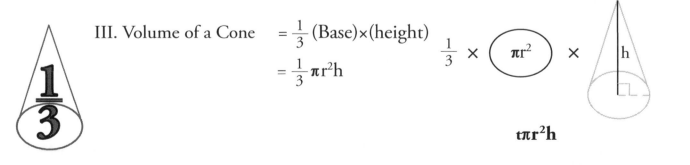

III. Volume of a Cone $\quad = \frac{1}{3}$ (Base)×(height)

$$= \frac{1}{3}\pi r^2 h$$

$\frac{1}{3}$ × πr^2 ×   h

tπr²h

Example 1.

Find the lateral area, total area and volume of the cone below:

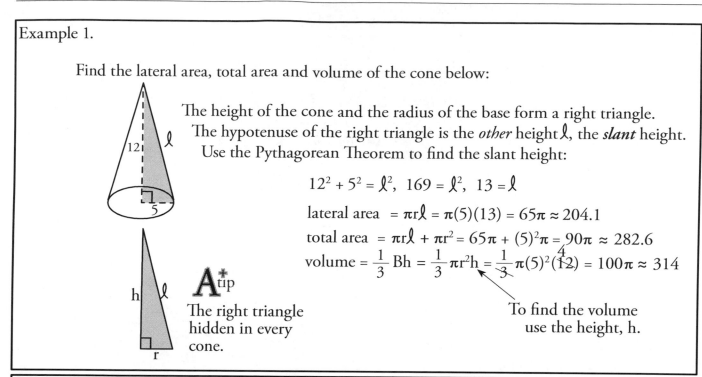

The height of the cone and the radius of the base form a right triangle. The hypotenuse of the right triangle is the *other* height ℓ, the **slant** height. Use the Pythagorean Theorem to find the slant height:

$$12^2 + 5^2 = \ell^2, \quad 169 = \ell^2, \quad 13 = \ell$$

lateral area $= \pi r \ell = \pi(5)(13) = 65\pi \approx 204.1$

total area $= \pi r \ell + \pi r^2 = 65\pi + (5)^2\pi = 90\pi \approx 282.6$

volume $= \dfrac{1}{3}Bh = \dfrac{1}{3}\pi r^2 h = \dfrac{1}{3}\pi(5)^2(12) = 100\pi \approx 314$

$\mathbf{A^{+}_{tip}}$

The right triangle hidden in every cone.

To find the volume use the height, h.

Example 2.

The total area of a cone is 90π. If the radius is 5, find the height and the slant height of the cone.

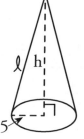

In the problem we are being given the answer to the total area formula, so set the answer equal to the formula:

$$\text{total area} = \pi r \ell + \pi r^2$$
$$90\pi = \pi r \ell + \pi r^2$$

Now substitute the radius, r = 5, into the equation:

$$90\pi = \pi(5)\ell + \pi(5)^2 = 5\pi\ell + 25\pi$$

Subtract 25π from both sides and then divide through by π:

$$\frac{65\pi}{\pi} = \frac{5\pi\ell}{\pi}, \quad 65 = 5\ell, \quad \ell = 13$$

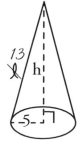

Add the value of ℓ to the original figure. Now we have the lengths of two sides of a right triangle so we can find the third side which is h, the height, using the Pythagorean Theorem:

$$h^2 + 5^2 = 13^2, \quad h = 12$$

Having found the height and the slant height, we could easily find the lateral area and volume of the cone if the problem had asked us to do so.
Did you notice that this is the same cone as the one in Example 1 above?

Solid Search

Write the letter of the matching figure. Be sure to check in the back of the book to see how you did.

Cone _____
Triangular pyramid _____
Square pyramid _____
Right Cylinder _____
Oblique prism _____
Right triangular prism _____
Right pentagonal prism _____
Cube _____
Right rectangular prism _____
Oblique cylinder _____

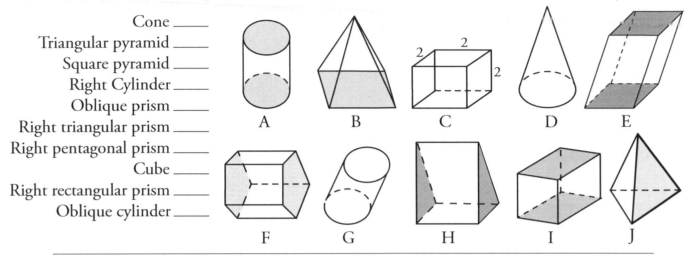

A   B   C   D   E

F   G   H   I   J

A Problem Combining Solids

Find the total area of the crayon at right.

Step 1. Study the figure. Notice that the crayon is made up of a cone and a cylinder.

Step 2. Studying the original figure, notice that the base of the cone and the top of the cylinder are not exposed and therefore are not part of the total area of the crayon.

Step 3. It's a good idea to work on each solid individually, so we'll start with the cone. The problem tells us the lengths of the slant height and the radius of the base. What the cone contributes to the total area of the crayon is only its lateral area.

lateral area = $\pi r \ell = \pi(5)(13) = 65\pi$

Step 4. Now consider the cylinder. The problem tells us the lengths of the height and the radius of the base. What the cylinder contributes to the total area of the crayon is its lateral area plus the area of its bottom base:

lateral area = $2\pi rh = 2\pi(5)(15) = 150\pi$
area of base (a circle) = $\pi r^2 = \pi(5)^2 = 25\pi$

Step 5. Adding together the contributions of the cone and the cylinder:

Area of crayon = $65\pi + 150\pi + 25\pi = 240\pi \approx 753.6$ ✓

If you were asked to find the volume of the crayon, (or how much wax was required to make a crayon), you would find the volume of each part and add them together.

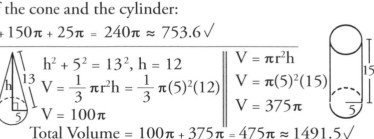

$h^2 + 5^2 = 13^2$, h = 12
$V = \frac{1}{3}\pi r^2 h = \frac{1}{3}\pi(5)^2(12)$
$V = 100\pi$

$V = \pi r^2 h$
$V = \pi(5)^2(15)$
$V = 375\pi$

Total Volume = $100\pi + 375\pi = 475\pi \approx 1491.5$ ✓

Spheres

A sphere is the collection of points in *space* that are equally far away from a given point called its center. Spheres are part of our everyday lives, for example tennis balls, soccer balls, volleyballs and basketballs and even the Earth itself. The terms that we use with circles are also used with spheres.

There are several types of sphere problems that you need to know how to do:

I. The Area of a Sphere — Since a sphere has no base, its "area" refers to its total surface area. Area tells us how much material it would require to make a cover for a bowling ball, or how much paint it would take to cover a spherical post top.

Area of a Sphere = $4\pi r^2$

Memory Tip!
Formulas for spheres include the number **4**.
$$A = \textcircled{4}\pi r^2$$
Since areas always come from "something times something", squaring the radius makes sense.

Example 1. Find the area of a tennis ball with radius = 2 inches.

$$A = 4\pi r^2 = 4\pi(2)^2 = 16\pi \approx 50.3 \checkmark$$

Example 2. The area of a basketball = 256π. Find its diameter.

In the problem we are given the answer to the area formula, so set the answer equal to the formula and work backwards to find the radius:

$$A = 4\pi r^2$$
$$256\pi = 4\pi r^2$$
$$\frac{\overset{64}{\cancel{256\pi}}}{4\pi} = \frac{4\pi r^2}{4\pi} \qquad \text{Divide both sides by } 4\pi.$$

$$64 = r^2, \text{ radius} = 8, \text{ so the diameter} = 2r = 16 \checkmark$$

II. The Volume of a Sphere

$$\text{Volume of a Sphere} = \frac{4}{3}\pi r^3$$

Memory Tips!
Formulas for spheres include the number 4. $V = \frac{4}{3}\pi r^3$ Since volumes always come from "something times something times something" cubing the radius makes sense.

Notice the 3 appears *two times* in the volume formula.

$$\frac{4}{3}\pi r^3$$

Example 1. Find the volume of a tennis ball with radius = 2 inches.

$$V = \frac{4}{3}\pi r^3 = \frac{4}{3}\pi(2)^3 = \frac{32}{3}\pi \approx 33.5 \checkmark$$

Example 2. The area of a basketball = 256. Find its volume.

In the problem we are given the answer to the area formula, so set the answer equal to the area (of a sphere) formula and work backwards to find the radius:

$$A = 4\pi r^2$$
$$256\pi = 4\pi r^2$$
$$\frac{\overset{64}{\cancel{256\pi}}}{\cancel{4\pi}} = \frac{\cancel{4\pi r^2}}{\cancel{4\pi}} \quad \text{divide both sides by } 4\pi$$

$64 = r^2$, radius = 8. Knowing the radius, we can find the volume:

$$V = \frac{4}{3}\pi r^3 = \frac{4}{3}\pi(8)^3 = \frac{2048}{3}\pi \approx 2143.6 \checkmark$$

III. Planes Passing Through Spheres

1. Imagine a solid wooden ball.

3. Study the larger piece. Its top will always be a circle. This circle is called the *circle of intersection.*

2. Imagine a plane (think of it as a sharp electric saw) making a clean cut through the ball. The ball would be divided into two pieces.

4. Draw a radius of the sphere to the edge of the top. Draw a radius of the top (the circle) to the same point. The distance from the cut to the sphere's center forms a right triangle with the two radii.

radius of circle
radius of sphere

A Close Look at Planes Passing Through Spheres

When a plane cuts through a sphere the intersection (the points shared by both the plane and the sphere) is the *circle of intersection*.

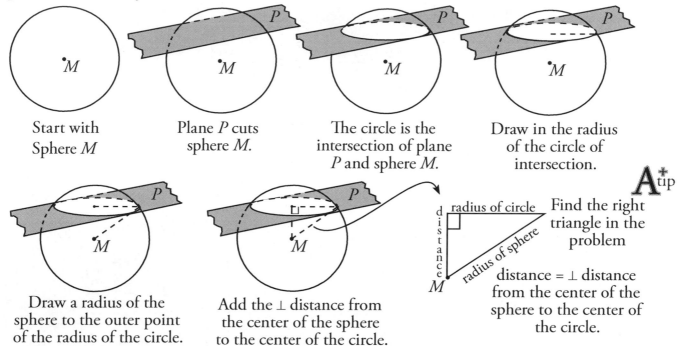

Start with Sphere *M*

Plane *P* cuts sphere *M*.

The circle is the intersection of plane *P* and sphere *M*.

Draw in the radius of the circle of intersection.

Draw a radius of the sphere to the outer point of the radius of the circle.

Add the ⊥ distance from the center of the sphere to the center of the circle.

A^+_{tip} Find the right triangle in the problem

distance = ⊥ distance from the center of the sphere to the center of the circle.

Example: A plane *P* cuts through sphere *M* in a circle that has radius 8. If plane *P* is 5 units away from the center of *M*, what is the radius of sphere *M*?

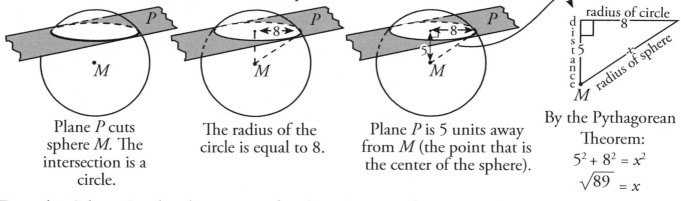

Plane *P* cuts sphere *M*. The intersection is a circle.

The radius of the circle is equal to 8.

Plane *P* is 5 units away from *M* (the point that is the center of the sphere).

By the Pythagorean Theorem:
$$5^2 + 8^2 = x^2$$
$$\sqrt{89} = x$$

Example: Sphere *G* with radius 13 is cut by plane *C*, 5 units from *G*. Find the area of the circle of intersection.

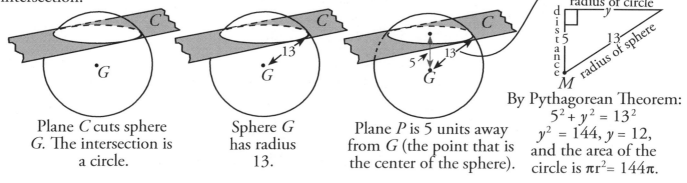

Plane *C* cuts sphere *G*. The intersection is a circle.

Sphere *G* has radius 13.

Plane *P* is 5 units away from *G* (the point that is the center of the sphere).

By Pythagorean Theorem:
$$5^2 + y^2 = 13^2$$
$y^2 = 144$, $y = 12$, and the area of the circle is $\pi r^2 = 144\pi$.

Example 1. A sphere has radius = 5. If a plane cuts the sphere 3 units from its center, what is the diameter ($2 \times r_c$) of the circle of intersection that is formed? Let r_c equal the radius of the circle of intersection. Then, by the Pythagorean Theorem:

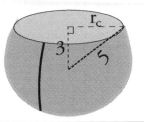

$$3^2 + (r_c)^2 = 5^2$$
$$9 + (r_c)^2 = 25$$
$$(r_c)^2 = 16, \ r_c = 4, \ \text{diameter} = 8 \checkmark$$

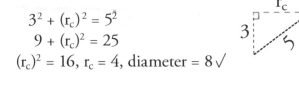

Example 2. A plane cuts a sphere 4 units from its center. If the area of the circle of intersection is 9π, what is the area of the sphere?

In this problem we are given the answer to the area formula, so set the answer equal to the area (of a *circle*) formula and work backwards to find the radius (r_c) of the circle. Since two different radiuses are involved, let r_c equal the area of the circle and r_s equal the area of the sphere:

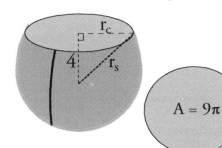

$A = 9\pi$

Looking at the top (the circle), from above the ball.

$$A = \pi(r_c)^2$$
$$9\pi = \pi(r_c)^2$$
$$\frac{9\pi}{\pi} = \frac{\pi(r_c)^2}{\pi} \quad \text{Divide both sides by } \pi$$
$$9 = (r_c)^2, \quad 3 = r_c$$

By Pythagorean Theorem
$$4^2 + 3^2 = (r_s)^2$$
$$16 + 9 = 25 = (r_s)^2$$
$$5 = r_s \checkmark$$

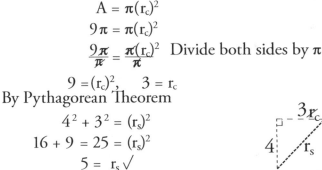

As with most circle and sphere problems, finding the radius is the key to the solution. We were asked for the area of the sphere:

$$A = 4\pi r^2 = 4\pi(5)^2 = 100\pi \approx 314 \checkmark$$

Example 3. If the radius of a tennis ball is 1.75 inches, how much empty space is in a can of tennis balls?

The air space equals the volume of the can (a cylinder) minus the volume of the 3 balls (spheres).

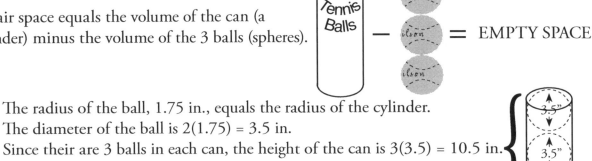

The radius of the ball, 1.75 in., equals the radius of the cylinder.
The diameter of the ball is 2(1.75) = 3.5 in.
Since their are 3 balls in each can, the height of the can is 3(3.5) = 10.5 in.

Let the empty space = $x = \pi r^2 h - 3(\frac{4}{3} \pi r^3)$

$$x = \pi(1.75)^2(10.5) - 3(\frac{4}{3}\pi(1.75)^3) = (32.15\pi - 21.43\pi) \approx 10.7\pi \approx 33.6 \ \text{in.}^3 \checkmark$$

Now You Try It

1. A plane cuts a sphere with a 10 inch radius and forms a circle of intersection with a 4 inch radius. How far away from the sphere's center does the plane cut the sphere? (Hint: Find the right triangle in the problem.)

2. Find the area and volume of the sphere in problem 1.

3. A basketball with an 8 inch radius fits perfectly in a box. Find the volume of the unused space in the box.

	Quick Solid Review
	Right Prisms: **2** Bases which are: 1) Congruent 2) Polygons 3) In parallel planes. Faces are rectangles. ⌐—⌐ I. Lateral Area (the area of the faces, the sum of the rectangles' areas). II. Total Area (the area of the faces *and* the 2 bases). III. Volume = Bh. (Capital B stands for the area of the base.)
	Regular Pyramids: **1** Base which is a regular (equal sides, equal angles) polygon. Faces are congruent isosceles triangles. *Pyramids* come to a *Point!* I. Lateral Area = ½ pℓ, ℓ is the slant height. II. Total Area = L.A. + 1B. III. Volume = ⅓ Bh
	Cylinders: **2** Bases which are congruent circles in parallel planes. I. Lateral Area = $2\pi rh$ II. Total Area = $2\pi rh + 2\pi r^2$ III. Volume = Bh = $\pi r^2 h$
	Cones: **1** Base which is a circle. I. Lateral Area = $\pi r\ell$ II. Total Area = $\pi r\ell + \pi r^2$ III. Volume = ⅓ $\pi r^2 h$
	Spheres: Remember the number **4** Area = $4\pi r^2$ Volume = $\frac{4}{3}\pi r^3$ Note the two "3"s. circle of intersection→ distance from center of sphere

Lengths, Areas and Volumes of Similar Solids — If two solids are similar, that is, if they have the same shape but perhaps, different sizes, they are connected mathematically. Here's an example:

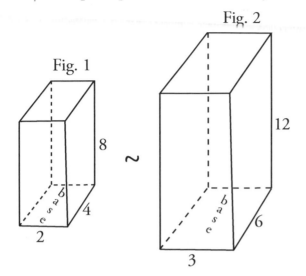

Fig. 2

Fig. 1

The two rectangular prisms are similar. This means the ratio of each pair of their corresponding sides is the same, that is, they are *in proportion* to one another:
$$\frac{2}{3} = \frac{4}{6} = \frac{8}{12}$$

Order Counts! The smaller figure appears first in the problem. Therefore, for this particular problem, the order is: **small to big**

The **Scale Factor** (SF) of two similar figures is the value of the common ratio of corresponding lengths, in this case: $\frac{2}{3}$ or "2 to 3" or 2:3

The 3 Levels Of Similarity

Level I — All Lengths (SF)

The *ratio* of *all* corresponding lengths of the two objects equals their scale factor, in our example: $\frac{2}{3}$

For example: The perimeter of the base of Fig. 1 = 2 + 4 + 2 + 4 = 12 $\frac{12}{18} = \frac{2}{3}$ √

The perimeter of the base of Fig. 2 = 3 + 6 + 3 + 6 = 18

In addition to the perimeters, other corresponding lengths whose ratios equal the scale factor include sides, diagonals, altitudes, and radii.

Level II — All Areas (SF)²

The *ratio* of all corresponding *areas* of the two objects equals their scale factor *squared*: $\frac{(2)^2}{(3)^2} = \frac{4}{9}$

For example: The area of the base of Fig. 1 = 2 × 4 = 8 $\frac{8}{18} = \frac{4}{9}$ √

The area of the base of Fig. 2 = 3 × 6 = 18

Other areas on Level II include the total areas, the lateral areas, and the areas of corresponding faces.

Level III — Volume (SF)³

The *ratio* of the *volumes* of the two figures equals their scale factor *cubed*, using our example:

$\frac{(2)^3}{(3)^3} = \frac{8}{27}$

For example: The volume of Fig. 1 = 2 × 4 × 8 = 64 $\frac{64}{216} = \frac{8}{27}$ √

The volume of Fig. 2 = 3 × 6 × 12 = 216

Example 1. The diagonal of the right face of Fig. 1 ≈ 8.9. Find x, the corresponding diagonal of Fig. 2. A diagonal is a length so it is on Level 1. Therefore, the *ratio* of the diagonals must equal the scale factor (SF) itself.

In creating the ratio, be consistent, small to big: $\frac{8.9}{x} = \frac{2}{3}$ Cross multiply and solve, $x \approx 13.4$ √

Example 2. The lateral area of Fig 2 on the previous page is 216, find the lateral area of Fig. 1.

A lateral area is a type of area so it must be on Level II. Therefore, the ratio of the two lateral areas must equal the scale factor squared, $(\text{SF})^2 = \dfrac{(2)^2}{(3)^2} = \dfrac{4}{9}$.

Let x equal the lateral area of Fig. 1.
Set the ratio up in the right order, small figure to big. Fig. 1 to Fig. 2.

$$\dfrac{4}{9} = \dfrac{x}{216} \quad \text{Cross multiply and solve.}$$

$$x = 96 \;\checkmark$$

To prove the above method works, find the lateral area of Fig. 1 by using the formula:

$$\text{L.A.} = ph = (2 + 4 + 2 + 4) \times 8 = 96 \;\checkmark$$

In fact, the ratio method works for pairs of similar objects of *any* shape:

___

Example 3. The radius of a basketball equals 8 in. and its volume $\approx 682.7\pi$ in³. If the radius of the soccer ball equals 6 in., find its volume.

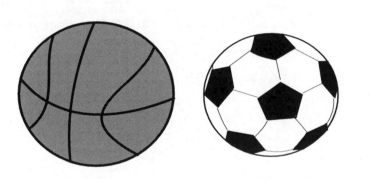

All spheres are similar, so we can use the ratio method. Since the radii are lengths, their ratio will give us the scale factor for these two solids:

$$\dfrac{8}{6} = \dfrac{4}{3} = \text{SF} \;\checkmark$$

The order that the balls are given in is big to small, so that's the order that the ratio must follow. Reducing the ratio makes it easier to work with.

The ratio of the volumes of similar objects equals their scale factor cubed: $(\text{SF})^3 = \dfrac{(4)^3}{(3)^3} = \dfrac{64}{27}$.
Let x equal the volume of the soccer ball and remember *order counts*:

$$\dfrac{\text{Volume of Basketball}}{\text{Volume of Soccer Ball}} = \dfrac{682.7\pi}{x} = \dfrac{64}{27} \qquad \begin{array}{l}\text{Cross multiply and solve:} \\ x = 288\pi \,\text{in}^3 \approx 904.3 \,\text{in}^3 \;\checkmark\end{array}$$

___

Example 4. What is the area of the basketball if the area of the volleyball is 144π?
Let y equal the area of the basketball. All areas are on Level II, so the ratio of the areas of the balls must equal their scale factor squared:

$$(\text{SF})^2 = \dfrac{(4)^2}{(3)^2} = \dfrac{16}{9}$$

$$\dfrac{\text{Area of Basketball}}{\text{Area of Soccer Ball}} = \dfrac{y}{144\pi} = \dfrac{16}{9} \qquad \begin{array}{l}\text{Cross multiply and solve:} \\ x \approx 256\pi \,\text{in}^2 \approx 803.8 \,\text{in}^2 \;\checkmark\end{array}$$

Example 5. The volumes of two similar pyramids are 343 un.^{3} and 216 un.^{3}, respectively.

a. Find the scale factor of the two figures and
b. Find the ratio of their lateral areas.

a. The problem gives the two volumes (Level III) of two similar solids. The ratio of the volumes equals the scale factor cubed:

$$\frac{\text{Volume of Pyramid I}}{\text{Volume of Pyramid II}} = \frac{343}{216} = (SF)^3$$

We need to find the scale factor (SF). We have its cube, so we need to work backwards to find the cube root:

$$\sqrt[3]{(SF)^3} = \sqrt[3]{\frac{343}{216}} = \frac{\sqrt[3]{343}}{\sqrt[3]{216}} = \frac{7}{6} = (SF)\ \checkmark$$

CALCULATOR NOTE: A cube root of a number is a number which when multiplied by itself and then by itself again equals the original number. For example, $4 \times 4 \times 4 = 64$, so 4 is the cube root of 64. To find a cube root using your calculator, find the power key. Here are three possibilities for how your power key might look: x^y, y^x, or $\wedge$. After you've found the power key, here's how to find a *cube* root: enter the base number, (use 64 for a trial), then press the power key, then enter 1/3 (if you have a fraction key) or .33333 (if you don't), then press the equal key. If you're doing it correctly, your calculator should display 4 or 3.9999.....

b. Knowing that the scale factor, (SF) = $\frac{7}{6}$ we can now find the ratio of the lateral areas (Level II) of the pyramids:

$$(SF)^2 = \frac{(7)^2}{(6)^2} = \frac{49}{36}\ \checkmark$$

A$^+_{\text{tip}}$ If the given information in the last problem had been the areas of the two solids and if the question had been to find the ratio of the volumes, the key to solving would have been to take the *square* root of the ratio of areas in order to find the scale factor (SF). The next step would have been to cube the scale factor to find the ratio of volumes. The harder problems of this type won't ask for the scale factor but will leave it up to you to figure out that finding the scale factor is the key to solving the problem.

Example — Two similar prisms have areas 64 un^2 and 121 un^2 respectively. If the volume of the smaller prism is 1,024 un^3, find the volume of the larger prism.

Step 1.

The ratio of the areas, $\frac{64}{121}$, equals $(SF)^2$.

Step 2. Find (SF):

$$SF = \sqrt{(SF)^2} = \sqrt{\frac{64}{121}} = \frac{\sqrt{64}}{\sqrt{121}} = \frac{8}{11}$$

Step 3. Find the ratio of the volumes, $(SF)^3$:

$$(SF)^3 = \frac{(8)^3}{(11)^3} = \frac{512}{1331}$$

Step 4. Let x = the volume of the larger prism:

$$\frac{1,024}{x} = \frac{512}{1331}$$

Cross multiply and solve: $x = 2{,}662\ \text{un}^3\ \checkmark$

Ways N͡ot to Do Similar Solid Problems

I. Setting up the ratio in the *wrong* order —

Example: The areas of two similar prisms are 225 and 169 respectively. Find their scale factor.

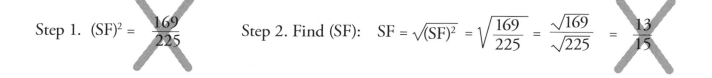

Step 1. $(SF)^2 = \dfrac{169}{225}$ ✗   Step 2. Find (SF): $SF = \sqrt{(SF)^2} = \sqrt{\dfrac{169}{225}} = \dfrac{\sqrt{169}}{\sqrt{225}} = \dfrac{13}{15}$ ✗

A⁺ₜᵢₚ When you're doing this kind of problem, before you give the answer, go back and re-read the question. This problem named the prisms in big to small order. The problem needs to be set up in the same order and the answer must be in the same order.

The correct answer is: $\dfrac{15}{13}$ ✓

II. Reducing fractions that do N͡ot reduce —

Example: The areas of two similar cylinders are 256 and 625 respectively. Find the ratio of their radii. (Radii are lengths so their ratio is the scale factor of the two figures)

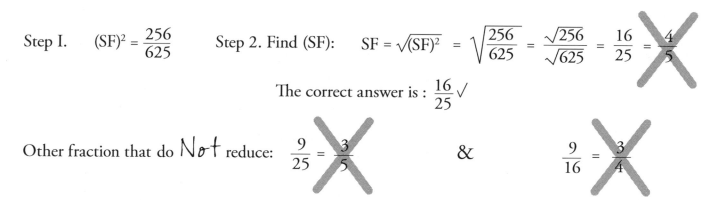

Step I. $(SF)^2 = \dfrac{256}{625}$ Step 2. Find (SF): $SF = \sqrt{(SF)^2} = \sqrt{\dfrac{256}{625}} = \dfrac{\sqrt{256}}{\sqrt{625}} = \dfrac{16}{25} = \dfrac{4}{5}$ ✗

The correct answer is : $\dfrac{16}{25}$ ✓

Other fraction that do N͡ot reduce: $\dfrac{9}{25} = \dfrac{3}{5}$ ✗   & $\dfrac{9}{16} = \dfrac{3}{4}$ ✗

Reducing and taking the square root are two different processes. Be sure to keep them separate.

CALCULATOR NOTE: Here's how to reduce proper fractions with your calculator. If you have a fraction key $a^{b/c}$, on your calculator, use it. Enter the numerator, then press the fraction key, then enter the denominator and then press the equal sign. Your calculator will display the reduced fraction if it reduces. If the original fraction is still shown, it does not reduce! To test an improper fraction invert it (flip it) and enter it. Remember to flip it back when you're done. If you do not have a fraction key, you should get a new calculator if you can. If you can't, just start looking for common divisors. And remember, no number can be divided by a number larger than one half itself.

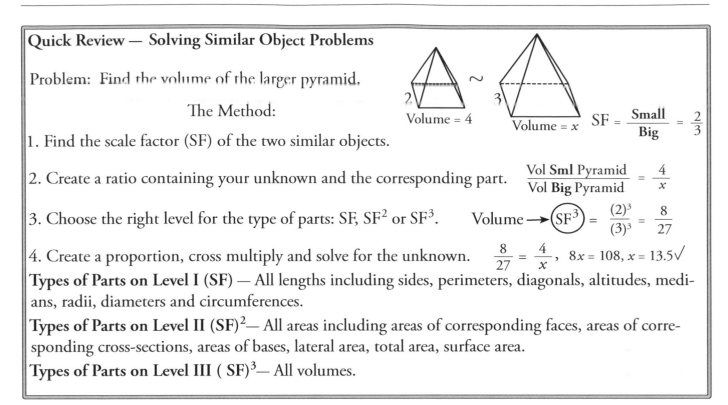

Quick Review — Solving Similar Object Problems

Problem: Find the volume of the larger pyramid.

The Method:

Volume = 4   Volume = x $SF = \dfrac{Small}{Big} = \dfrac{2}{3}$

1. Find the scale factor (SF) of the two similar objects.

2. Create a ratio containing your unknown and the corresponding part. $\dfrac{\text{Vol } \mathbf{Sml} \text{ Pyramid}}{\text{Vol } \mathbf{Big} \text{ Pyramid}} = \dfrac{4}{x}$

3. Choose the right level for the type of parts: SF, SF^2 or SF^3.   Volume → $\boxed{SF^3} = \dfrac{(2)^3}{(3)^3} = \dfrac{8}{27}$

4. Create a proportion, cross multiply and solve for the unknown. $\dfrac{8}{27} = \dfrac{4}{x}$, $8x = 108$, $x = 13.5\checkmark$

Types of Parts on Level I (SF) — All lengths including sides, perimeters, diagonals, altitudes, medians, radii, diameters and circumferences.

Types of Parts on Level II (SF)^2 — All areas including areas of corresponding faces, areas of corresponding cross-sections, areas of bases, lateral area, total area, surface area.

Types of Parts on Level III (SF)^3 — All volumes.

Now You Try It

1. The two cylinders at right are similar.
   In less than 30 seconds, find the **ratios of:**

    a. the heights of the cylinders

    b. the circumferences of the bases

    c. the lateral area of the cylinders

    d. the total area of the cylinders

    e. the volumes of the cylinders.

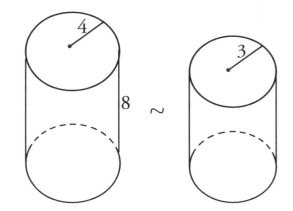

2. Find the height of the smaller cylinder.

3. The lateral area of the smaller cylinder is 36π. Using the ratio method, find the lateral area of the larger cylinder.

4. The volume of the larger cylinder is 128π. Using the ratio method, find the volume of the smaller cylinder.

14. Transformations

To *transform* or *map* a geometric figure means to change:

      1. its position, or
      2. its size, or
      3. its shape.

Transformation terms:

Preimage: The original figure (before any transformation).

Image: The transformed figure.
(Note that arrow ($\rightarrow$) and prime ($'$) notation indicate a transformation, or mapping.)

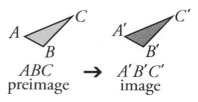

$ABC \rightarrow A'B'C'$
preimage   image

Isometry: A transformation in which the preimage and image are congruent.

Rigid Motion: A motion that preserves both length and angles. Isometries are rigid motions.

Translations

A **translation** (T) (or slide, or glide) maps all points in a figure the same distance in the same direction. Translations are isometries.

To find an image, plot each vertex based on the rule of the translation, then connect the vertices.

Example: Find the image of PQR under the translation:

$$(x, y) \rightarrow (x+7, y-1).$$
$$P(-6,5): (-6+7, 5-1) \rightarrow P'(1,4)$$
$$Q(-3,0): (-3+7, 0-1) \rightarrow Q'(4,-1)$$
$$R(-1,3): (-1+7, 3-1) \rightarrow R'(6,2)$$

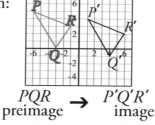

$PQR \rightarrow P'Q'R'$
preimage   image

Writing the rule of the translation when given the preimage and image:

(Note: You only need to calculate the change in **one** vertex.)

Given the figures ABC and $A'B'C'$:

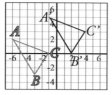

$A(-6,2) \rightarrow A'(-1,5)$

The horizontal (x) change is: $-1-(-6) = +5$

The vertical (y) change is: $5-2 = +3$

Therefore, the rule for the translation is: $(x,y) \rightarrow (x+5, y+3)$.

A **composition** of transformations means more than one. The mappings are done in order. The image of the first becomes the preimage of the second:

$$(x,y) \rightarrow (x-8, y+5)$$
$$(x,y) \rightarrow (x+2, y-1)$$

The composition of the two translations is: $(x, y): (-8+2, +5-1) \rightarrow (x-6, y+4)$

Reflections

Reflections, (R) or flips, map geometric objects across a line of reflection ($\overleftrightarrow{r}$).

Reflections are isometries.

Methods for mapping reflections:

1. If $\overleftrightarrow{r}$ is the x axis: Map each vertex of the object using this rule: $(x,y) \rightarrow (x,-y)$, then, connect the vertices of the image. Here's an example,

      Reflect **ABC** across the x axis.

Note:

1. Each point and its image are equidistant from $\overleftrightarrow{r}$, (the x axis in this case).

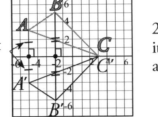

2. A point on $\overleftrightarrow{r}$ is its own image (**C** and C').

3. $\overleftrightarrow{r}$ (the x axis in this case) is the ⊥ bisector of the segment joining any point of the pre-image and its image, (for points not on $\overleftrightarrow{r}$).

2. If $\overleftrightarrow{r}$ is the y axis: Map each vertex of the object using this rule: $(x, y) \rightarrow (-x, y)$, then, connect the vertices of the image. Here's an example,

      Reflect **PRQ** across the y axis.

Note:

1. Each point and its image are equidistant from $\overleftrightarrow{r}$, (the y axis in this case) .

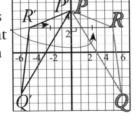

2. A point on $\overleftrightarrow{r}$ is its own image (**P** and P').

3. $\overleftrightarrow{r}$ (the y axis in this case) is the ⊥ bisector of the segment joining any point of the pre-image and its image, (for points not on $\overleftrightarrow{r}$).

3. Examples of an object (the gray polygon) reflected over various lines of reflection:

$\overleftrightarrow{r}$ is the line $x=1$ $\overleftrightarrow{r}$ is the line $y = 2$ $\overleftrightarrow{r}$ is the line $y=0$*   The lines $y = x$ and $y =-x$

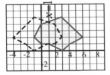   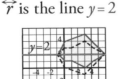   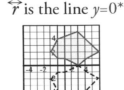   

(*y=0 is the equation of the x axis.)

Rotations

A Rotation (R) has:

        1. A center (a point),
        2. An angle (number of degrees), and
        3. A direction, *counterclockwise.
        (*Clockwise rotation indicates negative angle measure.)

The graph shows $R_{P,90}$, a 90° rotation(R) of $\triangle ABC$ around point P.

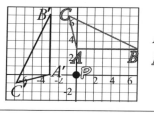

$R_{P,90}$ maps A to A', B to B' and C to C'.

Facts to know about rotations:

A rotation is an isometry.

An angle measuring: 90° is a quarter rotation, 180° is a half rotation, 360° is a full rotation.

Rotations that transform a figure to the same final position are said to be equal:

$$R_{P,-270} \text{ and } R_{P,90} \text{ are "equal".}$$

To graph the image of figure ABC, note that the distance from the center (P) to a pre-image point of the figure equals the distance from P to the corresponding image point:

$$PA = PA' \quad PB = PB' \quad PC = PC'$$

Rotations can also be made around a point on or in the interior region of a figure:

Rotating polygon $BCDE$ using point E as the center.

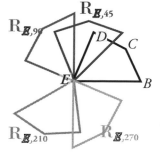

Plot vertex by vertex and connect the points.

Rotating a polygon using an interior point P as the center.

Common angles of rotation.

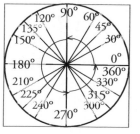

Dilations

Dilations are similarity mappings and are generally not congruence mappings (not isometries).

Given a dilation with center O and scale factor SF ($Do,_{SF}$):

To map point P of a pre-image to its image P':

If SF > 0, P' lies on $\overrightarrow{OP}$ and $\boxed{OP' = (SF) \cdot OP}$
(If the scale factor is positive, the new (image) point lies on the ray, whose endpoint is O and that passes through the pre-image point P. The length of the segment, OP' equals the product of the scale factor times the length of the segment from O to P.)

Example of $Do,_4$:   SF $= 4$ $\bullet P'$

$\bullet O$ $\bullet P$

If SF < 0, P' lies on the ray opposite $\overrightarrow{OP}$ and $\boxed{OP' = |SF| \cdot OP}$
(If the scale factor is negative, the new (image) point lies on the ray opposite the ray, whose endpoint is O and that passes through the pre-image point P. The length of the segment, OP' equals the product of the absolute value of the scale factor (make the negative number positive) times the length of the segment from O to P.)

Example of $Do,_{-3}$:   SF $= -3$

$\bullet O$ $\bullet P$

$\bullet P'$

If $|SF| > 1$ the dilation is an enlargement or "expansion".

If $|SF| = 1$ the dilation is an isometry.

If $|SF| < 1$ the dilation is a reduction or "contraction".

Here are some examples of dilation problems:, notice that you multiply each coordinate by the SF:
Find the coordinates of the image of
A(-3, 2), B(1, 4), C(2, -5) under the dilations:

$Do,_2$: A'(-6,4), B'(2,8), C'(4,-10)

$Do,_{-4}$: A'(12,-8), B'(-4,-16), C'(-8,20)

Polygons: A dilation $Do,_k$, maps any polygon to a similar polygon with area $= (k^2 \cdot$ (area preimage)).

Example: Map $\overline{RST}$
under $Do,2$

Area of a $\triangle = (½)bh$

$A_{\triangle RST} = (½)(6)(3) = 9$

$A_{\triangle R'S'T'} = (½)(12)(6) = 36$

$k=2$, $k^2=4$

$4 \times 9 = 36$ ✓

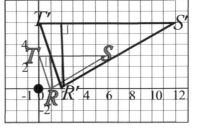

page 5

Now You Try It —

$$A\ \ P\ B\ \ \ \ C\ \ \ \ D\ \ \ \ E\ \ \ \ F\ \ \ \ G\ W\ H\ \ \ \ I$$
$$-2\ \underset{-1.2}{\blacktriangle}-1\ \ \ \ 0\ \ \ 1\ \ \ 2\ \ \ 3\ \ \ 4\underset{4.5}{\blacktriangle}5\ \ \ 6$$

Using the number line at right find:

1. $AE = |-2 - 2| = 4$ (or count)√   3. $EA = |2 - -2| = 4$ (or count)√

(Notice that $AE = EA$)

2. $WP = |4.5 - -1.2| = 5.7$√   4. $GA = |4 - -2| = 6$ (or count)√

5. The midpoint and the coordinate of the midpoint of $\overline{HB}$.

First find HB, (the length), $HB = |5 - -1| = 6$ (or just count). A midpoint divides a segment into two equal parts. Each part must be ½ of the total, in this problem, ½ of 6 or 3 units long. So the midpoint must be 3 units from either endpoint. Subtract 3 from 5 (or count 3 units back from 5), which equals 2. Since the number 2 is paired with point E, the midpoint of $\overline{HB}$ is E and 2 is the coordinate of the midpoint of $\overline{HB}$. √

page 6

Now You Try It — Give the definition of each symbol shown below.

1. $\overrightarrow{AB}$ Ray AB. √
2. AB The distance from point A to point B. AB is a number. √
3. $\overleftrightarrow{AB}$ Line AB. √
4. $\overline{AB}$ Segment AB √

page 11

Now You Try It —   1.

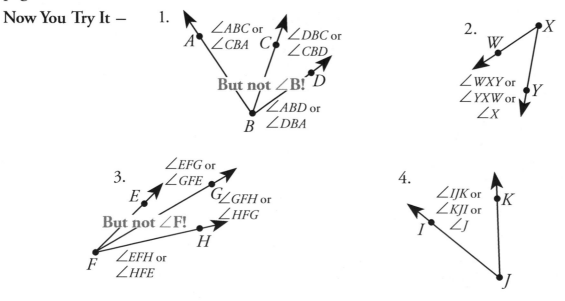

236

Now You Try It —

Pictures of angles do not always show the points through which the rays pass. In fact, there are many different ways in which your teacher and your textbook might draw and label angles. Several of these are shown in the measuring exercises below. Using a protractor:

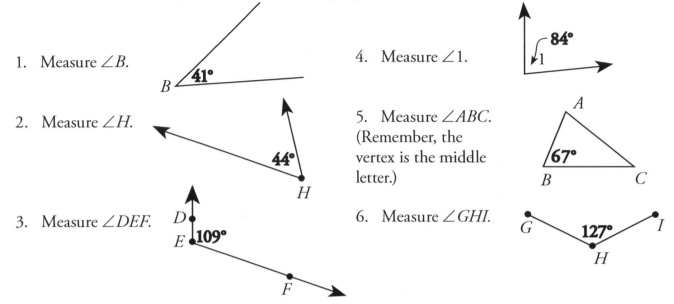

1. Measure ∠*B*.

2. Measure ∠*H*.

3. Measure ∠*DEF*.

4. Measure ∠1.

5. Measure ∠*ABC*. (Remember, the vertex is the middle letter.)

6. Measure ∠*GHI*.

And remember, always measure as carefully as you can, but these are still just approximations! (And your answers might be slightly different.)

Now You Try It —

Identify each of the following as an acute, right, obtuse or straight angle.

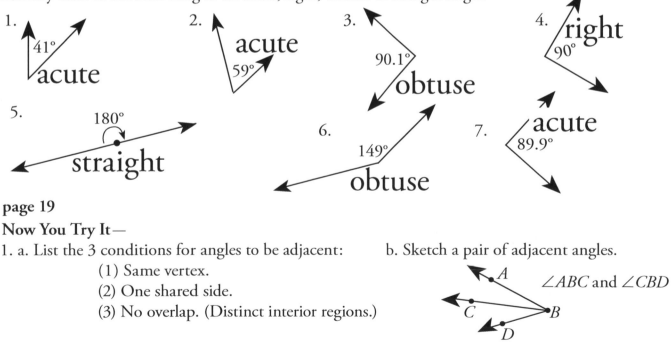

1. acute 41°

2. acute 59°

3. obtuse 90.1°

4. right 90°

5. straight 180°

6. obtuse 149°

7. acute 89.9°

Now You Try It—

1. a. List the 3 conditions for angles to be adjacent:

    (1) Same vertex.

    (2) One shared side.

    (3) No overlap. (Distinct interior regions.)

b. Sketch a pair of adjacent angles.

∠*ABC* and ∠*CBD*

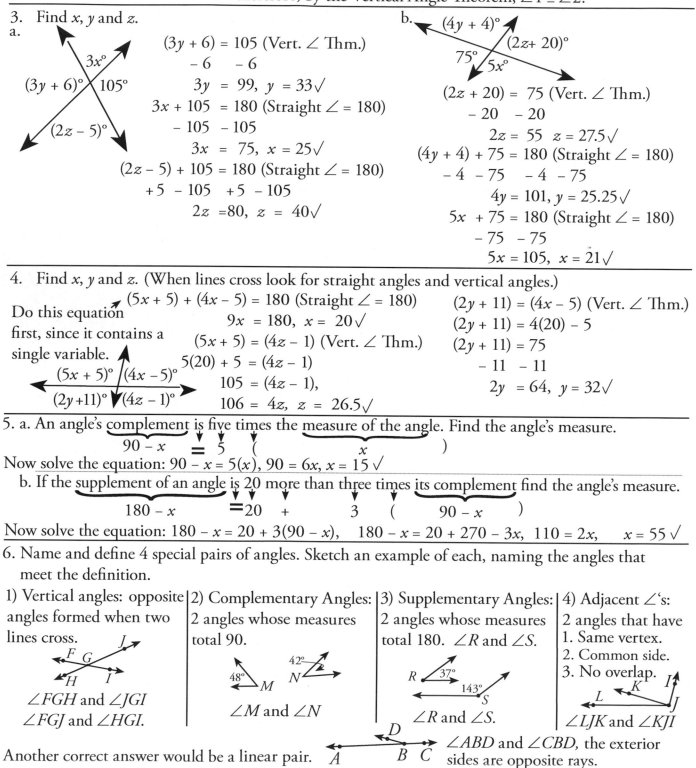

page 19 (continued) 2. Are ∠1 and ∠2 in the figure adjacent angles? If not why not? Do you know anything else about these two angles?

∠1 and ∠2 are not adjacent angles because they do not share a common side. However, ∠1 and ∠2 are vertical angles, and therefore, by the Vertical Angle Theorem, ∠1 ≅ ∠2.

3. Find x, y and z.

a.

$(3y + 6) = 105$ (Vert. ∠ Thm.)
   $- 6 \quad - 6$
   $3y = 99, \ y = 33 ✓$
$3x + 105 = 180$ (Straight ∠ = 180)
   $- 105 \ - 105$
   $3x = 75, \ x = 25 ✓$
$(2z - 5) + 105 = 180$ (Straight ∠ = 180)
   $+ 5 \ - 105 \ + 5 \ - 105$
   $2z = 80, \ z = 40 ✓$

b.

$(2z + 20) = 75$ (Vert. ∠ Thm.)
   $- 20 \quad - 20$
   $2z = 55 \ z = 27.5 ✓$
$(4y + 4) + 75 = 180$ (Straight ∠ = 180)
   $- 4 \ - 75 \quad - 4 \ - 75$
   $4y = 101, \ y = 25.25 ✓$
$5x + 75 = 180$ (Straight ∠ = 180)
   $- 75 \ - 75$
   $5x = 105, \ x = 21 ✓$

4. Find x, y and z. (When lines cross look for straight angles and vertical angles.)

Do this equation first, since it contains a single variable.

$(5x + 5) + (4x - 5) = 180$ (Straight ∠ = 180)
   $9x = 180, \ x = 20 ✓$
$(5x + 5) = (4z - 1)$ (Vert. ∠ Thm.)
$5(20) + 5 = (4z - 1)$
$105 = (4z - 1),$
$106 = 4z, \ z = 26.5 ✓$

$(2y + 11) = (4x - 5)$ (Vert. ∠ Thm.)
$(2y + 11) = 4(20) - 5$
$(2y + 11) = 75$
   $- 11 \ - 11$
   $2y = 64, \ y = 32 ✓$

5. a. An angle's complement is five times the measure of the angle. Find the angle's measure.

$90 - x = 5 (\ x \ )$

Now solve the equation: $90 - x = 5(x), \ 90 = 6x, \ x = 15 ✓$

b. If the supplement of an angle is 20 more than three times its complement find the angle's measure.

$180 - x = 20 + 3 (\ 90 - x \ )$

Now solve the equation: $180 - x = 20 + 3(90 - x), \quad 180 - x = 20 + 270 - 3x, \quad 110 = 2x, \quad x = 55 ✓$

6. Name and define 4 special pairs of angles. Sketch an example of each, naming the angles that meet the definition.

1) Vertical angles: opposite angles formed when two lines cross.

∠FGH and ∠JGI
∠FGJ and ∠HGI.

2) Complementary Angles: 2 angles whose measures total 90.

∠M and ∠N

3) Supplementary Angles: 2 angles whose measures total 180. ∠R and ∠S.

∠R and ∠S.

4) Adjacent ∠'s: 2 angles that have
1. Same vertex.
2. Common side.
3. No overlap.

∠LJK and ∠KJI

∠ABD and ∠CBD, the exterior sides are opposite rays.

Another correct answer would be a linear pair.

Now You Try It —

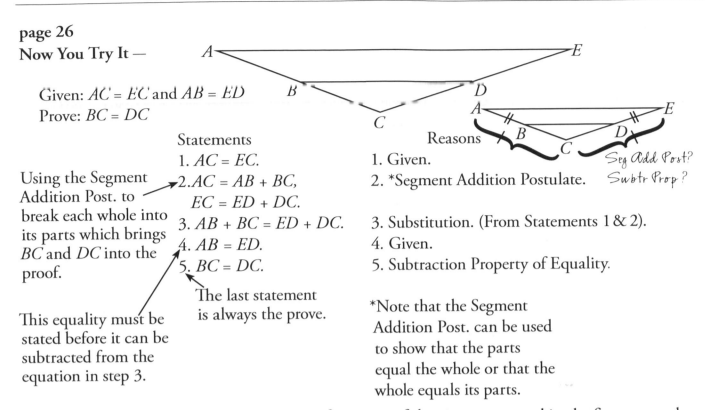

Given: $AC = EC$ and $AB = ED$
Prove: $BC = DC$

Statements	Reasons
1. $AC = EC$.	1. Given.
2. $AC = AB + BC$, $EC = ED + DC$.	2. *Segment Addition Postulate.
3. $AB + BC = ED + DC$.	3. Substitution. (From Statements 1 & 2).
4. $AB = ED$.	4. Given.
5. $BC = DC$.	5. Subtraction Property of Equality.

Using the Segment Addition Post. to break each whole into its parts which brings BC and DC into the proof.

This equality must be stated before it can be subtracted from the equation in step 3.

The last statement is always the prove.

*Note that the Segment Addition Post. can be used to show that the parts equal the whole or that the whole equals its parts.

When to put in the Given? In the above proof one part of the given was stated in the first step and the other part in the fourth. It was not necessary to do this but it made more sense because $AB = ED$ was not needed until the fourth step.

page 30
Now You Try It —

1. Given: $AB = CD$.
   Prove: $AC = BD$

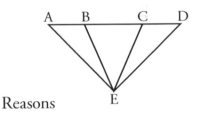

Statements	Reasons
1. $AB = CD$.	1. Given.
2. $BC = BC$.	2. Reflexive.
3. $AB + BC = CD + BC$.	3. Addition Property of Equality.
4. $AB + BC = AC$ $BC + CD = BD$.	4. Segment Addition Postulate.
5. $AB = CD$.	5. Substitution. (From Steps 3 & 4).

This equality must be stated before it can be added to the equation from step 1.

Adding the 2 equations from Steps 1 & 2 together.

Using the Segment Addition Post. to state that the sum of the parts equals the whole. Doing this brings AC and BD into the proof.

The last statement is always the prove.

Calling the same thing (from Step 3) by a different name (the one determined in Step 4).

2. Given: $m\angle AEB = m\angle DEC$
   Prove: $m\angle AEC = m\angle DEB$

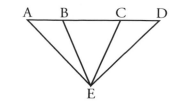

Note: This proof is logically identical to the previous proof.

This equation must be introduced before $m\angle BEC$ can be added to the equation from Step 1.

Adding the 2 equations from Steps 1 & 2 together.

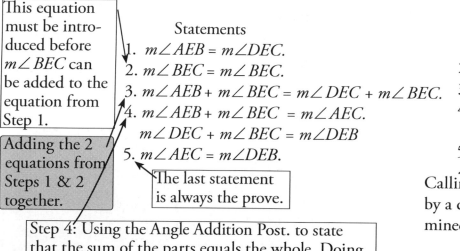

Statements	Reasons
1. $m\angle AEB = m\angle DEC$.	1. Given.
2. $m\angle BEC = m\angle BEC$.	2. Reflexive.
3. $m\angle AEB + m\angle BEC = m\angle DEC + m\angle BEC$.	3. Addition Property of Equality.
4. $m\angle AEB + m\angle BEC = m\angle AEC$.	4. *Angle* Addition Postulate.
$\quad m\angle DEC + m\angle BEC = m\angle DEB$	
5. $m\angle AEC = m\angle DEB$.	5. Substitution (Stmts. 2 & 3).

The last statement is always the prove.

Calling the same thing (from Step 3) by a different name (the one determined in Step 4).

Step 4: Using the Angle Addition Post. to state that the sum of the parts equals the whole. Doing this brings $m\angle AEC$ and $m\angle DEB$ into the proof.

3. Prove the following:

THEOREM: *If two lines are perpendicular, they form congruent adjacent angles.*
Create an appropriate figure and state the given and the prove.
The theorem at the top of the previous page and this one have a close connection. They are called "converses". Can you see what the connection is? **Answer:** The connection is that the given and the prove have been switched.

Given: $\overleftrightarrow{CE} \perp \overleftrightarrow{AD}$
Prove: $\angle ABC \cong \angle CBD$

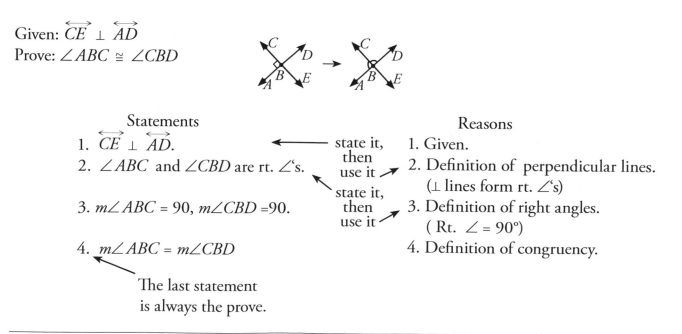

Statements	Reasons
1. $\overleftrightarrow{CE} \perp \overleftrightarrow{AD}$.	1. Given.
2. $\angle ABC$ and $\angle CBD$ are rt. $\angle$'s.	2. Definition of perpendicular lines. ($\perp$ lines form rt. $\angle$'s)
3. $m\angle ABC = 90$, $m\angle CBD = 90$.	3. Definition of right angles. (Rt. $\angle = 90°$)
4. $m\angle ABC = m\angle CBD$	4. Definition of congruency.

state it, then use it

state it, then use it

The last statement is always the prove.

1. List the four logical conditionals and using symbol shorthand give an example of each. State the two pairs which are logically connected.

If-then statement:	If p then q.	} These two conditionals are logically connected. Both are true or both are false.
Contrapositive:	If not q then not p.	
Converse:	If q then p.	} These two conditionals are logically connected. Both are true or both are false.
Inverse:	If not p then not q.	

2. Given the statement: If you are in a top band, then you are famous,
a. Form the contrapositive, converse and inverse and state the truth value of each of the 4 statements.

If-then statement:	If you are in a top band, then you are famous.	True
Contrapositive:	If you are not famous then you are not in a top band.	True
Converse:	If you are famous then you are in a top band.	False
Inverse:	If you are not in a top band, then you are not famous.	False

b. If your cousin Sally is not famous, what can you conclude about her?

Sally is not in a top band.

c. If your cousin Sam is famous, what can you conclude about him?

Nothing. Maybe he is in a top band or maybe he is famous for some other reason.

3. Given the statement: If $x^2 > 16$, then $x > 4$, form the contrapositive, the converse and the inverse and determine the truth value of each of the four statements.

If-then statement:	If $x^2 > 16$, then $x > 4$.	False (Counterexample, $(-5)^2 = 25$ and $-5 < 4$)
Contrapositive:	If $x \leq 4$ then $x^2 \leq 16$.	False (For example, $x = -5$, $x^2 = 25$)
Converse:	If $x > 4$ then $x^2 > 16$.	True
Inverse:	If $x^2 \leq 16$ then $x \leq 4$.	True.

4. Explain what are the requirements for a syllogism and make up an original example to demonstrate this logical form.
A syllogism is made up of two if-then statements in which the conclusion of the first statement is the hypothesis of the second statement.
The Law of Syllogism states that if both statements of a syllogism are true, then you can skip directly from the first hypothesis to the second conclusion.

$$p \rightarrow \text{\textcircled{q}} \quad \text{\textcircled{q}} \rightarrow r \qquad\qquad p \rightarrow r$$

Problem 4. continued on next page.

page 34 (continued) Example: If you had salad you had a healthy lunch. (Note, that the second statement happens to be false.)

If you had a healthy lunch you must have eaten chicken.

Notice that the conclusion of the first statement *is* the hypothesis of the second but the second statement is false and both statements must be true for the Law Of Syllogism to apply.

Example: If you can run the four hundred meters in under 50 seconds you are an extraordinary runner.

If you are an extraordinary runner you can qualify for a top track team.

Since the conclusion of the first statement is the hypothesis of the second *and* both statement are true The Law of Syllogism applies and we may state: If you can run the four hundred meters in under 50 seconds then you can qualify for a top track team. The Law of Syllogism applies in the following 2 examples:

Example: Tawnee is an Olympic gymnast.

Olympic gymnasts are excellent athletes.

Given that both statements are *true*, we can conclude that Tawnee is an excellent athlete.

Example: If $x > 5$ then $x^2 > 25$.   The Law of Syllogism can be represented symbolically as:

If $x^2 > 25$ then $x^4 > 625$.

$$\begin{array}{ccc} p & \rightarrow & \textcircled{q} \\ \textcircled{q} & \rightarrow & r \\ p & \rightarrow & r \end{array}$$

Since both statements are *true*, we can conclude that if $x > 5$ then $x^4 > 625$.

page 36   **Now You Try It —**

Given: $m\angle I \neq m\angle J$,

Prove: $\angle I$ and $\angle J$ are not both right angles. Do the proof indirectly and use a paragraph proof.

Proof: Assume temporarily that $\angle I$ and $\angle J$ are both right angles. Then, by the definition of right angles, $m\angle I = 90$ and $m\angle J = 90$. But this would mean that $m\angle I = m\angle J$ (Substitution). But this contradicts the given which is impossible. Therefore, our temporary assumption was wrong and $\angle I$ and $\angle J$ are not both right angles.

Here are two more examples of Indirect Proofs using theorems and definitions from later chapters. We have numbered the steps for clarity, but the numbers may be discarded and the proofs reformatted in paragraph form.

Example 1. Given: $m\angle 4 = 128°$, $m\angle 7 = 51°$

Prove: l is *not* parallel to m (the <u>negative</u> of the prove)

1. *Assume temporarily* that l is parallel to m,

2. If l is parallel to m, then PSSIS tells us that since $m\angle 4 = 128°$, $m\angle 6 = 52°$.

3. But if $m\angle 6 = 52°$, since vertical angles are equal, we also know that $m\angle 7 = 52°$.

4. But this contradicts the given, $m\angle 7 = 51°$.

5. Therefore, our temporary assumption must be false and the prove is true, l is *not* parallel to m.

Example 2. Given: $BD = BC = 6$, $AB = 7$, $m\angle 2 = m\angle 3$

Prove: $m\angle 1 \neq m\angle 4$ (the <u>negative</u> of the prove)

Proof:1. *Assume temporarily* that $m\angle 1 = m\angle 4.$

2. Since $DB = CB$, $m\angle 3 = m\angle 4$. (the Isosceles Triangle Theorem).

3. Since $m\angle 3 = m\angle 4$ (step 1) and $m\angle 2 = m\angle 3$ (given), $m\angle 2 = m\angle 4$ (Substitution).

4. But if $m\angle 1 = m\angle 4$ (our temporary assumption), and $m\angle 2 = m\angle 4$ (step 2)   then $m\angle 1 = m\angle 2$ (substitution).

5. But the converse of the Isosceles Triangle Theorem tells us that if $m\angle 1 = m\angle 2$ (step 3.) then $AB = BD$, which contradicts the given. Since the given is always true, our *temporary assumption*, $m\angle 1 = m\angle 4$, must be false and the prove, $m\angle 1 \neq m\angle 4$ must be true.

Now You Try It —

Using the above figure, classify the following pairs of angles
as corresponding angles, alternate interior angles, same side
interior angles or none of these. If the angles do form one
of the listed special pairs, name the lines and the transversal
that form the 2 angles.

1. $\angle 3$ and $\angle 7$: Corresp. $\angle$'s, the lines are k and l, the transversal is j.
2. $\angle 4$ and $\angle 5$: Alternate interior $\angle$'s, the lines are k and l, the transversal is j.
3. $\angle 15$ and $\angle 6$: Alternate exterior $\angle$'s, the lines are i and j, the transversal is l.
4. $\angle 2$ and $\angle 5$: Same side interior (or consecutive) $\angle$'s, the lines are k and l, the transversal is j.
5. $\angle 3$ and $\angle 10$: Alternate interior $\angle$'s, the lines are i and j, the transversal is k.
6. $\angle 11$ and $\angle 10$: None. (They *are* vertical angles but always answer the question that was asked.)
7. $\angle 7$ and $\angle 13$: Same side interior (or consecutive) $\angle$'s, the lines are i and j, the transversal is l.
8. $\angle 4$ and $\angle 13$: None, (no transversal is shared by both angles).
9. $\angle 11$ and $\angle 8$: None, (no transversal is shared by both angles).

page 42

Now You Try It —

1.)

$$7x - 5 = 4x + 31$$
$$\underline{-4x \qquad -4x} \quad \text{Subtract } 4x \text{ from each side.}$$
$$3x - 5 = 31 \qquad \text{Do the math.}$$
$$\underline{+5 \quad +5} \qquad\qquad \text{Add 5 to each side.}$$
$$\frac{3x}{3} = \frac{36}{3} \qquad \text{Divide by the coefficient of } x.$$
$$x = 12 \ \checkmark \qquad \text{Done! Because, we have a single}$$
positive x (that's what "isolated"
means) all by itself on one side.

2.)

$$3x + 24 = 4x - 8$$
$$\underline{-3x \qquad -3x} \qquad \text{Subtract } 3x \text{ from each side.}$$
$$24 = x - 8 \qquad \text{Do the math.}$$
$$\underline{+8 \quad +8} \qquad \text{Add 8 to each side.}$$
$$32 = x \ \checkmark \qquad \text{Done.}$$

3.)

$$4x + 7 = 5x - 10.5 \text{(The answer is not an integer)}$$
$$\underline{-4x \qquad -4x} \quad \text{Subtract } 4x \text{ from each side.}$$
$$7 = x - 10.5$$
$$\underline{+10.5 \qquad +10.5} \ \text{Add 10.5 to each side.}$$
$$17.5 = x \ \checkmark \qquad \text{Done,}$$

4.)

$$20x + 4 = 13x + 39$$
$$\underline{-13x \qquad -13x} \qquad \text{Subtract } 13x \text{ from each side.}$$
$$7x + 4 = 39$$
$$\underline{-4 \quad -4} \qquad \text{Subtract 4 from each side.}$$
$$\frac{7x}{7} = \frac{35}{7} \qquad \text{Divide by the coefficient of } x.$$
$$x = 5 \checkmark$$

Now you try it — Solve for x:

1. $10x + 5 = 180 - (4x - 7)$

   $10x + 5 = 180 - 1(4x) - 1(-7)$ Distribute the formerly invisible negative one.

   $10x + 5 = 180 - 4x + 7$ $(-)(+) = -$ $(-)(-) = +$

   $+4x \;\; -5 = \qquad\; +4x - 5$ Add $4x$ to each side, subtract 5 from each side. Do the math.

   $\dfrac{14x}{14} = \dfrac{182}{14}$ Divide each side by the coefficient of x.

   $\qquad x = 13 \checkmark$ Done.

2. $180 - (2x + 3) = 6x - 3$

   $180 - 1(2x) - 1(3) = 6x - 3$ Distribute the formerly invisible negative one.

   $180 - 2x - 3 = \;\; 6x - 3$ $(-)(+) = -$ $(-)(+) = -$

   $+2x + 3 = +2x + 3$ Add $2x$ to each side, add 3 to each side. Do the math.

   $\dfrac{180}{8} = \dfrac{8x}{8}$ Divide both sides by the coefficient of x.

   $22.5 = x \checkmark$ Done.

Now You Try It   See:

1. Solve for a, b and c.   Think

In a problem like this one, sometimes it's helpful to extend the lines (see Figure 2). Now it's clearer that the figure has (2 pairs of) parallel lines cut by transversals. The arrowheads should tell you to think of

P ^{CC}
P ^{AIAC}
P ^{SSIS} $\checkmark$
P ^{AEAC}

Fig. 1   Fig. 2

the 4 methods and of the 4, PSSIS is the right method for this problem because it tells us that each pair of consecutive angles is supplementary. Now, which equation to create? Remember, in order to be solvable, one equation can have at most, one variable. The only pair of expressions that share the same variable are $4a$ and $8a - 48$, so that's the equation to start with:

$$4a + (8a - 48) = 180$$
$$12a - 48 = 180, \; 12a = 228, \; a = 19 \checkmark$$

Knowing that $a = 19$, we can solve for the remaining variables:

$$3b + 5 + 4a = 180, \; 3b + 5 + 4(19) = 180$$
$$3b + 5 + 76 = 180, \; 3b + 81 = 180, \; 3b = 99, \; b = 33 \checkmark$$
$$c + 3b + 5 = 180, \; c + 3(33) + 5 = 180 = 76, \; c = 76 \checkmark$$

2. Solve for *u*, *v* and *w*.

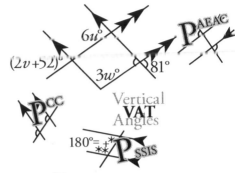

To solve this problem, we need to create an equation with a single variable. This means that we should form the equation using the 81°angle and one other expression based on one of the 4 methods. Studying the problem, there are several ways to solve, however, it is probably easiest to use PAEAC:

$$6u = 81, \quad u = 13.5 \checkmark$$

PCC tells us: $(2v + 52)° = 6u = 81$(from above), so

$$2v = 81 - 52 = 29, \quad v = 14.5 \checkmark$$

We can then use the Vertical Angle Theorem and PSSIS,

$$3w + 81 = 180, \quad w = 33 \checkmark$$

page 45
Now You Try It:

Given line $p \| q$

1. Prove $\angle 2$ is supplementary to $\angle 5$.

None of the 4 methods deal with angles in the relative positions of $\angle 2$ and $\angle 5$ so we need to use other methods to get started. Study the figure. The Vertical Angle Theorem (VAT) tells us that $m\angle 2 = m\angle 3$. PSSIS tells us that $\angle 3$ and $\angle 5$ are supplementary, but this is the same as saying $\angle 2$ is supplementary to $\angle 5$. As you do a proof mark up your sketch. Your "notes" help you.

Proof:

Statements	Reasons
1. $p \| q$.	1. Given.
2. $m\angle 2 = m\angle 3$.	2. VAT.
3. $\angle 3$ is supplementary to $\angle 5$.	3. PSSIS.
4. $m\angle 3 + m\angle 5 = 180$.	4. Definition of supplementary.
5. $m\angle 2 + m\angle 5 = 180$.	5. Substitution Property (Statements 2 & 4).
6. $\angle 2$ is supplementary to $\angle 5$.	6. Definition of supplementary.

Note: We can't substitute right after Step 3 because substitution is a property of equality and congruence. You can't use substitution in other types of statements.

2. Prove $m\angle 1 = m\angle 8$ *without* using PAEAC

Only PAEAC deals with angles in the relative positions of $\angle 1$ and $\angle 8$ so we need to use other methods to get started. Study the figure. PCC tells us that $m\angle 1 = m\angle 5$ and the Vertical Angle Theorem (VAT) tells us that $m\angle 5 = m\angle 8$. Since both steps involve equations, we can use the Transitive Property with no intermediate steps.

Proof:

Statements	Reasons
1. $p \| q$.	1. Given.
2. $m\angle 1 = m\angle 5$.	2. PCC.
3. $m\angle 5 = m\angle 8$.	3. VAT.
4. $m\angle 1 = m\angle 8$.	4. Transitive Property (Statements 2 & 3).

page 48
Find the measures of *a* through *f*.

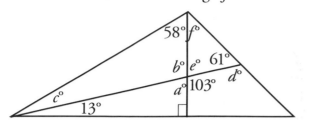

Letter	Reason	Equation	Answer
a	Straight ∠ = 180	(180 –103)	77
b	VAT	—	103
c	Angles of a △ = 180	(180-103-58)	19
d	Straight ∠ = 180	(180 – 61)	119
e	VAT	—	77
f	Angles of a △ = 180	(180 - 61 -77)	42

page 53 Now You Try It —

Find the measures of ∠*SRQ*, ∠*SRT*, ∠Q and ∠S.
(Hint: Form a "system" of equations with *u* and *v*.)

When a problem **A⁺tip** gives a hint, *always* take it.

①Since a straight ∠ = 180:
$(2u – 6) + (5v – 3) = 180$
$2u + 5v = 189$

②By the Ext. ∠ Theorem:
$(2u – 6) = (u +5) + (v – 4)$
$2u – u – v = 5 – 4 + 6$
$u – v = 7$, now isolate *u*
$u = v + 7$

③Now substitute:
$2u + 5v = 189$ [The equation from step 1.]
$2(v + 7) + 5v = 189$, $2v + 14 + 5v = 189$, $7v + 14 = 189$, $v = 25$.
Now that we know $v = 25$, substitute 25 into the simplest equation in order to find *u*: $u = 25 + 7 = 32$.

Note: Some students might find this problem challenging. Learning to solve "systems of equations" is covered in first year Algebra courses. Some of the methods used to solve include substitution, (which we used here), addition and subtraction, multiplication and a combination of the above.

A "system of equations," simply means more than one equation, solved together.

Knowing *u* and *v*, it's easy to substitute their values into the expressions for the various angles:

$$m∠SRQ = 5v – 3 = 5(25) – 3 = 122√   ∠SRT = 2u – 6 = 2(32) – 6 = 58√$$

$$∠Q = u +5 = (32)+5 = 37√, ∠S = v – 4 = (25)– 4 = 21√$$

page 60

Now You Try It – Each pair of triangles shown below is congruent. In each problem you are supposed to find an additional pair of equal parts by using previously introduced theorems or postulates. Then, name the postulate or theorem which proves that the pair of triangles is congruent.

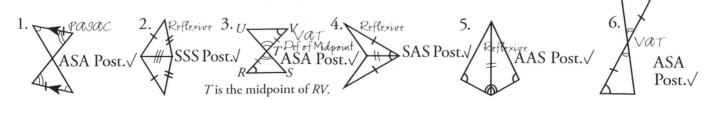

1. *PAJAC* ASA Post.√   2. Reflexive SSS Post.√   3. VAT Def of Midpoint ASA Post.√ *T* is the midpoint of *RV*.   4. Reflexive SAS Post.√   5. Reflexive AAS Post.√   6. VAT ASA Post.√

page 61

Now You Try It

Given: $\overline{AB} \parallel \overline{DE}$ and C is the midpoint of $\overline{AE}$ Prove: $\triangle ACB \cong \triangle ECD$

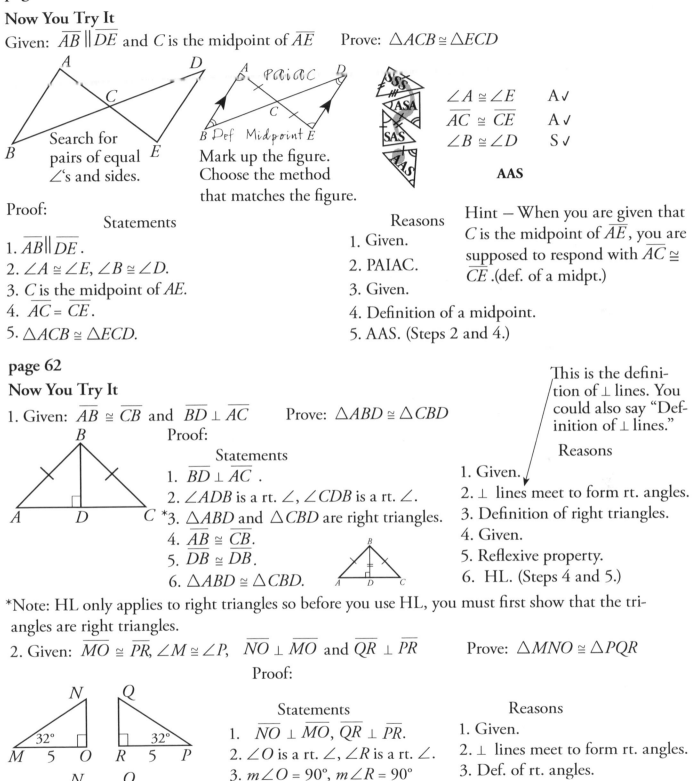

Search for pairs of equal ∠'s and sides.

Mark up the figure. Choose the method that matches the figure.

$\angle A \cong \angle E$ A ✓
$\overline{AC} \cong \overline{CE}$ A ✓
$\angle B \cong \angle D$ S ✓

AAS

Proof:

Statements	Reasons
1. $\overline{AB} \parallel \overline{DE}$.	1. Given.
2. $\angle A \cong \angle E$, $\angle B \cong \angle D$.	2. PAIAC.
3. C is the midpoint of AE.	3. Given.
4. $\overline{AC} = \overline{CE}$.	4. Definition of a midpoint.
5. $\triangle ACB \cong \triangle ECD$.	5. AAS. (Steps 2 and 4.)

Hint – When you are given that C is the midpoint of $\overline{AE}$, you are supposed to respond with $\overline{AC} \cong \overline{CE}$.(def. of a midpt.)

page 62

Now You Try It

1. Given: $\overline{AB} \cong \overline{CB}$ and $\overline{BD} \perp \overline{AC}$ Prove: $\triangle ABD \cong \triangle CBD$

Proof:

This is the definition of ⊥ lines. You could also say "Definition of ⊥ lines."

Statements	Reasons
1. $\overline{BD} \perp \overline{AC}$.	1. Given.
2. $\angle ADB$ is a rt. $\angle$, $\angle CDB$ is a rt. $\angle$.	2. ⊥ lines meet to form rt. angles.
*3. $\triangle ABD$ and $\triangle CBD$ are right triangles.	3. Definition of right triangles.
4. $\overline{AB} \cong \overline{CB}$.	4. Given.
5. $\overline{DB} \cong \overline{DB}$.	5. Reflexive property.
6. $\triangle ABD \cong \triangle CBD$.	6. HL. (Steps 4 and 5.)

*Note: HL only applies to right triangles so before you use HL, you must first show that the triangles are right triangles.

2. Given: $\overline{MO} \cong \overline{PR}$, $\angle M \cong \angle P$, $\overline{NO} \perp \overline{MO}$ and $\overline{QR} \perp \overline{PR}$ Prove: $\triangle MNO \cong \triangle PQR$

Proof:

Statements	Reasons
1. $\overline{NO} \perp \overline{MO}$, $\overline{QR} \perp \overline{PR}$.	1. Given.
2. $\angle O$ is a rt. $\angle$, $\angle R$ is a rt. $\angle$.	2. ⊥ lines meet to form rt. angles.
3. $m\angle O = 90°$, $m\angle R = 90°$	3. Def. of rt. angles.
4. $\angle O \cong \angle R$.	4. Definition of congruency.
5. $\angle M \cong \angle P$, $\overline{MO} \cong \overline{PR}$.	5. Given.
6. $\triangle MNO \cong \triangle PQR$.	6. ASA. (Steps 4 and 5.)

Now You Try It

Every proof on this page can be done using congruent triangles and/or the definition of congruence.

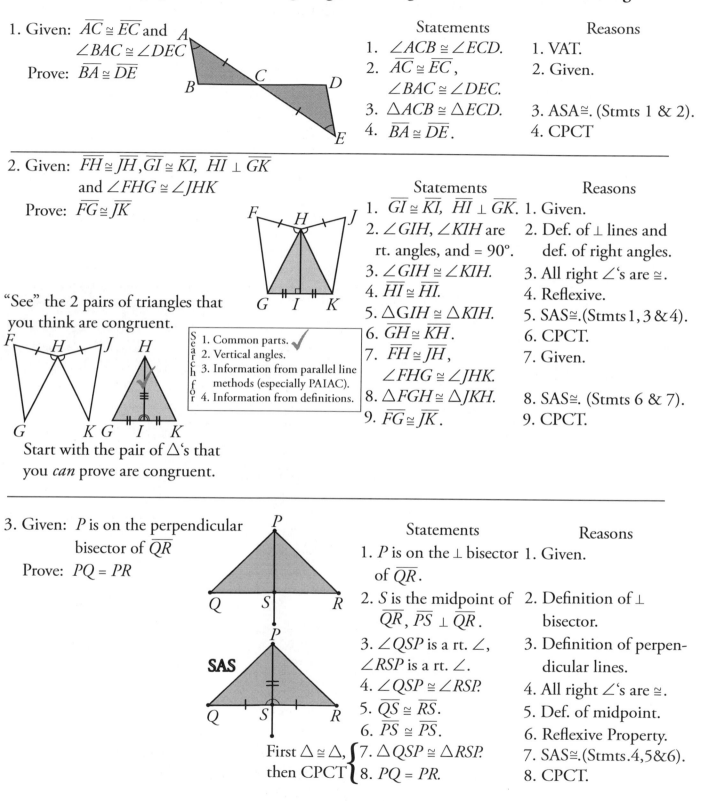

1. Given: $\overline{AC} \cong \overline{EC}$ and
$\angle BAC \cong \angle DEC$
   Prove: $\overline{BA} \cong \overline{DE}$

Statements	Reasons
1. $\angle ACB \cong \angle ECD$.	1. VAT.
2. $\overline{AC} \cong \overline{EC}$, $\angle BAC \cong \angle DEC$.	2. Given.
3. $\triangle ACB \cong \triangle ECD$.	3. ASA≅. (Stmts 1 & 2).
4. $\overline{BA} \cong \overline{DE}$.	4. CPCT

2. Given: $\overline{FH} \cong \overline{JH}, \overline{GI} \cong \overline{KI}, \ \overline{HI} \perp \overline{GK}$
and $\angle FHG \cong \angle JHK$
   Prove: $\overline{FG} \cong \overline{JK}$

"See" the 2 pairs of triangles that you think are congruent.

> **Search for**
> 1. Common parts. ✓
> 2. Vertical angles.
> 3. Information from parallel line methods (especially PAIAC).
> 4. Information from definitions.

Start with the pair of △'s that you *can* prove are congruent.

Statements	Reasons
1. $\overline{GI} \cong \overline{KI}$, $\overline{HI} \perp \overline{GK}$.	1. Given.
2. $\angle GIH, \angle KIH$ are rt. angles, and = 90°.	2. Def. of $\perp$ lines and def. of right angles.
3. $\angle GIH \cong \angle KIH$.	3. All right $\angle$'s are $\cong$.
4. $\overline{HI} \cong \overline{HI}$.	4. Reflexive.
5. $\triangle GIH \cong \triangle KIH$.	5. SAS≅.(Stmts 1, 3 & 4).
6. $\overline{GH} \cong \overline{KH}$.	6. CPCT.
7. $\overline{FH} \cong \overline{JH}$, $\angle FHG \cong \angle JHK$.	7. Given.
8. $\triangle FGH \cong \triangle JKH$.	8. SAS≅. (Stmts 6 & 7).
9. $\overline{FG} \cong \overline{JK}$.	9. CPCT.

3. Given: P is on the perpendicular bisector of $\overline{QR}$
   Prove: $PQ = PR$

SAS

First △ ≅ △, then CPCT {

Statements	Reasons
1. P is on the $\perp$ bisector of $\overline{QR}$.	1. Given.
2. S is the midpoint of $\overline{QR}$, $\overline{PS} \perp \overline{QR}$.	2. Definition of $\perp$ bisector.
3. $\angle QSP$ is a rt. $\angle$, $\angle RSP$ is a rt. $\angle$.	3. Definition of perpendicular lines.
4. $\angle QSP \cong \angle RSP$.	4. All right $\angle$'s are $\cong$.
5. $\overline{QS} \cong \overline{RS}$.	5. Def. of midpoint.
6. $\overline{PS} \cong \overline{PS}$.	6. Reflexive Property.
7. $\triangle QSP \cong \triangle RSP$.	7. SAS≅.(Stmts.4,5&6).
8. $PQ = PR$.	8. CPCT.

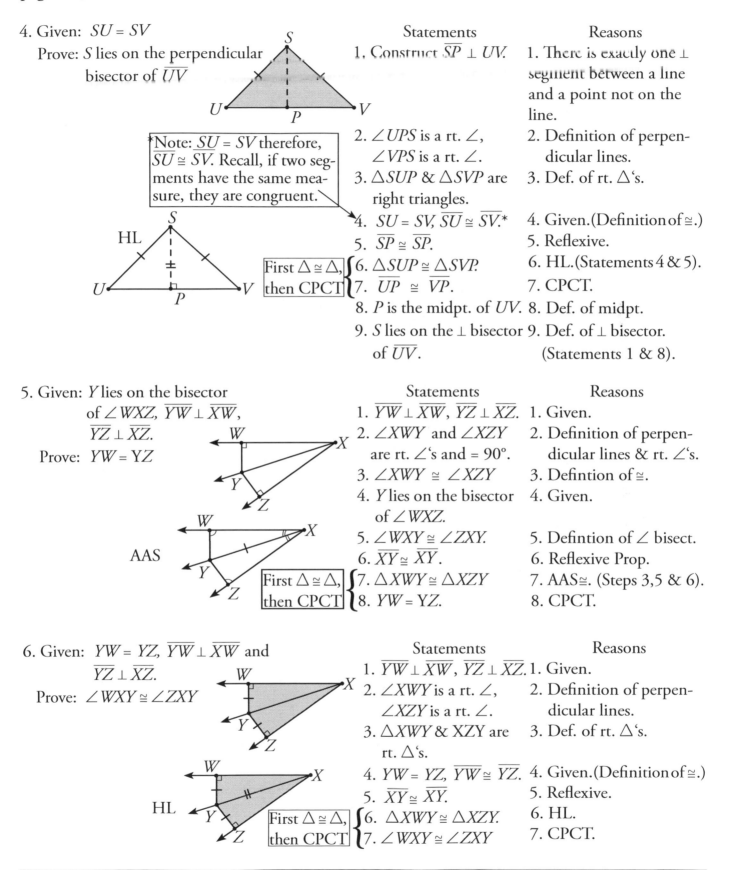

4. Given: $SU = SV$
Prove: S lies on the perpendicular bisector of $\overline{UV}$

*Note: $SU = SV$ therefore, $\overline{SU} \cong \overline{SV}$. Recall, if two segments have the same measure, they are congruent.

HL

First △ ≅ △, then CPCT

Statements	Reasons
1. Construct $\overline{SP} \perp UV$.	1. There is exactly one ⊥ segment between a line and a point not on the line.
2. $\angle UPS$ is a rt. $\angle$, $\angle VPS$ is a rt. $\angle$.	2. Definition of perpendicular lines.
3. $\triangle SUP$ & $\triangle SVP$ are right triangles.	3. Def. of rt. △'s.
4. $SU = SV$, $\overline{SU} \cong \overline{SV}$.*	4. Given.(Definition of ≅.)
5. $\overline{SP} \cong \overline{SP}$.	5. Reflexive.
6. $\triangle SUP \cong \triangle SVP$.	6. HL.(Statements 4 & 5).
7. $\overline{UP} \cong \overline{VP}$.	7. CPCT.
8. P is the midpt. of UV.	8. Def. of midpt.
9. S lies on the ⊥ bisector of $\overline{UV}$.	9. Def. of ⊥ bisector. (Statements 1 & 8).

5. Given: Y lies on the bisector of $\angle WXZ$, $\overline{YW} \perp \overline{XW}$, $\overline{YZ} \perp \overline{XZ}$.
Prove: $YW = YZ$

AAS

First △ ≅ △, then CPCT

Statements	Reasons
1. $\overline{YW} \perp \overline{XW}$, $\overline{YZ} \perp \overline{XZ}$.	1. Given.
2. $\angle XWY$ and $\angle XZY$ are rt. $\angle$'s and = 90°.	2. Definition of perpendicular lines & rt. $\angle$'s.
3. $\angle XWY \cong \angle XZY$	3. Defintion of ≅.
4. Y lies on the bisector of $\angle WXZ$.	4. Given.
5. $\angle WXY \cong \angle ZXY$.	5. Defintion of $\angle$ bisect.
6. $\overline{XY} \cong \overline{XY}$.	6. Reflexive Prop.
7. $\triangle XWY \cong \triangle XZY$	7. AAS≅. (Steps 3,5 & 6).
8. $YW = YZ$.	8. CPCT.

6. Given: $YW = YZ$, $\overline{YW} \perp \overline{XW}$ and $\overline{YZ} \perp \overline{XZ}$.
Prove: $\angle WXY \cong \angle ZXY$

HL

First △ ≅ △, then CPCT

Statements	Reasons
1. $\overline{YW} \perp \overline{XW}$, $\overline{YZ} \perp \overline{XZ}$.	1. Given.
2. $\angle XWY$ is a rt. $\angle$, $\angle XZY$ is a rt. $\angle$.	2. Definition of perpendicular lines.
3. $\triangle XWY$ & XZY are rt. △'s.	3. Def. of rt. △'s.
4. $YW = YZ$, $\overline{YW} \cong \overline{YZ}$.	4. Given.(Definition of ≅.)
5. $\overline{XY} \cong \overline{XY}$.	5. Reflexive.
6. $\triangle XWY \cong \triangle XZY$.	6. HL.
7. $\angle WXY \cong \angle ZXY$	7. CPCT.

Polygon Chart page 70

Number of Sides	Number of Triangles	Calculation	Total of Interior Angles
3	1	1 x 180	180
4	2	2 x 180	360
5	3	3 x 180	540
6	4	4 x 180	720
7	5	5 × 180	900
8	6	6 × 180	1080
9	7	7 × 180	1260
10	8	8 × 180	1440

page 71 Now You Try It

1. Find the total sum of the measure of the interior angles of a 100-gon.

Use the formula, $(n - 2)(180) = (\overset{100}{\cancel{n}} - 2)(180) = (98)(180) = 17,640° \checkmark$

2. Find the measure of the smallest and largest angle of polygon $PQRSTUVW$.

Step 1. Counting the sides of the polygon, there are eight, which means that the total measure of the interior angles is:

$$(n - 2)(180) = (\overset{8}{\cancel{n}} - 2)(180) = (6)(180) = 1080°$$

Step 2. Now set the total of the values of each angle equal to 1080°:

$$(10x - 7)° + 106° + (14x - 13)° + 12x° + (9x+1)° + 11x° + (12x+13)° + (13x+8)° = 1080°$$

$$81x + 108 = 1080, \qquad 81x = 972, \qquad x = 12$$

Step 3. Go back and read the question and answer the questions that were asked: The smallest angle is $\angle Q$ which equals 106°, the largest is $\angle W$ which equals 164°. $\checkmark$

page 72

Now You Try It — If the sum of the interior angles of a regular polygon is 1260, find the measure of one exterior angle of the polygon.

This problem is giving the *answer* to the formula $(n - 2)(180)$ so set the formula equal to the answer:

$$(n - 2)(180) = 1260°$$

Carefully distribute each term of the expression $(n - 2)$:

$$n(180) - 2(180) = 1260$$
$$180n - 360 = 1260$$

Balance the equation:

$$+ 360 \quad + 360$$

Divide by the coefficient of n:

$$\frac{180n}{180} = \frac{1620}{180} \qquad n = 9$$

Using the compact formula $\frac{(n - 2)180}{n}$ for the measure of one interior angle, $\frac{(9 - 2)180}{9} = 140°. \checkmark$

Now, draw one corner of the polygon:

Since a straight angle equals 180°, the exterior angle must equal $180 - 40 = 40$. $\checkmark$

250

page 74
Now You Try It

1. If an exterior angle of a regular polygon is 72°, what is the total measure of the interior angles of the polygon?

Since the exterior angles of a regular polygon are equal, use the formula:

$$\frac{360°}{\text{the measure of one exterior angle of a } \textit{regular} \text{ polygon}} = n \qquad \frac{360°}{72°} = 5$$

Since n = 5, the total measure of the interior angles = $(\overset{5}{\cancel{n}} - 2)(180) = 540°$ ✓

2. If the interior angles of a regular polygon total 1620°, find the measure of one exterior angle.

The problem is giving the answer to the formula (n – 2)180 so set the formula equal to the answer:

$$(n-2)180 = 1620.$$
Now distribute: $180n - 360 = 1620,$
$$180n = 1980, \quad n = 11$$

Use the formula:

$$\frac{360°}{n} = \frac{\text{the measure of one exterior}}{\text{angle of a } \textit{regular} \text{ polygon}} \qquad \frac{360°}{n} = \frac{360°}{11} \approx 32.7° ✓$$

3. Polygon *ABCDEFGH* is regular. Find the measure of ∠*I*.

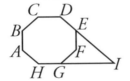

There are lots of ways to do this problem, here's one of them:
Polygon *ABCDEFGH* is a regular octagon.

Each exterior angle measures: $\dfrac{360°}{n} = \dfrac{360°}{8} = 45°$

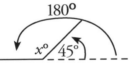

Sketching one vertex of the octagon helps you to see that each interior angle equals 180° – 45° = 135°. Now study the original figure. Polygon *EFGI* is a quadrilateral so its interior angles total (n – 2)180 = 360°:

$$\angle FEI + \angle IGF + (360° - \angle EFG) + \angle I = 360°$$

Now substitute: $\underset{45°}{\angle \cancel{FEI}} + \underset{45°}{\angle \cancel{IGF}} + (360° - \underset{135°}{\angle \cancel{EFG}}) + \angle I = 360°$

$45° + 45° + (225°) + \angle I = 360°, \ \angle I = 360° - 315° = 45°$ ✓

page 75
Now You Try It
Given ▱ *GHIJ* at right, add the correct measurements for the following:

1. $m\angle GHI = m\angle GJI = 133$ (⌓) ✓, 2. $m\angle HIJ = 180° - 133° = 47°$ (consecutive angles of ▱ are supplementary), 3. $m\angle HGJ = m\angle HIJ = 47°$ (⌓),
4. $\overline{HG} \cong \overline{IJ} = 5$ (⌓), 5. $\overline{HI} \cong \overline{GJ} = 8$ (⌓),
6. $\overline{GI} = 6 + 6 = 12$ (⌓). ✓

Now You Try It

Supply the reasons to prove the diagonals of a parallelogram bisect each other.
Given: ▱ ABCD   Prove: $\overline{BD}$ bisects $\overline{AC}$ and $\overline{AC}$ bisects $\overline{BD}$

Proof:

Statements	Reasons
1. ▱ ABCD.	1. Given.
2. ∠AEB ≅ ∠DEC.	2. VAT.
3. $\overline{AB} \cong \overline{DC}$.	3. Opposite sides of a ▱ are ≅ .
4. $\overline{BC} \parallel \overline{AD}$; $\overline{AB} \parallel \overline{DC}$.	4. Definition of a ▱.
5. ∠DBA ≅ ∠CDB.	5. PAIAC.
6. △AEB ≅ △CED.	6. AAS.
7. $\overline{AE} = \overline{EC}$; $\overline{BE} = \overline{ED}$.	7. CPCT.
8. E is the midpoint of $\overline{AC}$; E is the midpoint of $\overline{BD}$.	8. Definition of a midpoint.
9. $\overline{BD}$ bisects $\overline{AC}$; $\overline{AC}$ bisects $\overline{BD}$.	9. Definition of segment bisector.

Now You Try It

THEOREM: *If both diagonals of a quadrilateral bisect each other, the quadrilateral is a parallelogram.*

Given: Quadrilateral ABCD with $\overline{AE} \cong \overline{EC}$ and $\overline{BE} \cong \overline{ED}$
Prove: ABCD is a parallelogram.

Proof:

Statements	Reasons
1. $\overline{AE} \cong \overline{EC}$, $\overline{BE} \cong \overline{ED}$.	1. Given.
2. ∠BEA ≅ ∠CED; ∠AED ≅ ∠BEC.	2. VAT.
3. △AEB ≅ △CED; △BEC ≅ △DEA.	3. SAS.
4. $\overline{AB} \cong \overline{DC}$, $\overline{AD} \cong \overline{BC}$.	4. CPCT.
5. ABCD is a parallelogram.	5. A Quad. with 2 prs. of ≅ opp. sides is a ▱.

THEOREM: *If one pair of opposite sides of a quadrilateral are congruent and parallel, the quadrilateral is a parallelogram.*

Given: Quadrilateral MNOP with $\overline{MN} \cong \overline{PO}$ and $\overline{MN} \parallel \overline{PO}$ Prove: MNOP is a parallelogram.

Proof:

Statements	Reasons
1. Construct $\overline{MO}$.	1. Two points determine a line.
2. $\overline{MN} \parallel \overline{PO}$.	2. Given.
3. ∠NMO ≅ ∠POM.	3. PAIAC.
4. $\overline{MN} \cong \overline{PO}$.	4. Given.
5. $\overline{MO} \cong \overline{MO}$.	5. Reflexive Property.
6. △MNO ≅ △OPM.	6. SAS≅.
7. $\overline{NO} \cong \overline{MP}$.	7. CPCT.
8. MNOP is a parallelogram.	8. A Quad. with 2 prs. of ≅ opp. sides is a ▱.

Now You Try It

For problems 1 – 4, find any parallelograms in the drawings. Be sure to thoroughly explain your conclusions.

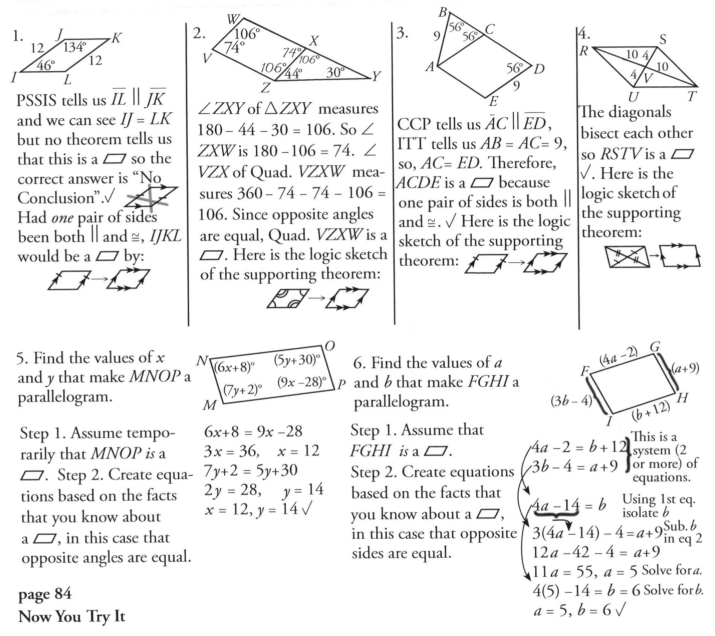

1.

PSSIS tells us $\overline{IL} \parallel \overline{JK}$ and we can see $IJ = LK$ but no theorem tells us that this is a ▱ so the correct answer is "No Conclusion". √ Had *one* pair of sides been both ∥ and ≅, *IJKL* would be a ▱ by:

2.

∠*ZXY* of △*ZXY* measures $180 - 44 - 30 = 106$. So ∠ *ZXW* is $180 - 106 = 74$. ∠ *VZX* of Quad. *VZXW* measures $360 - 74 - 74 - 106 = 106$. Since opposite angles are equal, Quad. *VZXW* is a ▱. Here is the logic sketch of the supporting theorem:

3.

CCP tells us $\overline{AC} \parallel \overline{ED}$, ITT tells us $AB = AC = 9$, so, $AC = ED$. Therefore, *ACDE* is a ▱ because one pair of sides is both ∥ and ≅. √ Here is the logic sketch of the supporting theorem:

4.

The diagonals bisect each other so *RSTV* is a ▱ √. Here is the logic sketch of the supporting theorem:

---

5. Find the values of x and y that make *MNOP* a parallelogram.

Step 1. Assume temporarily that *MNOP is* a ▱. Step 2. Create equations based on the facts that you know about a ▱, in this case that opposite angles are equal.

$6x + 8 = 9x - 28$
$3x = 36, \quad x = 12$
$7y + 2 = 5y + 30$
$2y = 28, \quad y = 14$
$x = 12, y = 14$ √

6. Find the values of a and b that make *FGHI* a parallelogram.

Step 1. Assume that *FGHI is* a ▱.
Step 2. Create equations based on the facts that you know about a ▱, in this case that opposite sides are equal.

$4a - 2 = b + 12$ This is a system (2
$3b - 4 = a + 9$ or more) of equations.

$4a - 14 = b$ Using 1st eq. isolate b
$3(4a - 14) - 4 = a + 9$ Sub. b in eq 2
$12a - 42 - 4 = a + 9$
$11a = 55, a = 5$ Solve for a.
$4(5) - 14 = b = 6$ Solve for b.
$a = 5, b = 6$ √

Now You Try It

1. Find the perimeter of ▱*ABCD*, $m\angle AED$ and $m\angle EAD$.

ABCD is a rhombus because $\overline{BA} \cong \overline{AD}$. (A ▱ with 2 consecutive ≅ sides is a rhombus.) Therefore, the perimeter is $4 \times 10 = 40$. √

Diagonals of a rhombus are ⊥ so, $m\angle AED = 90°$ √ (def. of ⊥).

$m\angle BAD = 68°$ ($\angle ABC = 2 \times 56° = 112°$, so $\angle BAD = 180° - 112° = 68°$ (Consec. ∠'s of ▱ are suppl.),

$\angle EAD = 68° \div 2 = 34°$.√ (Diagonals of a rhombus bisect the ∠'s.)

2. a. Find x, y and z in $\square$ $FGHI$. b. $\triangle GEF \cong$?

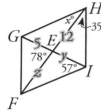

3. Given $\square$ $JKLM$, find NM.

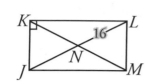

$\square$ $FGHI$ is not a rhombus ($m\angle GEF = 78$), nor is it a rectangle ($\overline{GE} \not\cong \overline{EH}$) so only the regular properties of parallelograms can be used to solve the problem: $y = 5$, $z = 12$ (diagonals of a $\square$ bisect each other), $m\angle FEI = 102$ ($180 - 78$), $m\angle EFI = 21$ ($180 - 102 - 57$, $3\angle$'s of a $\triangle = 180°$), so $x = 21$, (PAIAC). $\checkmark$ b. $\triangle GEF \cong \triangle IEH$ (Since $GE = IE$, $\angle GEF = \angle IEH$, $\overline{EF} \cong \overline{EH}$.)

$JKLM$ is a rectangle (a $\square$ with 1 rt. $\angle$ is a rectangle). $\overline{JN} \cong \overline{NL}$ (diagonals of a $\square$ bisect each other), so $JL =$ 32. Diagonals of a rectangle are $\cong$, so $KM = 32$, and $NM = 16$. (A second proof: $\triangle JLM$ is a right triangle, and N is the midpoint of JL, so $NM = 16$ because the midpoint of the hypotenuse is equidistant from each vertex.)

page 87 Now You Try It

4. Find the perimeter of square $MNOP$, $m\angle MQN$, $m\angle POQ$ and name any triangles $\cong \triangle OQP$.

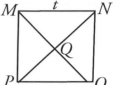

A square has four $\cong$ sides and four $\cong$ (rt.) angles. The length of one side of the square is t. In this case, we cannot solve for (find the value of) t, and that means the answer must be "in terms of" t; the perimeter is $4 \times t$ or, $4t$. $\checkmark$ To find $m\angle MQN$, recall that the diagonals of a square (since a square is a rhombus) are $\perp$, so $m\angle MQN = 90$ $\checkmark$ (def. of $\perp$ lines). The diagonals of a square (since it's a rhombus) bisect the vertices, so $m\angle POQ = 90 \div 2 = 45 \checkmark$. Since order counts, there are 7 triangles $\cong$ to $\triangle OQP$: $\triangle PQO$, $\triangle OQN$, $\triangle NQO$, $\triangle MQN \triangle NQM$, $\triangle MQP$, $\triangle PQM$. $\checkmark$

1. Is $ABCD$ a trapezoid?

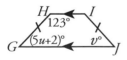

To be a trapezoid, two sides (and only two) must be $\parallel$ which would then mean that by PSSIS, two adjacent angles (two angles that are next to each other) must be supplementary. But $101 + 78 = 179$, and $78 + 93 = 171$. So no, $ABCD$ is *not* a trapezoid. $\checkmark$

2. Given $\triangle$ $MNOP$ with bases $\overline{NO}$ and $\overline{MP}$, find x and y.

$$N\overline{\underset{(5x+4)°\ (4x+7)°}{\overset{(9x+8)°\ \ y}{}}}O$$

Since $\overline{NO}$ and $\overline{MP}$ are bases, they must be $\parallel$. Therefore, by PSSIS, $\angle ONM$ and $\angle PMN$ must be supplementary, so set their sum equal to 180: $9x + 8 + 5x + 4 = 180$, $14x + 12 = 180$
$$14x = 168, \quad x = 12. \checkmark$$
For the same reason, $\angle MPO$ and $\angle NOP$ must also be supplementary. Substitute 12 for x in the equation:
$$4(12) + 7 + y = 180, \quad 55 + y = 180, \quad y = 125. \checkmark$$

3. Given $GHIJ$ find u and v.

This is an isosceles $\triangle$ so by the Isosceles Trapezoid Theorem, each pair of base angles is equal. But first, we need to find the value of either u or v. $\angle IHG$ and $\angle JGH$ are supplementary by PSSIS so set their sum equal to 180: $123 + 5u + 2 = 180$,
$$125 + 5u = 180, 5u = 55. \ u = 11. \checkmark$$
Now substitute: $u = 11$ into $(5u + 2)$ and set $= v$. $5(11) + 2 = v$.
$$v = 57 \checkmark$$

page 87 (Continued)

4. Given △ *UVWX* with median $\overline{YZ}$, find *a*, *b* and *c*.

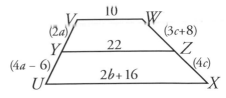

5. Given △ *QRST* find *x*.

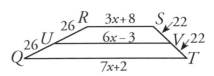

6. Given △ *IJKL* with median $\overline{MN}$, find *z*.

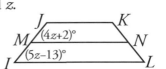

Finding *a* and *c*. Since $\overline{YZ}$ is the median of the △, *Y* and *Z* are midpoints of the sides. This means:

$$4a - 6 = 2a,$$
$$2a = 6, \ a = 3 \checkmark$$

and, $4c = 3c + 8$

$$c = 8 \checkmark$$

Because $\overline{QU} = \overline{UR}$ and $\overline{SV} = \overline{VT}$, $\overline{UV}$ is a median.

Finding *b*:

$$\text{median} = \frac{\text{base 1 + base 2}}{2}$$

$$22 = \frac{10 + (2b + 16)}{2}$$

$$2\left(22 = \frac{10 + 2b + 16}{\cancel{2}}\right)\cancel{2}$$

$$44 = 2b + 26, \ 18 = 2b, \ 9 = b. \checkmark$$

$$\text{median} = \frac{\text{base 1 + base 2}}{2}$$

$$6x - 3 = \frac{(3x + 8) + (7x + 2)}{2}$$

$$2\left(6x - 3 = \frac{10x + 10}{\cancel{2}}\right)\cancel{2}$$

$$12x - 6 = 10x + 10, \ 2x = 16, \ x = 8 \checkmark$$

The median of a △ is parallel to both bases. PCC tells us that ∠*JMN* ≅ ∠*MIL* so set the corresponding expressions equal: $5z - 13 = 4z + 2, \quad z = 15. \checkmark$

Top of page 88 — Now You Try It

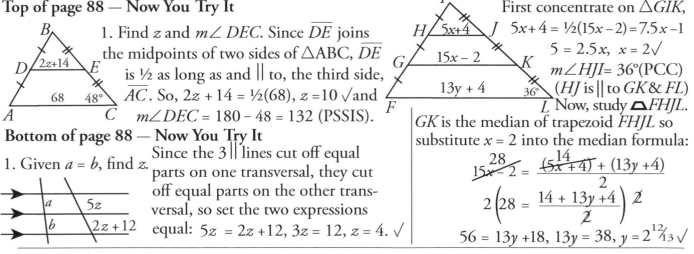

1. Find *z* and *m*∠ *DEC*. Since $\overline{DE}$ joins the midpoints of two sides of △ABC, $\overline{DE}$ is ½ as long as and ∥ to, the third side, $\overline{AC}$. So, $2z + 14 = \frac{1}{2}(68), \ z = 10 \checkmark$ and $m∠DEC = 180 - 48 = 132$ (PSSIS).

Bottom of page 88 — Now You Try It

1. Given *a* = *b*, find *z*.

Since the 3 ∥ lines cut off equal parts on one transversal, they cut off equal parts on the other transversal, so set the two expressions equal: $5z = 2z + 12, \ 3z = 12, \ z = 4. \checkmark$

2. Solve for *x*, *y* and *m*∠ *HJI*. First concentrate on △*GIK*,

$$5x + 4 = \frac{1}{2}(15x - 2) = 7.5x - 1$$
$$5 = 2.5x, \ x = 2 \checkmark$$
$$m∠HJI = 36°(\text{PCC})$$
$$(HJ \text{ is } \| \text{ to } GK \& FL)$$

Now, study △*FHJL*. *GK* is the median of trapezoid *FHJL* so substitute *x* = 2 into the median formula:

$$15x - 2 = \frac{(5x + 4) + (13y + 4)}{2}$$
$$2\left(28 = \frac{14 + 13y + 4}{\cancel{2}}\right)\cancel{2}$$
$$56 = 13y + 18, \ 13y = 38, \ y = 2\tfrac{12}{13} \checkmark$$

2. The three lines in the figure are equally far apart. Solve for *x* and *y*. (Hint: Think about which segment equals which segment.)

Parallel lines that are equally far apart, cut off equal segments on every transversal. Therefore, we can form these equations:

(1) $\quad 2y + 5 = 4x - 1$
(2) $\quad 3y - 7 = 2x + 4$ } This is a system (2 or more) of equations

Working on equation (1), isolate *y*: $2y = 4x - 6, \ y = \boxed{2x - 3}$

Now substitute into equation (2): $3(2x - 3) - 7 = 2x + 4$

Distribute: $6x - 9 - 7 = 2x + 4$
$$4x = 20, \ x = 5 \checkmark$$

Substitute *x* = 5 into equation (1): $2y + 5 = 4(5) - 1 = 19, \ y = 7 \checkmark$

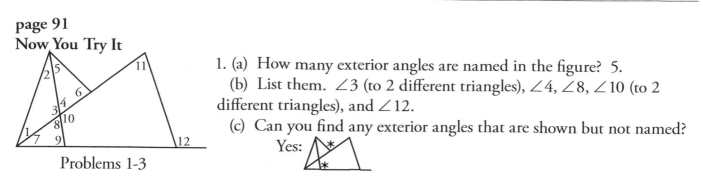

page 91
Now You Try It

Problems 1-3

1. (a) How many exterior angles are named in the figure? 5.
(b) List them. $\angle 3$ (to 2 different triangles), $\angle 4$, $\angle 8$, $\angle 10$ (to 2 different triangles), and $\angle 12$.
(c) Can you find any exterior angles that are shown but not named?
Yes:

2. Using the Exterior Angle Inequality Theorem, list as many inequalities as you can.
$m\angle 3 > m\angle 5$, $m\angle 3 > m\angle 6$, $m\angle 3 > m\angle 7$, $m\angle 3 > m\angle 9$, $m\angle 4 > m\angle 1$, $m\angle 4 > m\angle 2$, $m\angle 8 > m\angle 1$, $m\angle 8 > m\angle 2$ (Notice that $m\angle 8 = m\angle 4$, by VAT), $m\angle 10 > m\angle 7$, $m\angle 10 > m\angle 9$, $m\angle 10 > m\angle 5$, $m\angle 10 > m\angle 6$, (Notice that $m\angle 10 = m\angle 3$ by VAT), $m\angle 12 > m\angle 7$, $m\angle 12 > m\angle 11$.

3. If $m\angle 3 > m\angle 4$ (never assume that it is based on the drawing) what else can you conclude?
Since $m\angle 3 > m\angle 4$ and $m\angle 4 > m\angle 1$ and $m\angle 4 > m\angle 2$, then $m\angle 3 > m\angle 1$ and $m\angle 3 > m\angle 2$ as well. We could not have known this otherwise.

page 92
Now You Try It

1. Can you make a triangle with sides equal to
a) 4, 4, and 8? No, $4 + 4 = 8 \not> 8$. "Equal to" isn't good enough. The sum must be larger.
b) 4.01, 1 and 3? No. $1 + 3 = 4 \not> 4.01$. Remember always add the two *smaller* numbers together.
c) 2, 7.$\overline{9}$ and 6? Yes, $2 + 6 = 8$ and $8 > 7.\overline{9}$ (ever so slightly bigger, but that's enough).

2. Three sides of a triangle are 5, 5 and x. Find numbers a and b such that $a < x < b$.

$\underline{a} < x < \underline{b}$ $\underline{5-5} < x < \underline{5+5}$ $\underline{0} < x < \underline{10}$ $a = 0^*$, $b = 10\checkmark$

*But physical quantities are positive, how can $a = 0$? Because x will not never equal 0. Zero is simply its lower limit. But x may get as close to 0 as we please. For example, x could equal .0000000001, or perhaps ½ of that amount, or ¼ of that amount and so on.

Explanation of Triangle Inequality Problem Type 2
Here's an example: Given that the lengths of three sides of a triangle are 3, 4, and x. Find a and b such that $a < x < b$. Looking for a, the lower limit, we know by the Triangle Inequality Theorem, when x is added to 3 (the lessor of the *given* lengths) the sum must be greater than four, that is $x + 3 > 4$. To solve the inequality, subtract 3 from each side, $x > 1$, or $1 < x$ This gives us a, the lower limit. Looking for the upper limit, since $3 + 4 = 7$, x has to stay smaller than 7, that is $x < 7$. And this gives us the upper limit, b. Putting this all together, we have: $1 < x < 7$.
But looking back at the explanation, you can why this is the same as the method:

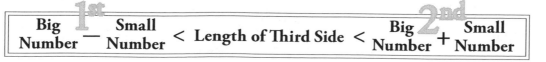

Big Number − Small Number	< Length of Third Side <	Big Number + Small Number

page 95
Now You Try It

Given: $\overline{BD} \cong \overline{BC}$, list everything you can discover about the angles, sides and triangles in the figure below:

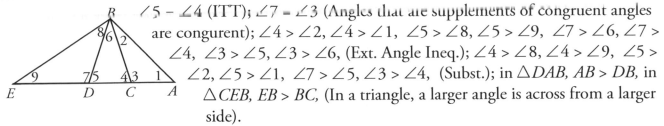

$\angle 5 = \angle 4$ (ITT); $\angle 7 = \angle 3$ (Angles that are supplements of congruent angles are congruent); $\angle 4 > \angle 2$, $\angle 4 > \angle 1$, $\angle 5 > \angle 8$, $\angle 5 > \angle 9$, $\angle 7 > \angle 6$, $\angle 7 > \angle 4$, $\angle 3 > \angle 5$, $\angle 3 > \angle 6$, (Ext. Angle Ineq.); $\angle 4 > \angle 8$, $\angle 4 > \angle 9$, $\angle 5 > \angle 2$, $\angle 5 > \angle 1$, $\angle 7 > \angle 5$, $\angle 3 > \angle 4$, (Subst.); in $\triangle DAB$, $AB > DB$, in $\triangle CEB$, $EB > BC$, (In a triangle, a larger angle is across from a larger side).

page 99
Now You Try It

Answers to Inequalities with Two Triangles:

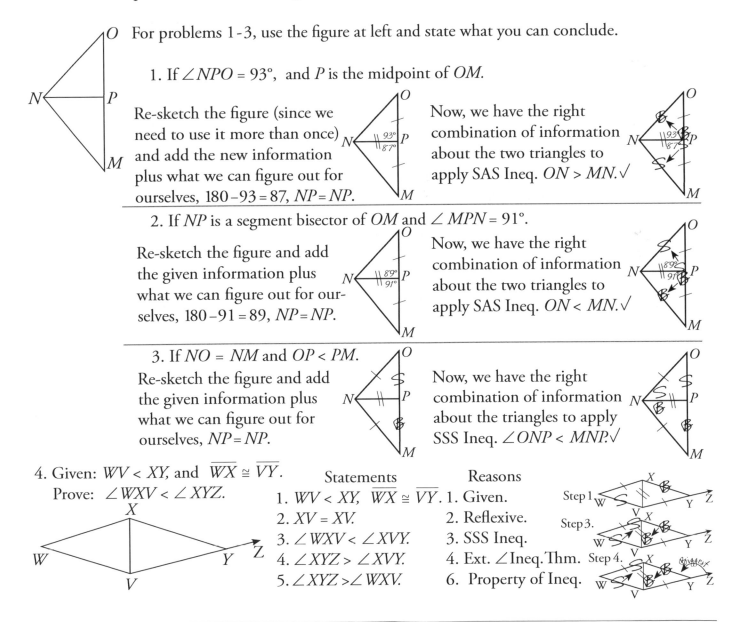

For problems 1-3, use the figure at left and state what you can conclude.

1. If $\angle NPO = 93°$, and P is the midpoint of OM.

Re-sketch the figure (since we need to use it more than once) and add the new information plus what we can figure out for ourselves, $180 - 93 = 87$, $NP = NP$.

Now, we have the right combination of information about the two triangles to apply SAS Ineq. $ON > MN.\checkmark$

2. If NP is a segment bisector of OM and $\angle MPN = 91°$.

Re-sketch the figure and add the given information plus what we can figure out for our-selves, $180 - 91 = 89$, $NP = NP$.

Now, we have the right combination of information about the two triangles to apply SAS Ineq. $ON < MN.\checkmark$

3. If $NO = NM$ and $OP < PM$.

Re-sketch the figure and add the given information plus what we can figure out for ourselves, $NP = NP$.

Now, we have the right combination of information about the triangles to apply SSS Ineq. $\angle ONP < MNP.\checkmark$

4. Given: $WV < XY$, and $\overline{WX} \cong \overline{VY}$.
   Prove: $\angle WXV < \angle XYZ$.

Statements	Reasons
1. $WV < XY$, $\overline{WX} \cong \overline{VY}$.	1. Given.
2. $XV = XV$.	2. Reflexive.
3. $\angle WXV < \angle XVY$.	3. SSS Ineq.
4. $\angle XYZ > \angle XVY$.	4. Ext. $\angle$ Ineq. Thm.
5. $\angle XYZ > \angle WXV$.	6. Property of Ineq.

page 101
Now You Try It

1. Name the 6 properties of proportion and give an example of each.

If $\dfrac{w}{x} = \dfrac{y}{z}$ and $\dfrac{3}{4} = \dfrac{6}{8}$

(1) Cross multiply: $wz = yx$ and $(3)(8) = (4)(6)$ (2) Flip the ratios: $\dfrac{x}{w} = \dfrac{z}{y}$ and $\dfrac{4}{3} = \dfrac{8}{6}$

(3) Swap the means $\dfrac{w}{y} = \dfrac{x}{z}$ and $\dfrac{3}{6} = \dfrac{4}{8}$ (4) Swap the extremes: $\dfrac{z}{x} = \dfrac{y}{w}$ and $\dfrac{8}{4} = \dfrac{6}{3}$

(5) Bring up the denominator $\dfrac{w+x}{x} = \dfrac{y+z}{z}$ (6) Add them up: $\dfrac{w+y}{x+z} = \dfrac{w}{x} = \dfrac{y}{z}$

$\dfrac{3+4}{4} = \dfrac{6+8}{8}$ $\dfrac{3+6}{4+8} = \dfrac{9}{12} = \dfrac{3}{4} = \dfrac{6}{8}$

2. If $\boxed{\dfrac{g}{h} = \dfrac{i}{j}}$ which of the following are equivalent proportions?

a.) $\dfrac{j}{h} = \dfrac{i}{g}$ What happened? g and j got switched. Since swapping the extremes is one of the properties, the answer is yes, this is an equivalent proportion.

b.) $\dfrac{h}{g} = \dfrac{j}{i}$ What happened? Compare b to the original proportion; each ratio has been flipped. Since flipping (inverting) the ratios is one of the properties, the answer is yes, this is an equivalent proportion.

c.) $\dfrac{g}{h} = \dfrac{j}{i}$ What happened? Compare c to the **original** proportion; the right ratio has been flipped (inverted), but not the left. None of the properties of proportion allow only one side to be flipped. Conclusion: No, this is not an equivalent proportion.

3. If $\dfrac{w}{x} = \dfrac{y}{z}$ then $\dfrac{z}{y} = ?$ Compare to the original, the ratios were flipped and then both the means and extremes were swapped, so $\dfrac{z}{y} = \dfrac{x}{w}$. $\checkmark$

4. If $\dfrac{a}{b} = \dfrac{c}{d}$ then, $\dfrac{b}{a+b} = ?$ Compare to the original, the denominator of the left side got "brought up" and then the ratio was flipped (inverted) so do exactly the same to the right side: $\dfrac{b}{a+b} = \dfrac{d}{c+d}$. $\checkmark$

page 105   Now You Try It

1. Write four congruencies and the extended proportion given by this similarity: $MNOP \sim QRST$.

$\angle M \cong \angle Q \quad \angle N \cong \angle R \quad \angle O \cong \angle S \quad \angle P \cong \angle T$

$\dfrac{MN}{QR} = \dfrac{NO}{RS} = \dfrac{OP}{ST} = \dfrac{PM}{TQ}$

Order counts!

$MNOP \sim QRST$

page 105 (continued)

2. The two quadrilaterals shown are similar.

a. Name two similar quadrilaterals.

Answer: Any similarity where the corresponding angles are matched up as shown below, is correct.

$GHIJ \sim BADC.$ (Match up the corresponding angles)

b. *CBAD* is similar to what quadrilateral?

$JGHI$

c. What is the scale factor of the two figures?

Answer: Studying the two figures, the problem gives a length of two corresponding sides, $\overline{BA}$ and $\overline{GH}$, the sides between the 125° and 31° angles in each figure. Since the figures were given in small to big order, that is the order for *this* problem, small to big. The scale factor is $\dfrac{8}{12}$ which reduces to $\dfrac{2}{3}$.

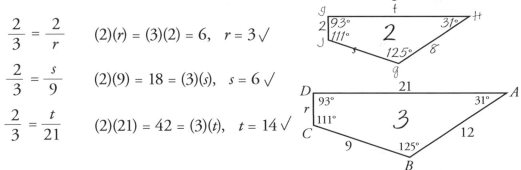

Re-sketched $GHIJ$ so that it is oriented like $BADC$.

Always write the scale factor on the figures. It helps you to keep track of the scale factor and the order of the problem.

d. Find *r*, *s*, and *t*.

$$\dfrac{2}{3} = \dfrac{2}{r} \qquad (2)(r) = (3)(2) = 6, \quad r = 3 \checkmark$$

$$\dfrac{2}{3} = \dfrac{s}{9} \qquad (2)(9) = 18 = (3)(s), \quad s = 6 \checkmark$$

$$\dfrac{2}{3} = \dfrac{t}{21} \qquad (2)(21) = 42 = (3)(t), \quad t = 14 \checkmark$$

page 109

Now You Try It

In problems 1-4, decide if two triangles are similar. If they are similar, explain why and name the similarity. If not, explain why not.

1.

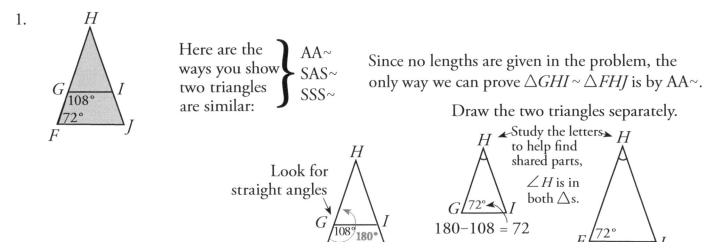

Here are the ways you show two triangles are similar: } AA~
SAS~
SSS~

Since no lengths are given in the problem, the only way we can prove $\triangle GHI \sim \triangle FHJ$ is by AA~.

Draw the two triangles separately.

Look for straight angles

Study the letters to help find shared parts, $\angle H$ is in both $\triangle$s.

$180 - 108 = 72$

$\angle G \cong \angle F$ and $\angle H \cong \angle H$
$\triangle GHI \sim \triangle FHJ$ by AA~. $\checkmark$

2.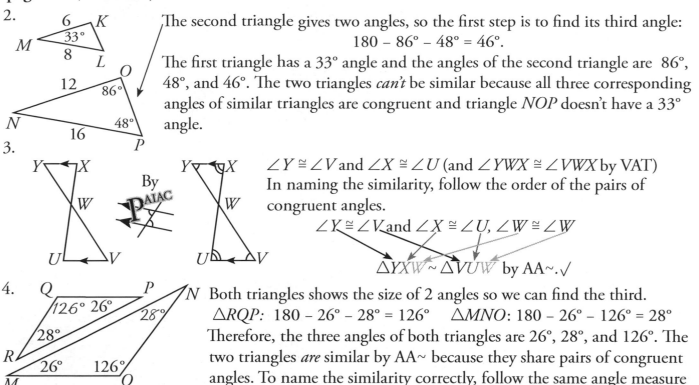

The second triangle gives two angles, so the first step is to find its third angle:
$$180 - 86° - 48° = 46°.$$
The first triangle has a 33° angle and the angles of the second triangle are 86°, 48°, and 46°. The two triangles *can't* be similar because all three corresponding angles of similar triangles are congruent and triangle *NOP* doesn't have a 33° angle.

3.

$\angle Y \cong \angle V$ and $\angle X \cong \angle U$ (and $\angle YWX \cong \angle VWX$ by VAT)
In naming the similarity, follow the order of the pairs of congruent angles.
$$\angle Y \cong \angle V \text{ and } \angle X \cong \angle U, \angle W \cong \angle W$$
$$\triangle YXW \sim \triangle VUW \text{ by AA} \sim . \checkmark$$

By PAIAC

4.

Both triangles shows the size of 2 angles so we can find the third.
$\triangle RQP$: $180 - 26° - 28° = 126°$ $\triangle MNO$: $180 - 26° - 126° = 28°$
Therefore, the three angles of both triangles are 26°, 28°, and 126°. The two triangles *are* similar by AA~ because they share pairs of congruent angles. To name the similarity correctly, follow the same angle measure path in each triangle: 26°, 28°, 126° $\triangle PRQ \sim \triangle MNO.\checkmark$

5. Using the figures below, name two similar triangles and find *g* and *i*.

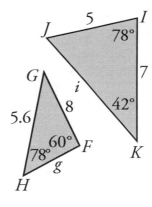

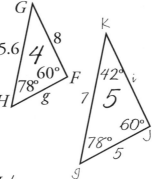

The problem is telling us that the 2 triangles are similar so the first thing to do is to re-orient the second triangle (triangle *IJK*) so both triangles are turned the same way. Do this by drawing a larger copy of *FGH*. You know where the angles go (in the same relative locations as in $\triangle FGH$). Then carefully transfer the letters and other numbers by referring to the original figure.

Now it's easy to name 2 similar triangles: $\triangle FGH \sim \triangle JKI \checkmark$
Find the scale factor by forming the ratio of two corresponding sides that have length: $\dfrac{5.6}{7} = .8 = \dfrac{4}{5}$ $\dfrac{4}{5} = \dfrac{g}{5}$ $g = 4\checkmark$ $\dfrac{4}{5} = \dfrac{8}{i}$ $i = 10\checkmark$

6. Using the figure below, prove:

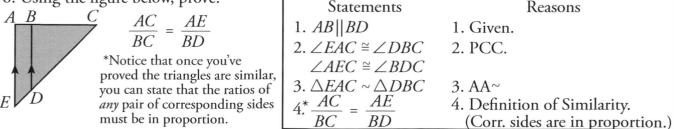

$$\frac{AC}{BC} = \frac{AE}{BD}$$

*Notice that once you've proved the triangles are similar, you can state that the ratios of *any* pair of corresponding sides must be in proportion.*

Statements	Reasons
1. $AB \| BD$	1. Given.
2. $\angle EAC \cong \angle DBC$ $\angle AEC \cong \angle BDC$	2. PCC.
3. $\triangle EAC \sim \triangle DBC$	3. AA~
4.* $\dfrac{AC}{BC} = \dfrac{AE}{BD}$	4. Definition of Similarity. (Corr. sides are in proportion.)

page 112

Now You Try It — Complete the indicated operation and if necessary, simplify (finish) the
  following radical expressions.

Answers:

1. $(3)(2\sqrt{27}) = 6\sqrt{27} = 6\sqrt{(3)(3)(3)} = (6)(3)(\sqrt{3}) = 18\sqrt{3}$ ✓

2. $(4\sqrt{8})(2\sqrt{6}) = 8\sqrt{48} = 8\sqrt{(3)(4)(4)} = (8)(4)(\sqrt{3}) = 32\sqrt{3}$ ✓

3. $\dfrac{3}{4\sqrt{3}}\left(\dfrac{\sqrt{3}}{\sqrt{3}}\right) = \dfrac{3\sqrt{3}}{(4)(3)} = \dfrac{\sqrt{3}}{4}$ ✓

4. $\sqrt{\dfrac{4}{15}} = \dfrac{\sqrt{4}}{\sqrt{15}} = \dfrac{\sqrt{4}}{\sqrt{15}}\left(\dfrac{\sqrt{15}}{\sqrt{15}}\right) = \left(\dfrac{\sqrt{60}}{15}\right) = \dfrac{\sqrt{(2)(2)(3)(5)}}{15} = \dfrac{2\sqrt{15}}{15}$ ✓

Note: You *can't* cancel $\dfrac{\sqrt{\cancel{15}}}{\cancel{15}}$
Think of the radical sign as
protecting its contents.

page 114

Now You Try It

Given right $\triangle GHL$ with altitude $\overline{HJ}$ drawn from the right angle, name three pairs of similar triangles.
Remember, order counts!

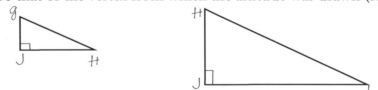

What this problem is testing is if you know how to name the similarity; that means,
listing the angles of each triangle in the correct (corresponding) order.
  We know the three triangles are similar so they all have the same shape. Draw
two smaller triangles that are the same shape as the original triangle
(GHL) and that are oriented (turned) like *GHL*. And then *carefully*
place each letter on the correct vertex, beginning with the right angle, and
then the acute angle from which the altitude was *not* drawn (i.e. not *H*). The third
letter will be that of the vertex from which the altitude was drawn (*H* in this example).

Now it's easy to name three similar triangles: 1. $\triangle GHL$
                  2. $\triangle GJH$
                  3. $\triangle HJL$

There are lots of other correct answers, for example, $\triangle GLH \sim \triangle GHJ \sim \triangle HLJ$.

page 117
Now You Try It

1. Based on the figure below,
   fill in the blanks:

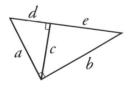

a is the geometric mean between __d__ and $(d+e)$.
b is the geometric mean between __e__ and $(d+e)$.
c is the geometric mean between __e__ and __d__.

page 117 (continued)

2. Based on the figure, find *AC*, *AB*, *DB* and *CB*:

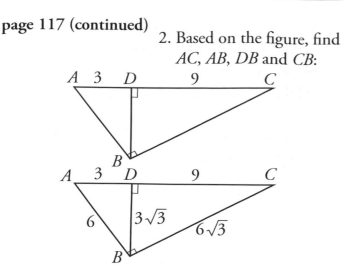

$AC = AD + DC = 3 + 9 = 12$ ✓

Finding *AB*: $\dfrac{3}{AB} = \dfrac{AB}{12}$

$(AB)^2 = 36,\ AB = 6$ ✓

Finding *DB*: $\dfrac{3}{DB} = \dfrac{DB}{9}$

$(DB)^2 = 27,\ DB = \sqrt{(3)(3)(3)} = 3\sqrt{3}\ \approx 5.2$ ✓

Finding *BC*: $\dfrac{9}{BC} = \dfrac{BC}{12}$

$(BC)^2 = 108,\ BC = \sqrt{(3)(6)(6)} = 6\sqrt{3} \approx 10.4$ ✓

page 119 Now You Try It   1. Find *z* in the right triangle below. Answer: In the problem, you are given that this is a right triangle so you know that the lengths of the sides of the triangle are related by the Pythagorean Theorem:

$2z + 2$ ⟍ $3z - 2$

⟍ z

$(\text{Leg1})^2 + (\text{Leg2})^2 = (\text{Hypotenuse})^2$

$(2z + 2)^2 + z^2 \qquad = (3z - 2)^2$ Foil the binomials.

$(4z^2 + 8z + 4) + z^2 = 9z^2 - 12z + 4$

$4z^2 - 20z = 0,\quad 4z(z - 5) = 0$

$4z = 0,\ z \neq 0$ (impossible), $(z - 5) = 0,\ z = 5$ ✓

2. Is triangle *PQR* a right triangle? Answer: If the lengths of a triangle are not related by the formula: $(\text{Leg 1})^2 + (\text{Leg 2})^2 = (\text{Leg 3})^2$ then the triangle is not a right triangle*. When testing to see, always set the largest side of the triangle to be Leg 3:

$3^2 + 2^2 \overset{?}{=} 4^2,\ 9 + 4 \neq 16$, so triangle *PQR* is *not* a right triangle.

(*Note, this is the contrapositive of the Pythagorean Theorem.)

Q 4 *R*
3
P 2

page 121   Now You Try It — Given: $a^2 + b^2 < c^2$ Prove: $\triangle ABC$ is obtuse.

Hints: Construct right $\triangle TUV$, with right $\angle V$ and legs *a* and *b*. Now recall the inequality theorems!

Statements	Reasons
1. $a^2 + b^2 < c^2$.	1. Given.
2. $a^2 + b^2 = v^2$.	2. Pythagorean Theorem.
3. $v^2 < c^2,\ v < c$.	3. Substitution.
4. $\angle C > \angle V$.	4. SSS Ineq. Thm.
5. $\angle C$ & $\triangle ABC$ are obtuse.	5. Def. of Obtuse $\angle$'s &. $\triangle$'s.

page 122 Now You Try It   1. Is triangle *BCD* acute, right or obtuse? Prove it.

To use the three theorems, we need to know the lengths of all three sides. Since triangle *ABC* is a right triangle, we can use the Pythagorean Theorem to find *CB*:

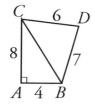

$8^2 + 4^2 = (CB)^2,\quad 80 = (CB)^2$

$\sqrt{80} = \sqrt{(5)(4)(4)} = 4\sqrt{5} \approx 8.94 \approx CB$

Knowing all three sides we can now form a test equation to test the triangle:

$6^2 + 7^2\ ?\ (4\sqrt{5})^2$ Since $4\sqrt{5}$ is the largest leg, we placed it by itself on the right.

Left side: $6^2 + 7^2 = 36 + 49 = 85$ Right side: $(4\sqrt{5})^2 = (4)^2\,(\sqrt{5})^2 = (16)(5) = 80$

Since $85 > 80$, that is, the left side of the test equation is larger than the right side, we know the triangle is acute. ✓

1. Find the length of a diagonal of the square below.

A diagonal divides a square into two congruent 45°–45°–90° triangles.

Draw the 45°–45°–90° model so that it is oriented like the problem.

Since $\sqrt{6}$ is the length of a leg, the hypotenuse of the triangle (which is the diagonal of the square) = 's

$\sqrt{6}\sqrt{2} = \sqrt{12} = 2\sqrt{3} \approx 3.46\,\checkmark$

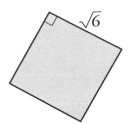

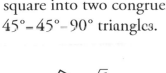

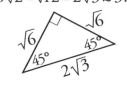

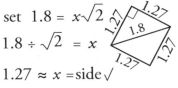

2. Find the length of a side and the perimeter of the square below.

The diagonal divides the square into two congruent 45°–45°–90° triangles with hypotenuse = 1.8

Draw the 45°–45°–90° model so that it is oriented like the problem.

Since 1.8 is the length of the diagonal (the hypotenuse of each triangle),

set $1.8 = x\sqrt{2}$

$1.8 \div \sqrt{2} = x$

$1.27 \approx x = \text{side}\,\checkmark$

Perimeter $\approx 4(1.27)$

$\approx 5.08\,\checkmark$

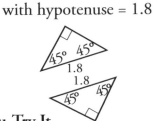

Find a.

Answer: To solve a right triangle problem we need one side and one acute angle. At first glance, it doesn't look like we have enough information about either triangle to get started. But the lower triangle is isosceles, and since it's a right triangle, w must equal 45.

① Draw the 45°–45°–90° model so that it is oriented like the lower triangle

② Comparing the lower triangle to the model, the hypotenuse of the lower triangle must equal $5\sqrt{2}$.

Lower Triangle

③ But the hypotenuse of the lower triangle is the left side of the upper triangle, and since $w = 45$, $2w = 90$. Now we know that the upper triangle is a 30°–60°–90° triangle.

Upper Triangle

④ Draw the 30°–60°–90° model so that it is oriented like the upper triangle in the problem.

⑤ Comparing the upper triangle to the model, $5\sqrt{2}$ is in the same position as $x\sqrt{3}$ so set them equal.

$5\sqrt{2} = x\sqrt{3}$

$\dfrac{5\sqrt{2}}{\sqrt{3}} = \dfrac{x\sqrt{3}}{\sqrt{3}}$ Divide by $\sqrt{3}$

$\dfrac{5\sqrt{2}}{\sqrt{3}}\left(\dfrac{\sqrt{3}}{\sqrt{3}}\right) = x$ Rationalize denomin. $(\sqrt{3}\sqrt{3} = 3)$

$x = \dfrac{5\sqrt{6}}{3} \approx 4.08$ so $2x = a = \dfrac{10\sqrt{6}}{3} \approx 8.16\,\checkmark$

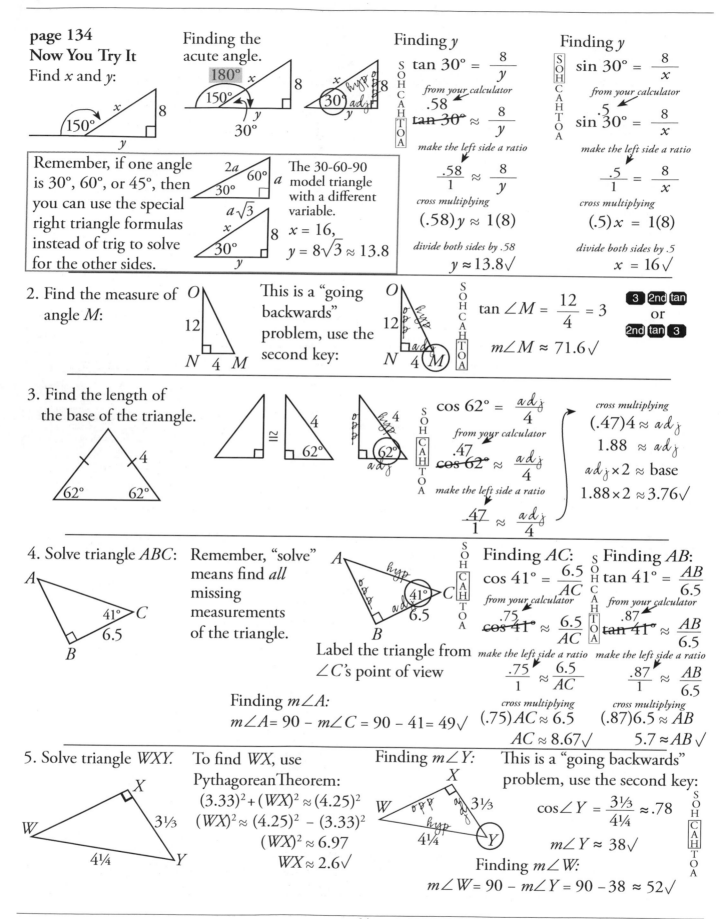

page 134
Now You Try It
Find x and y:

Finding the acute angle.

Finding y

$$\tan 30° = \frac{8}{y}$$

from your calculator

$$\tan 30° \approx \frac{8}{y}$$ .58

make the left side a ratio

$$\frac{.58}{1} \approx \frac{8}{y}$$

cross multiplying

$$(.58)y \approx 1(8)$$

divide both sides by .58

$$y \approx 13.8 \checkmark$$

Finding y

$$\sin 30° = \frac{8}{x}$$

from your calculator

$$\sin 30° = \frac{8}{x}$$ .5

make the left side a ratio

$$\frac{.5}{1} = \frac{8}{x}$$

cross multiplying

$$(.5)x = 1(8)$$

divide both sides by .5

$$x = 16 \checkmark$$

Remember, if one angle is 30°, 60°, or 45°, then you can use the special right triangle formulas instead of trig to solve for the other sides.

The 30-60-90 model triangle with a different variable.
$x = 16,$
$y = 8\sqrt{3} \approx 13.8$

2. Find the measure of angle M:

This is a "going backwards" problem, use the second key:

$$\tan \angle M = \frac{12}{4} = 3$$

3 2nd tan or 2nd tan 3

$$m\angle M \approx 71.6 \checkmark$$

3. Find the length of the base of the triangle.

$$\cos 62° = \frac{adj}{4}$$

from your calculator

$$\cos 62° \approx \frac{adj}{4}$$ .47

make the left side a ratio

$$\frac{.47}{1} \approx \frac{adj}{4}$$

cross multiplying

$$(.47)4 \approx adj$$
$$1.88 \approx adj$$
$$adj \times 2 \approx \text{base}$$
$$1.88 \times 2 \approx 3.76 \checkmark$$

4. Solve triangle ABC:

Remember, "solve" means find *all* missing measurements of the triangle.

Label the triangle from $\angle C$'s point of view

Finding $m\angle A$:
$$m\angle A = 90 - m\angle C = 90 - 41 = 49 \checkmark$$

Finding AC:

$$\cos 41° = \frac{6.5}{AC}$$

from your calculator

$$\cos 41° \approx \frac{6.5}{AC}$$ .75

make the left side a ratio

$$\frac{.75}{1} \approx \frac{6.5}{AC}$$

cross multiplying

$$(.75)AC \approx 6.5$$
$$AC \approx 8.67 \checkmark$$

Finding AB:

$$\tan 41° = \frac{AB}{6.5}$$

from your calculator

$$\tan 41° \approx \frac{AB}{6.5}$$ .87

make the left side a ratio

$$\frac{.87}{1} \approx \frac{AB}{6.5}$$

cross multiplying

$$(.87)6.5 \approx AB$$
$$5.7 \approx AB \checkmark$$

5. Solve triangle WXY.

To find WX, use Pythagorean Theorem:
$$(3.33)^2 + (WX)^2 \approx (4.25)^2$$
$$(WX)^2 \approx (4.25)^2 - (3.33)^2$$
$$(WX)^2 \approx 6.97$$
$$WX \approx 2.6 \checkmark$$

Finding $m\angle Y$:

This is a "going backwards" problem, use the second key:

$$\cos \angle Y = \frac{3\frac{1}{3}}{4\frac{1}{4}} \approx .78$$

$$m\angle Y \approx 38 \checkmark$$

Finding $m\angle W$:
$$m\angle W = 90 - m\angle Y = 90 - 38 \approx 52 \checkmark$$

Now You Try It

In problems 1-5, name a method that could be used to solve each triangle, then solve the triangle.

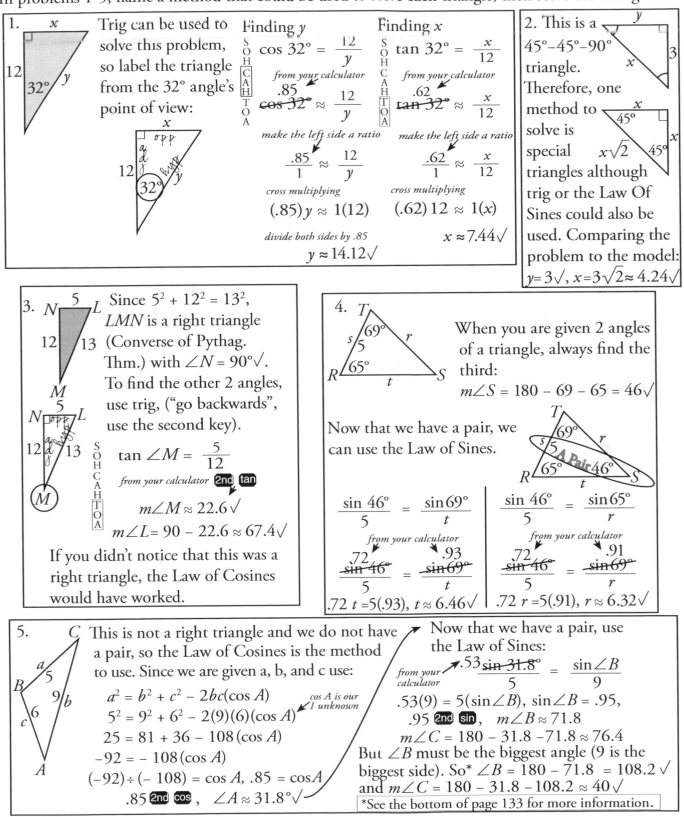

1. Trig can be used to solve this problem, so label the triangle from the 32° angle's point of view:

Finding y

S O H
C A H
T O A

$\cos 32° = \dfrac{12}{y}$

from your calculator

$\cancel{\cos 32°} \approx \dfrac{.85}{} \dfrac{12}{y}$

make the left side a ratio

$\dfrac{.85}{1} \approx \dfrac{12}{y}$

cross multiplying

$(.85)y \approx 1(12)$

divide both sides by .85

$y \approx 14.12 \checkmark$

Finding x

S O H
C A H
T O A

$\tan 32° = \dfrac{x}{12}$

from your calculator

$\cancel{\tan 32°} \approx \dfrac{.62}{} \dfrac{x}{12}$

make the left side a ratio

$\dfrac{.62}{1} \approx \dfrac{x}{12}$

cross multiplying

$(.62)12 \approx 1(x)$

$x \approx 7.44 \checkmark$

2. This is a 45°–45°–90° triangle. Therefore, one method to solve is special triangles although trig or the Law Of Sines could also be used. Comparing the problem to the model:

$y = 3\checkmark$, $x = 3\sqrt{2} \approx 4.24\checkmark$

3. Since $5^2 + 12^2 = 13^2$, LMN is a right triangle (Converse of Pythag. Thm.) with $\angle N = 90°\checkmark$. To find the other 2 angles, use trig, ("go backwards", use the second key).

S O H
C A H
T O A

$\tan \angle M = \dfrac{5}{12}$

from your calculator **2nd** **tan**

$m\angle M \approx 22.6 \checkmark$

$m\angle L = 90 - 22.6 \approx 67.4 \checkmark$

If you didn't notice that this was a right triangle, the Law of Cosines would have worked.

4. When you are given 2 angles of a triangle, always find the third:

$m\angle S = 180 - 69 - 65 = 46\checkmark$

Now that we have a pair, we can use the Law of Sines.

A Pair

$\dfrac{\sin 46°}{5} = \dfrac{\sin 69°}{t}$

from your calculator

$\dfrac{.72}{\cancel{\sin 46°}} = \dfrac{.93}{\cancel{\sin 69°}}$
$\dfrac{}{5} \quad \dfrac{}{t}$

$.72 t = 5(.93), t \approx 6.46\checkmark$

$\dfrac{\sin 46°}{5} = \dfrac{\sin 65°}{r}$

from your calculator

$\dfrac{.72}{\cancel{\sin 46°}} = \dfrac{.91}{\cancel{\sin 69°}}$
$\dfrac{}{5} \quad \dfrac{}{r}$

$.72 r = 5(.91), r \approx 6.32\checkmark$

5. This is not a right triangle and we do not have a pair, so the Law of Cosines is the method to use. Since we are given a, b, and c use:

$a^2 = b^2 + c^2 - 2bc(\cos A)$ ← *cos A is our 1 unknown*

$5^2 = 9^2 + 6^2 - 2(9)(6)(\cos A)$

$25 = 81 + 36 - 108(\cos A)$

$-92 = -108(\cos A)$

$(-92) \div (-108) = \cos A, .85 = \cos A$

$.85$ **2nd** **cos** , $\angle A \approx 31.8°\checkmark$

Now that we have a pair, use the Law of Sines:

from your calculator

$\dfrac{.53 \cancel{\sin 31.8°}}{5} = \dfrac{\sin \angle B}{9}$

$.53(9) = 5(\sin \angle B), \sin \angle B = .95,$

$.95$ **2nd** **sin** , $m\angle B \approx 71.8$

$m\angle C = 180 - 31.8 - 71.8 \approx 76.4$

But $\angle B$ must be the biggest angle (9 is the biggest side). So* $\angle B = 180 - 71.8 = 108.2 \checkmark$ and $m\angle C = 180 - 31.8 - 108.2 \approx 40 \checkmark$

See the bottom of page 133 for more information.

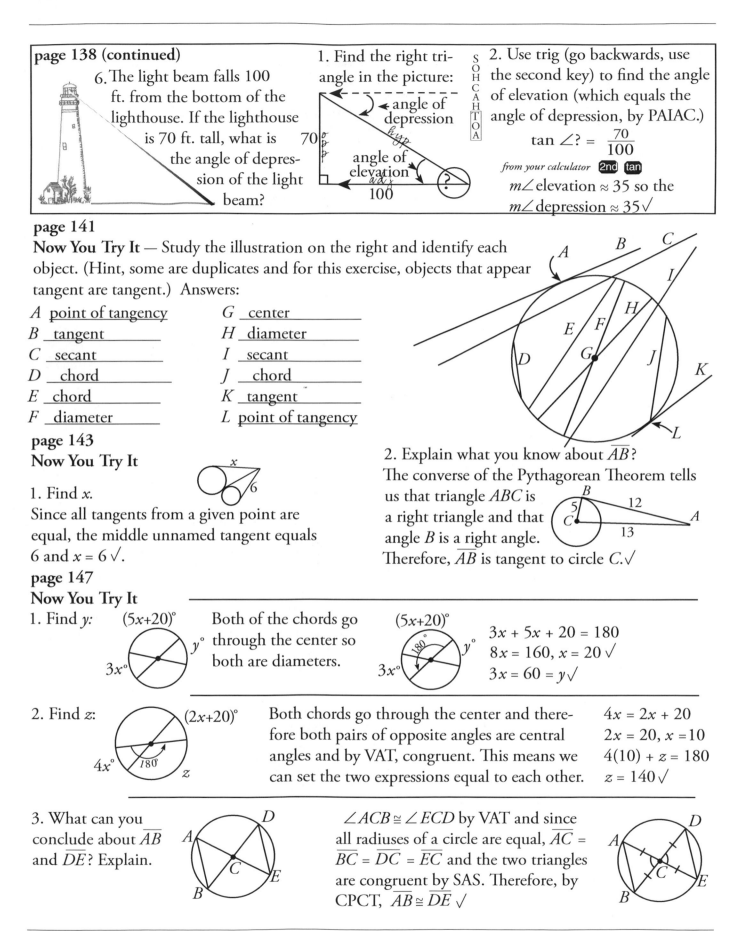

page 138 (continued)

6. The light beam falls 100 ft. from the bottom of the lighthouse. If the lighthouse is 70 ft. tall, what is the angle of depression of the light beam?

1. Find the right triangle in the picture:

angle of depression

angle of elevation

70

100

S O H C A H T O A

2. Use trig (go backwards, use the second key) to find the angle of elevation (which equals the angle of depression, by PAIAC.)

$$\tan \angle? = \frac{70}{100}$$

from your calculator **2nd** **tan**

$m\angle$ elevation ≈ 35 so the
$m\angle$ depression $\approx 35\sqrt{}$

page 141

Now You Try It — Study the illustration on the right and identify each object. (Hint, some are duplicates and for this exercise, objects that appear tangent are tangent.) Answers:

A point of tangency
B tangent
C secant
D chord
E chord
F diameter

G center
H diameter
I secant
J chord
K tangent
L point of tangency

page 143

Now You Try It

1. Find *x*.

Since all tangents from a given point are equal, the middle unnamed tangent equals 6 and $x = 6\sqrt{}$.

2. Explain what you know about $\overline{AB}$?
The converse of the Pythagorean Theorem tells us that triangle *ABC* is a right triangle and that angle *B* is a right angle.
Therefore, $\overline{AB}$ is tangent to circle *C*.$\sqrt{}$

page 147

Now You Try It

1. Find *y*: $(5x+20)°$ $3x°$

Both of the chords go through the center so both are diameters.

$(5x+20)°$ $3x°$

$3x + 5x + 20 = 180$
$8x = 160, x = 20\sqrt{}$
$3x = 60 = y\sqrt{}$

2. Find *z*: $(2x+20)°$ $4x°$ 180 *z*

Both chords go through the center and therefore both pairs of opposite angles are central angles and by VAT, congruent. This means we can set the two expressions equal to each other.

$4x = 2x + 20$
$2x = 20, x = 10$
$4(10) + z = 180$
$z = 140\sqrt{}$

3. What can you conclude about $\overline{AB}$ and $\overline{DE}$? Explain.

$\angle ACB \cong \angle ECD$ by VAT and since all radiuses of a circle are equal, $\overline{AC} = \overline{BC} = \overline{DC} = \overline{EC}$ and the two triangles are congruent by SAS. Therefore, by CPCT, $\overline{AB} \cong \overline{DE}$ $\sqrt{}$

page 150 — Now You Try It

Theorem — In the same or congruent circles, congruent chords are equally distant from the center of the circle(s).

Using the drawings as hints, prove the above theorem for a single circle.

Prove: In a ⊙ ≅ chords are equidistant from the center.

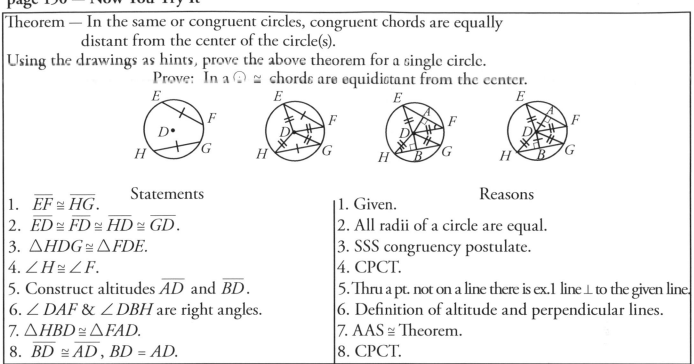

Statements	Reasons
1. $\overline{EF} \cong \overline{HG}$.	1. Given.
2. $\overline{ED} \cong \overline{FD} \cong \overline{HD} \cong \overline{GD}$.	2. All radii of a circle are equal.
3. $\triangle HDG \cong \triangle FDE$.	3. SSS congruency postulate.
4. $\angle H \cong \angle F$.	4. CPCT.
5. Construct altitudes $\overline{AD}$ and $\overline{BD}$.	5. Thru a pt. not on a line there is ex.1 line ⊥ to the given line.
6. $\angle DAF$ & $\angle DBH$ are right angles.	6. Definition of altitude and perpendicular lines.
7. $\triangle HBD \cong \triangle FAD$.	7. AAS ≅ Theorem.
8. $\overline{BD} \cong \overline{AD}$, $BD = AD$.	8. CPCT.

page 152 — Now You Try It

1. Find the length of a diameter of circle C.

Let x = the radius

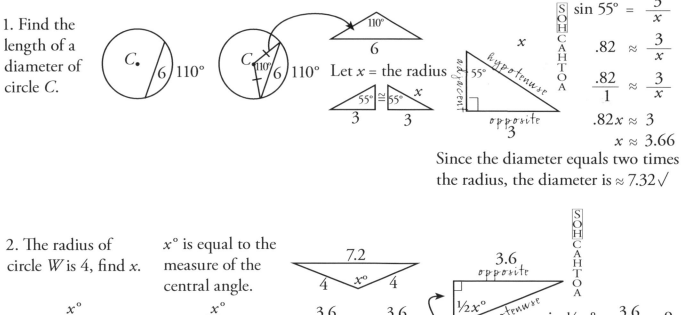

$$\sin 55° = \frac{3}{x}$$

$$.82 \approx \frac{3}{x}$$

$$\frac{.82}{1} \approx \frac{3}{x}$$

$$.82x \approx 3$$

$$x \approx 3.66$$

Since the diameter equals two times the radius, the diameter is $\approx 7.32 \checkmark$

2. The radius of circle W is 4, find x.

$x°$ is equal to the measure of the central angle.

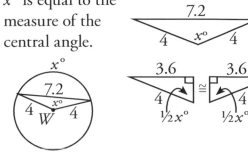

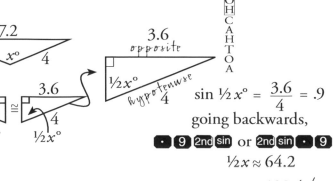

$$\sin \tfrac{1}{2}x° = \frac{3.6}{4} = .9$$

going backwards,

⚫ 9 2nd sin or 2nd sin ⚫ 9

$$\tfrac{1}{2}x \approx 64.2$$

$$x \approx 128.4 \checkmark$$

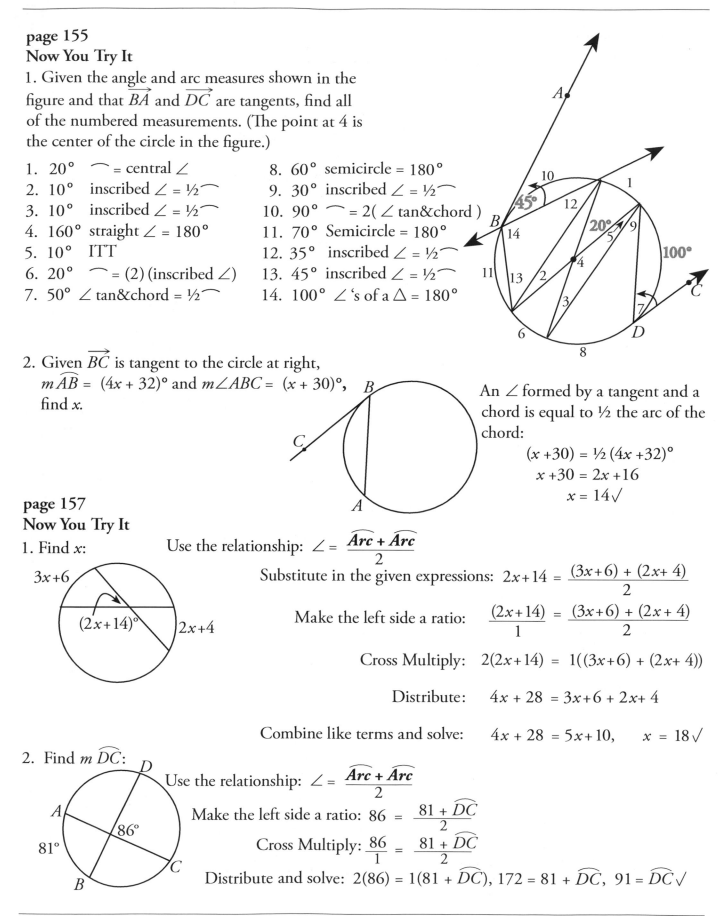

page 155
Now You Try It

1. Given the angle and arc measures shown in the figure and that $\vec{BA}$ and $\vec{DC}$ are tangents, find all of the numbered measurements. (The point at 4 is the center of the circle in the figure.)

1. 20° ⌒ = central ∠
2. 10° inscribed ∠ = ½ ⌒
3. 10° inscribed ∠ = ½ ⌒
4. 160° straight ∠ = 180°
5. 10° ITT
6. 20° ⌒ = (2)(inscribed ∠)
7. 50° ∠ tan&chord = ½ ⌒

8. 60° semicircle = 180°
9. 30° inscribed ∠ = ½ ⌒
10. 90° ⌒ = 2(∠ tan&chord)
11. 70° Semicircle = 180°
12. 35° inscribed ∠ = ½ ⌒
13. 45° inscribed ∠ = ½ ⌒
14. 100° ∠'s of a △ = 180°

2. Given $\vec{BC}$ is tangent to the circle at right, $m\,\widehat{AB} = (4x + 32)°$ and $m\angle ABC = (x + 30)°$, find x.

An ∠ formed by a tangent and a chord is equal to ½ the arc of the chord:

$$(x + 30) = \tfrac{1}{2}(4x + 32)°$$
$$x + 30 = 2x + 16$$
$$x = 14 \checkmark$$

page 157
Now You Try It

1. Find x:

Use the relationship: $\angle = \dfrac{\widehat{Arc} + \widehat{Arc}}{2}$

Substitute in the given expressions: $2x + 14 = \dfrac{(3x+6) + (2x+4)}{2}$

Make the left side a ratio: $\dfrac{(2x+14)}{1} = \dfrac{(3x+6) + (2x+4)}{2}$

Cross Multiply: $2(2x+14) = 1((3x+6) + (2x+4))$

Distribute: $4x + 28 = 3x + 6 + 2x + 4$

Combine like terms and solve: $4x + 28 = 5x + 10$, $\quad x = 18 \checkmark$

2. Find $m\,\widehat{DC}$:

Use the relationship: $\angle = \dfrac{\widehat{Arc} + \widehat{Arc}}{2}$

Make the left side a ratio: $86 = \dfrac{81 + \widehat{DC}}{2}$

Cross Multiply: $\dfrac{86}{1} = \dfrac{81 + \widehat{DC}}{2}$

Distribute and solve: $2(86) = 1(81 + \widehat{DC})$, $172 = 81 + \widehat{DC}$, $91 = \widehat{DC} \checkmark$

page 160 — Now You Try It

1. Find the value of x in the figure on the right:

Use the formula: $\angle = \dfrac{\overset{\frown}{Big} - \overset{\frown}{Sml}}{2}$

$4x - 2 = \dfrac{(8x+20) - (2x+4)}{2}$

Form a proportion: $\dfrac{(4x-2)}{1} = \dfrac{(8x+20) - (2x+4)}{2}$

Cross multiply: $2(4x-2) = 1((8x+20) - (2x+4))$

$8x - 4 = (8x+20) - (2x+4)$ Now, carefully distribute the invisible negative one.

$8x - 4 = 8x + 20 - 2x - 4 = 6x + 16$

$2x = 20, \quad x = 10 \checkmark$

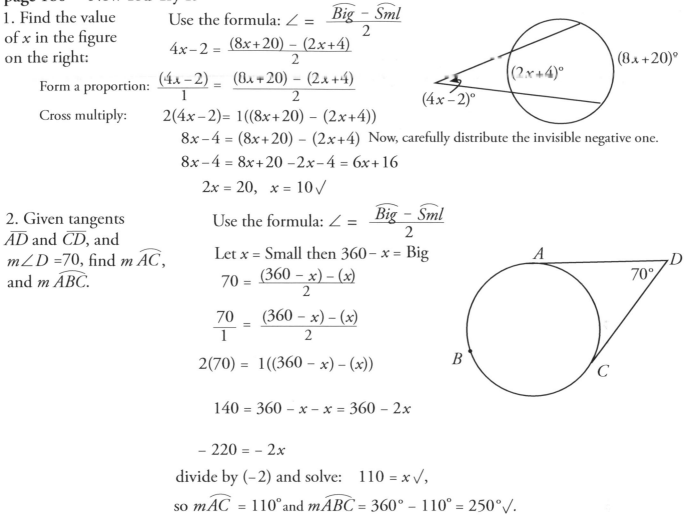

2. Given tangents $\overline{AD}$ and $\overline{CD}$, and $m\angle D = 70$, find $m\overset{\frown}{AC}$, and $m\overset{\frown}{ABC}$.

Use the formula: $\angle = \dfrac{\overset{\frown}{Big} - \overset{\frown}{Sml}}{2}$

Let x = Small then $360 - x$ = Big

$70 = \dfrac{(360 - x) - (x)}{2}$

$\dfrac{70}{1} = \dfrac{(360 - x) - (x)}{2}$

$2(70) = 1((360 - x) - (x))$

$140 = 360 - x - x = 360 - 2x$

$-220 = -2x$

divide by (-2) and solve: $\quad 110 = x \checkmark,$

so $m\overset{\frown}{AC} = 110°$ and $m\overset{\frown}{ABC} = 360° - 110° = 250° \checkmark.$

page 162 — Now You Try It

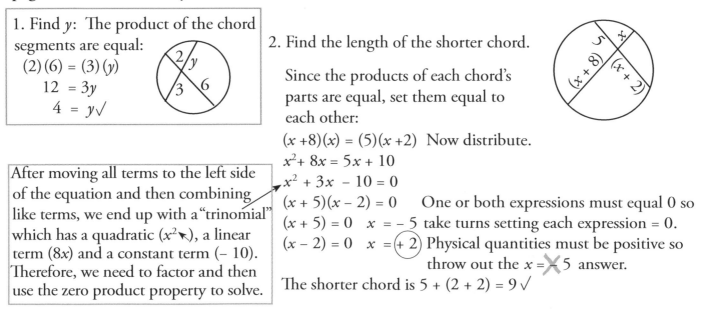

1. Find y: The product of the chord segments are equal:

$(2)(6) = (3)(y)$

$12 = 3y$

$4 = y \checkmark$

After moving all terms to the left side of the equation and then combining like terms, we end up with a "trinomial" which has a quadratic (x^2), a linear term ($8x$) and a constant term (-10). Therefore, we need to factor and then use the zero product property to solve.

2. Find the length of the shorter chord.

Since the products of each chord's parts are equal, set them equal to each other:

$(x+8)(x) = (5)(x+2)$ Now distribute.

$x^2 + 8x = 5x + 10$

$x^2 + 3x - 10 = 0$

$(x+5)(x-2) = 0$ One or both expressions must equal 0 so

$(x+5) = 0 \quad x = -5$ take turns setting each expression = 0.

$(x-2) = 0 \quad x = +2$ Physical quantities must be positive so throw out the $x = -5$ answer.

The shorter chord is $5 + (2+2) = 9 \checkmark$

269

1. Find x:

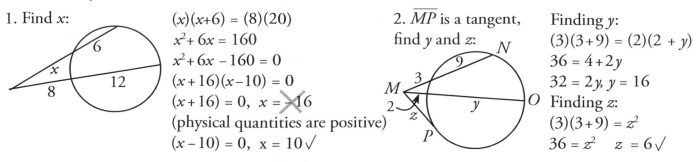

$(x)(x+6) = (8)(20)$
$x^2 + 6x = 160$
$x^2 + 6x - 160 = 0$
$(x + 16)(x - 10) = 0$
$(x + 16) = 0, \; x = \cancel{-16}$
(physical quantities are positive)
$(x - 10) = 0, \; x = 10 \checkmark$

2. $\overline{MP}$ is a tangent, find y and z:

Finding y:
$(3)(3+9) = (2)(2 + y)$
$36 = 4 + 2y$
$32 = 2y, \; y = 16$
Finding z:
$(3)(3+9) = z^2$
$36 = z^2 \quad z = 6 \checkmark$

3. Find AB:

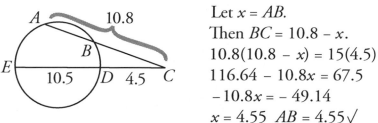

Let $x = AB$.
Then $BC = 10.8 - x$.
$10.8(10.8 - x) = 15(4.5)$
$116.64 - 10.8x = 67.5$
$-10.8x = -49.14$
$x = 4.55 \quad AB = 4.55 \checkmark$

1. Find the perimeter of square $UVWX$.
Let y equal the length of one side. By the Pythagorean Theorem,
$$y^2 + y^2 = 2y^2 = 14^2 = 196$$
$$\frac{\cancel{2}y^2}{\cancel{2}} = \frac{\cancel{196}^{\,98}}{\cancel{2}}$$
$$y^2 = 98$$
$$y = 7\sqrt{2} \;\approx 9.9, \text{ perimeter} \approx (4)(9.9) = 39.6 \checkmark$$

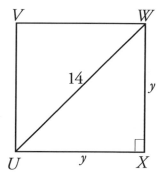

2. The area of a rectangle is 180. The ratio of the base to the height is 4:5.
Find the perimeter and the diagonal of the rectangle.

Draw a rectangle that roughly matches the given information.
The ratio is 4:5, so you must supply the x's, $4x : 5x$.

Now, mark up the drawing, showing all the information that you know.

180 | *5x*
4x

Since A = bh
$180 = (4x)(5x) = 20x^2$, ÷ both sides by 20, then $9 = x^2$, $3 = x$.
The perimeter equals $4x + 5x + 4x + 5x = 18x = (18)(3) = 54 \checkmark$
By the Pythagorean Theorem, the diagonal squared equals $(12)^2 + (15)^2 = 369$
The diagonal $= \sqrt{369} \approx 19.2 \checkmark$

3. Find the area of the figure below:

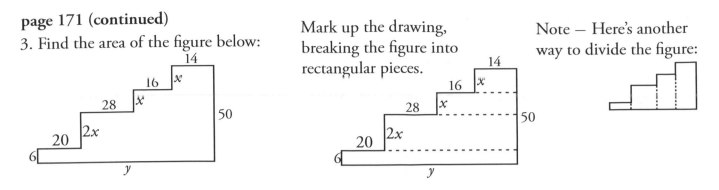

Mark up the drawing, breaking the figure into rectangular pieces.

Note — Here's another way to divide the figure:

Find the missing dimensions, find x: $50 = x + x + 2x + 6$, $50 = 4x + 6$, $44 = 4x$, $x = 11$

find y: $y = 20 + 28 + 16 + 14 = 78$, $y = 78$

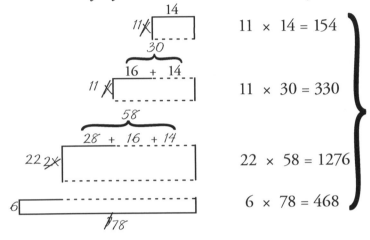

$11 \times 14 = 154$

$11 \times 30 = 330$

$22 \times 58 = 1276$

$6 \times 78 = 468$

Area = $154 + 330 + 1276 + 468 = 2228$ √

4. Find the area of paralellogram *GHIJ*

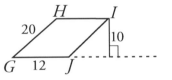

A = (base) ✗ (height to *that* base) = (12)(10) = 120 √

A = (20)(12) = 240   A = (20)(10) = 200

5. Given parallelogram *ABCD*, find *x*.

Answer: Area = (base) ✗ (height to *that* base)

Area = (20)(6) = 120

Now, since the area of the parallelogram doesn't change,

Area = 120 = (12)(x) The other base times the height to ***that*** base

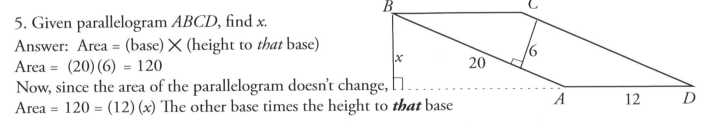

$\frac{\overset{10}{\cancel{120}}}{\cancel{12}} = \frac{(\cancel{12})(x)}{\cancel{12}}$ $x = 10$ √

6. Find the area of parallelogram *PQRS*.

Answer: Opposite sides of a parallelogram are equal so *PS* = 18 and *QP* = 11. Finding *QM* would allow us to find the area of the parallelogram. Using trig to find QM:

$$\sin 70° = \frac{opp}{11} \qquad \frac{.94}{1} = \frac{QM}{11}$$
$$QM = (.94)(11) = 10.34$$
$$\text{Area} = (18)(10.34) = 186.12 \checkmark$$

page 179

Now You Try It

1. The measure of each side of the equilateral triangle on the right is 5. Find the area of the triangle.

By Pythagorean Theorem:
$$x^2 + (2.5)^2 = 5^2$$
$$x^2 = 25 - 6.25 = 18.75$$
$$x \approx 4.33$$

$$A = \frac{(4.33)(2.5)}{2} \approx 5.4$$

There are two right triangles within the equilateral triangle, so the total area equals 2× 5.4 = 10.8 √

Alternatively, when we know the lengths of all 3 sides of a triangle, we can use Heron's formula:

Semiperimeter = S = $\frac{5 + 5 + 5}{2} = \frac{15}{2} = 7.5$ A = $\sqrt{7.5(7.5 - 5)(7.5 - 5)(7.5 - 5)} \approx \sqrt{117.2} \approx 10.8$ √

2. Find the area of the rhombus to the right. (Hint, draw the diagonals, which divide the rhombus into four 4 ≅ rt △, then use trigonometry.)

$$\cos 23° = \frac{adj}{6} \qquad \frac{.92}{1} \approx \frac{adj}{6} \qquad adj \approx 5.52$$

$$\sin 23° = \frac{opp}{6} \qquad \frac{.39}{1} \approx \frac{opp}{6} \qquad opp \approx 2.34$$

First Way:
Area of 1 right triangle ≈ ½(5.52)(2.34) ≈ 6.46

There are 4 right triangles so:
$$A \approx (4)(6.46) \approx 25.8 \checkmark$$

Second Way: Once we find the opposite and adjacent sides of the right triangles, we are also finding the lengths of the diagonals and so we can then use the special formula for the area of a rhombus.

Area of a rhombus = $\frac{(d_1)(d_2)}{2}$

$$A \approx \frac{(11.04)(4.68)}{2} = 25.83 \checkmark$$

(The .03 difference in answers is a rounding error.)

page 179 (continued)

3. Find the area of triangle *ABC* on the right.

7.98 is the altitude to the base which measures 13.6, so the area of the triangle is :

A = ½ [(base) × (height **to** *that* **base**)]

A = ½ (13.6)(7.98) = 54.264 √

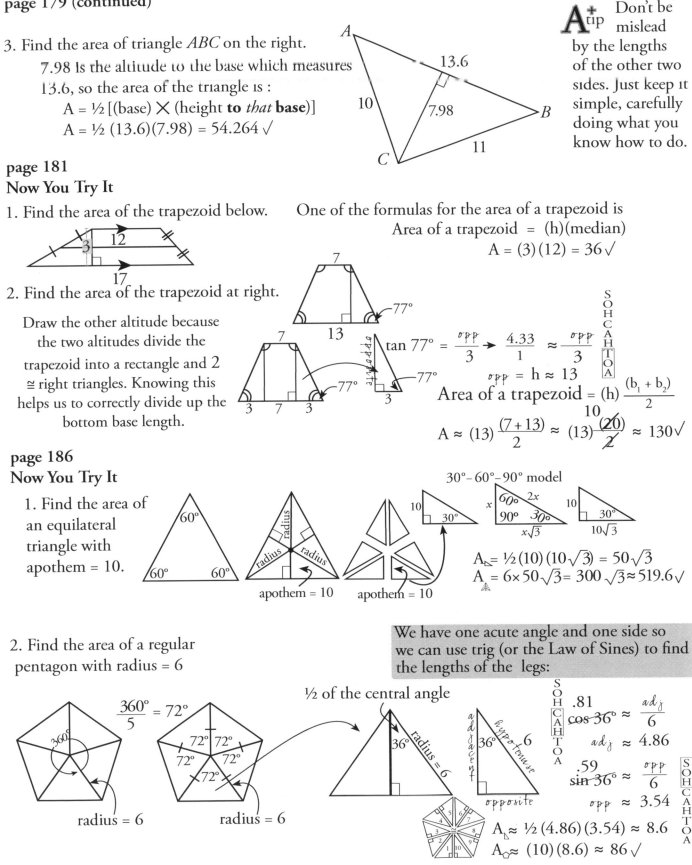

A^+_{tip} Don't be mislead by the lengths of the other two sides. Just keep it simple, carefully doing what you know how to do.

page 181
Now You Try It

1. Find the area of the trapezoid below.

One of the formulas for the area of a trapezoid is

Area of a trapezoid = (h)(median)

A = (3)(12) = 36 √

2. Find the area of the trapezoid at right.

Draw the other altitude because the two altitudes divide the trapezoid into a rectangle and 2 ≅ right triangles. Knowing this helps us to correctly divide up the bottom base length.

$\tan 77° = \frac{opp}{3} \rightarrow \frac{4.33}{1} \approx \frac{opp}{3}$

$opp = h \approx 13$

Area of a trapezoid = (h) $\frac{(b_1 + b_2)}{2}$

A ≈ (13) $\frac{(7+13)}{2}$ ≈ (13) $\frac{(20)}{2}$ ≈ 130 √

page 186
Now You Try It

1. Find the area of an equilateral triangle with apothem = 10.

30°–60°–90° model

$A_\triangle$ = ½ (10) (10√3) = 50√3

A = 6 × 50√3 = 300 √3 ≈ 519.6 √

2. Find the area of a regular pentagon with radius = 6

$\frac{360°}{5}$ = 72°

½ of the central angle

We have one acute angle and one side so we can use trig (or the Law of Sines) to find the lengths of the legs:

$\cos 36° \approx \frac{adj}{6}$ → .81

$adj \approx 4.86$

$\sin 36° \approx \frac{opp}{6}$ → .59

$opp \approx 3.54$

$A_\triangle \approx$ ½ (4.86)(3.54) ≈ 8.6

$A_\circ \approx$ (10)(8.6) ≈ 86 √

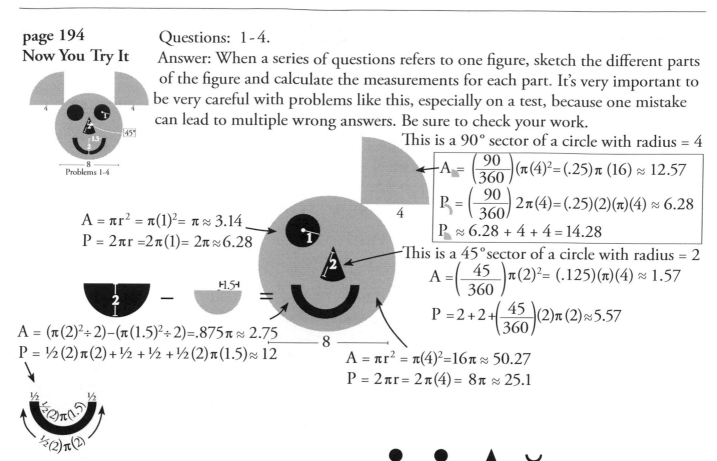

Problems 1-4

Questions: 1-4.

Answer: When a series of questions refers to one figure, sketch the different parts of the figure and calculate the measurements for each part. It's very important to be very careful with problems like this, especially on a test, because one mistake can lead to multiple wrong answers. Be sure to check your work.

This is a 90° sector of a circle with radius = 4

$$A = \left(\frac{90}{360}\right)(\pi(4)^2 = (.25)\pi(16) \approx 12.57$$

$$P = \left(\frac{90}{360}\right)2\pi(4) = (.25)(2)(\pi)(4) \approx 6.28$$

$$P \approx 6.28 + 4 + 4 = 14.28$$

$$A = \pi r^2 = \pi(1)^2 = \pi \approx 3.14$$
$$P = 2\pi r = 2\pi(1) = 2\pi \approx 6.28$$

This is a 45° sector of a circle with radius = 2

$$A = \left(\frac{45}{360}\right)\pi(2)^2 = (.125)(\pi)(4) \approx 1.57$$

$$P = 2 + 2 + \left(\frac{45}{360}\right)(2)\pi(2) \approx 5.57$$

$$A = (\pi(2)^2 \div 2) - (\pi(1.5)^2 \div 2) = .875\pi \approx 2.75$$
$$P = \tfrac{1}{2}(2)\pi(2) + \tfrac{1}{2} + \tfrac{1}{2} + \tfrac{1}{2}(2)\pi(1.5) \approx 12$$

$$A = \pi r^2 = \pi(4)^2 = 16\pi \approx 50.27$$
$$P = 2\pi r = 2\pi(4) = 8\pi \approx 25.1$$

1. Find the total area of the black parts of the figure: $3.14 + 3.14 + 1.57 + 2.75 \approx 10.6 \checkmark$

2. Find the total area of the gray parts of the figure: $50.27 - 10.6 + 12.57 + 12.57 = 64.84 \checkmark$

3. Find the total length of the perimeters of the black parts of the figure:
$12 + 6.28 + 6.28 + 5.57 = 30.13 \checkmark$

4. Find the total length of the perimeters (including all inner and outer edges) of the gray areas:
$30.13 + 25.1 + 14.28 + 14.28 = 83.79 \checkmark$

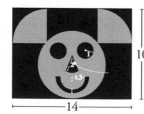

5. The figure is placed on a 14 × 10 rectangular black board. Find the total of the perimeters (including all inner and outer edges) of the black areas of the figure including the board: $27.77 + 25.1 + 14.28 + 14.28 + 48 = 129.43 \checkmark$

6. Find the total area of the visible black portions of the figure including those on the black board:
$2.75 + 3.14 + 3.14 + 1.57 + 140 - (50.27 + 12.57 + 12.57) = 75.19 \checkmark$

7a. What is the area of the black part of the figure:

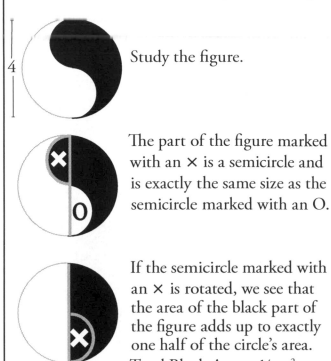

Study the figure.

The part of the figure marked with an × is a semicircle and is exactly the same size as the semicircle marked with an O.

If the semicircle marked with an × is rotated, we see that the area of the black part of the figure adds up to exactly one half of the circle's area.
Total Black Area = $\frac{1}{2}\pi r^2$
Note that r = $(\frac{1}{2})(4) = 2$.
$\frac{1}{2}\pi(2)^2 = 2\pi \approx 6.28\checkmark$

7b. What is the perimeter of the black part of the figure including inside and outside edges?

To start, we have ½ of the circumference of the outer circle: $C = 2\pi r = 2\pi(2) = 4\pi$ so $(\frac{1}{2})C = (\frac{1}{2})4\pi = \boxed{2\pi}$

For the inner edges, we have ½ the circumference of the upper small black circular area (×) plus ½ the circumference of the lower small white circular area (O).

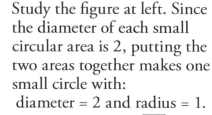

Study the figure at left. Since the diameter of each small circular area is 2, putting the two areas together makes one small circle with:
diameter = 2 and radius = 1.
$C = 2\pi(1) = \boxed{2\pi}$
Total = $2\pi + 2\pi = 4\pi \approx 12.6\checkmark$

8. Find the area of the circle segment shown in the figure on the right.

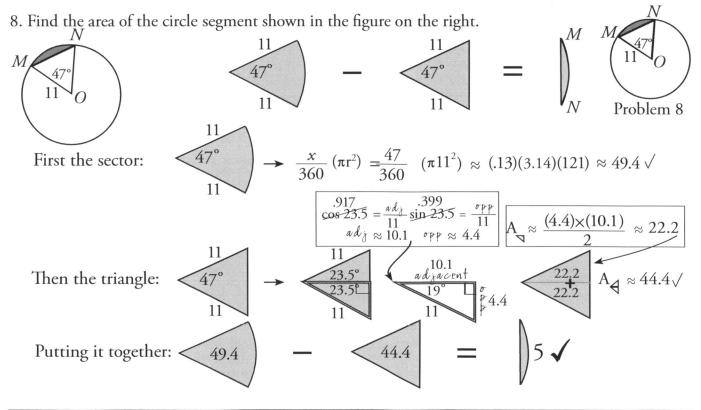

First the sector:

$\frac{x}{360}(\pi r^2) = \frac{47}{360}(\pi 11^2) \approx (.13)(3.14)(121) \approx 49.4\checkmark$

$\cos 23.5 = \frac{adj}{11} \quad \sin 23.5 = \frac{opp}{11}$
$.917 \qquad\qquad .399$
$adj \approx 10.1 \quad opp \approx 4.4$

$A \approx \frac{(4.4)\times(10.1)}{2} \approx 22.2$

Then the triangle:

$A \approx 44.4\checkmark$

Putting it together: 49.4 − 44.4 = 5 ✓

page 194 (continued) 9. Find the area of the gray portion of the figure. Answer: the two congruent circles intersect at their centers so the distance between the centers equals 6.

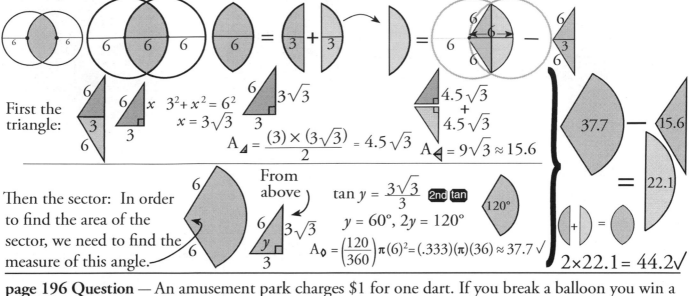

First the triangle:

$$3^2 + x^2 = 6^2$$
$$x = 3\sqrt{3}$$

$$A_\triangle = \frac{(3) \times (3\sqrt{3})}{2} = 4.5\sqrt{3} \qquad A_\triangleleft = 9\sqrt{3} \approx 15.6$$

Then the sector: In order to find the area of the sector, we need to find the measure of this angle.

From above

$$\tan y = \frac{3\sqrt{3}}{3} \quad \boxed{2nd}\,\boxed{tan}$$
$$y = 60°, \ 2y = 120°$$
$$A_\lozenge = \left(\frac{120}{360}\right)\pi(6)^2 = (.333)(\pi)(36) \approx 37.7 \ \checkmark$$

$$2 \times 22.1 = 44.2\sqrt{}$$

page 196 Question — An amusement park charges $1 for one dart. If you break a balloon you win a $3 stuffed animal. Is it a good bet? What is the probability of not breaking a balloon? Hint: The total of all probabilities add up to 1. Explain your answers. Answers: Unless you are an expert at darts it's a bad bet. Here's why. You have a .18 (random) chance of breaking a balloon which means you'll win about once every 5½ games at a cost of $5.50 for a $3 prize. (In real games, the odds are *much* worse.) Since the sum of all probabilities is 1, the probability of not hitting a balloon is: 1 − .18 = .82 $\checkmark$

page 199 — Now You Try It 1. What is the probability of hitting a diamond on the dart board on the right? The diamonds are congruent, each side is equal to 5" and the shorter diagonal is equal to 6".

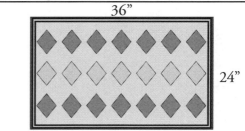

$$\text{Probability} = \frac{\text{Area of a win}}{\text{Area of whole figure}}$$

First the numerator, find the area of one diamond, then multiply by the number of diamonds:

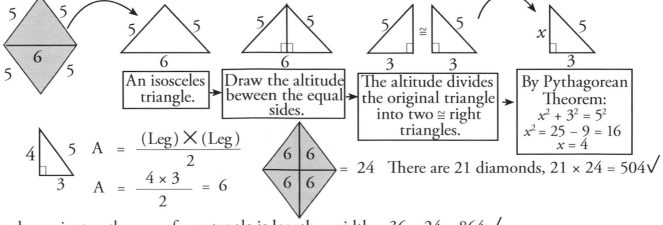

An isosceles triangle.

Draw the altitude beween the equal sides.

The altitude divides the original triangle into two ≅ right triangles.

By Pythagorean Theorem:
$$x^2 + 3^2 = 5^2$$
$$x^2 = 25 - 9 = 16$$
$$x = 4$$

$$A = \frac{(\text{Leg}) \times (\text{Leg})}{2}$$
$$A = \frac{4 \times 3}{2} = 6$$

= 24   There are 21 diamonds, 21 × 24 = 504$\checkmark$

The denominator, the area of a rectangle is length × width = 36 × 24 = 864 $\checkmark$

$$\text{Probability} = \frac{504}{864} \approx .583\sqrt{}$$

2. Study the archery target on the right, with the measurements as indicated. Assuming you are a beginning archer and only arrows that hit the target count, find the probability of:

a. hitting the bulls eye and

b. hitting either white area.

b.

a.

$$\frac{\bigcirc}{\bigcirc} = \frac{\pi(4)^2}{\pi(8)^2} = \frac{1}{4}\ \checkmark$$

$$\frac{\bigcirc - \bullet + \bigcirc}{\bigcirc} = \frac{\pi(8)^2 - \pi(6)^2 + \pi(4)^2}{\pi(8)^2}$$

$$= \frac{44}{64} = \frac{11}{16}\ \checkmark$$

3. The ceremony is 3 hours long. This means the ceremony is 3 × 60 minutes per hour = 180 minutes long. Tiger will be at the community center for 15 ÷ 180 = 1/12 of the time. Probability = 1/12 ✓

4. The local pizza place has a special which includes a large pizza with or without extra cheese and the choice of one meat topping and one vegetable topping. There are 3 kinds of meat, pepperoni, sausage and ham; and there are 4 different kinds of vegetables, green peppers, onions, mushrooms and pineapple. You want to buy the special and need to decide what to choose.

    a. How many different kinds of special pizzas are there to choose from?

        2 × 3 × 4 = 24. ✓

    b. If you always choose pepperoni for the meat choice, how many types of pizza can you chose from?

        1 × 2 × 4 = 8. ✓

    c. If the pizza place decides to offer thin and thick crusts, how would the answer to the above questions change?

        Each answer would be doubled. ✓

page 206
Now You Try It

1. Find the lateral area, total area and volume of the square prism on the right. Prisms are named by their bases so we know that the shaded regions must be the square bases and therefore, that the unlabeled sides of each base must measure 5.

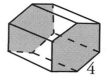

The lateral area is made up of the four rectangular faces each of which measures:

$$12 \times 5 = 60, \text{ so the lateral area is } 4 \times 60 = 240. \checkmark$$

The total area is the lateral area plus the area of the 2 bases, each of which is:

$$5 \times 5 = 25, \text{ so the total area is } 240 + (2)25 = 290. \checkmark$$

The volume equals Bh, where B is the area of a base and h is the distance between the bases:

$$25 \times 12 = 300. \checkmark$$

2. The volume of the regular hexagonal prism on the right = 300 un³. Find its lateral area and total area.

To find the lateral area, we need h, the height of the prism. Since the volume = Bh, if we calculate B we can then "work backwards" and solve for h. The problem states that the prism's base is a regular (equal sides, equal angles) hexagon with sides equal to 4. This is enough information to allow us to calculate its area:

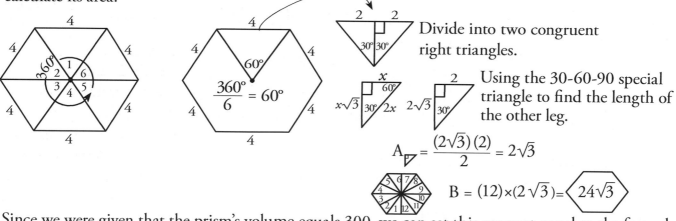

Divide into two congruent right triangles.

Using the 30-60-90 special triangle to find the length of the other leg.

$$A_{\triangledown} = \frac{(2\sqrt{3})(2)}{2} = 2\sqrt{3}$$

$$B = (12) \times (2\sqrt{3}) = 24\sqrt{3}$$

Since we were given that the prism's volume equals 300, we can set this amount equal to the formula for volume, substituting $24\sqrt{3}$ for B and then solve for h, the height of the prism:

$$300 = (24\sqrt{3})(h)$$
$$h \approx 7.2$$

Having found h, we can now calculate the lateral area of the prism. Lateral area is the sum of the six rectangular faces, each of which is (length × width) $7.2 \times 4 = 28.8$,

$$\text{L.A.} \approx 6 \times 28.8 = 172.8 \text{ un}^2 \checkmark$$

We could also have used the formula for lateral area: L.A. = ph = (6 × 4) × 7.2 = 172.8

Total area is the sum of the lateral area plus the area of the 2 bases:

$$T \approx 172.8 + (2)(24\sqrt{3}) \approx 255.9 \text{ un}^2 \checkmark$$

3. The total area of the triangular prism on the right is 768 cm². Find the volume of the prism.

'The total area is equal to the lateral area plus the area of the two bases.

The problem states 768 = Total Area = L.A. + 2 B = (13)(18) + (13)(18) + (10)(18) + 2B

$$768 = 648 + 2B$$
$$2B = 120, \ B = 60$$

Since V = Bh, the volume is (60)(18) = 1080 cm³ √

4. The volume of a cube is 125 un³. Find the total area of the cube.

Answer: Since this is a cube, V = s³ (where s = the length of every edge). The cube root* of 125 is 5, which means each side of the figure measures 5. One face is 5² = 25 un². There are 6 faces so the total area equals 6 × 25 = 150 un². √ *See the calculator note on page 226 to learn how to find cube roots.

page 213 — Now You Try It

1. A regular square pyramid fits perfectly in a cube with edges equal to 4. What is the volume of the unused space in the solid?

All edges of a cube are equal. Since the pyramid fits perfectly inside the cube, the pyramid's base edges and height are equal to 4.

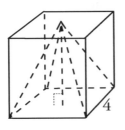

Notice that the pyramid's volume is equal to ⅓ of the cube's volume so the unused space *must* equal ⅔ of the total volume of the cube. Could we have simply found the volume of the cube (s³) and multiplied by ⅔? Absolutely.

— = Solution

B = (4)(4) = 16
V = Bh
V = (16)(4) = 64

B = (4)(4) = 16
V = ⅓ Bh
V = ⅓ (16)(4) ≈ 21.33

≈ 42.67 √

2. A regular triangular pyramid has height = 10. If the apothem of the base equals 5, find the volume of the pyramid.

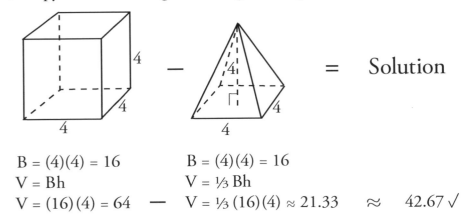

base of the pyramid

apothem = 5

Compare to the model 30-60-90 triangle.

$$A = \frac{(5\sqrt{3}) \times (5)}{2} = 12.5\sqrt{3}$$

B = 6 × 12.5√3 = 75√3

V = ⅓ Bh = ⅓ (75√3)(10) ≈ 433 √

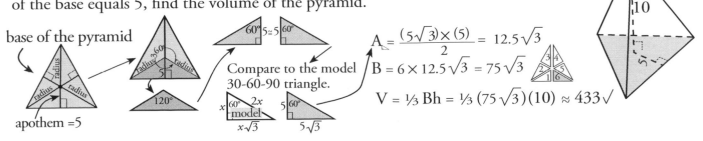

page 213 (continued)

3. The storage tank pictured on the right is a square pyramid. The tank has an upper
reserve tank as shown in the figure. Find the volume of the upper reserve tank.

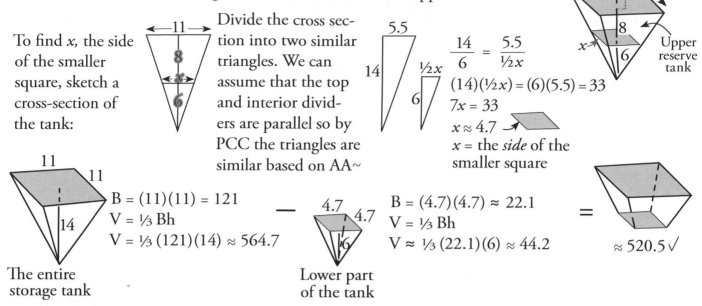

To find *x*, the side
of the smaller
square, sketch a
cross-section of
the tank:

Divide the cross sec-
tion into two similar
triangles. We can
assume that the top
and interior divid-
ers are parallel so by
PCC the triangles are
similar based on AA~

$\dfrac{14}{6} = \dfrac{5.5}{\frac{1}{2}x}$

$(14)(\tfrac{1}{2}x) = (6)(5.5) = 33$

$7x = 33$

$x \approx 4.7$

x = the *side* of the
smaller square

Upper
reserve
tank

B = (11)(11) = 121
V = ⅓ Bh
V = ⅓ (121)(14) ≈ 564.7

The entire
storage tank

B = (4.7)(4.7) ≈ 22.1
V = ⅓ Bh
V ≈ ⅓ (22.1)(6) ≈ 44.2

Lower part
of the tank

≈ 520.5 √

page 218
Now You Try It

1. If the radius of a cylinder is halved, what is the effect on the size of the volume of the cylinder?
Start with the formula for the volume of a cylinder: $V = \pi r^2 h$.
Substitute in the modified variable, in this case, replace r with ½r:

$$V = \pi(\tfrac{1}{2}r)^2 h = \pi(\tfrac{1}{2})^2 r^2 h = \pi(\tfrac{1}{4})r^2 h = \tfrac{1}{4}\pi r^2 h \qquad \text{(Math note: } (\tfrac{1}{2})^2 = (\tfrac{1}{2})(\tfrac{1}{2}) = \tfrac{1}{4})$$

Comparing the solution, $\tfrac{1}{4}\pi r^2 h$, to the original formula, $\pi r^2 h$, we see that when the radius is halved,
the volume is quartered (multiplied by ¼ which is the same as ÷4). √

2. The lateral area of a cylinder is 100π and the height is 5. Find the circumference of the base of the
cylinder. Answer: Since the lateral area is given, set that value equal to the formula for lateral area.

$$100\pi = 2\pi rh = 2\pi r(5) = 10\pi r.$$

$$\dfrac{100\cancel{\pi}}{\cancel{\pi}} = \dfrac{10\cancel{\pi}r}{\cancel{\pi}} \text{ Since both terms have } \pi \text{ as a factor, } \div \text{ by } \pi.$$

$$100 = 10r$$
$$r = 10$$

The formula for circumference is $2\pi r = 2\pi(10) = 20\pi \approx 62.8$ √

3. Find the total area and the volume of the cylinder in problem number 2.

Total area = $2\pi rh + 2\pi r^2$ and since we know that r = 10 and h = 5, the total area is:

$$2\pi(10)(5) + 2\pi(10)^2 = 100\pi + 200\pi = 300\pi \approx 942 \checkmark$$
$$V = \pi r^2 h = \pi(10)^2 5 = \pi(10)^2 5 = 500\pi \approx 1570 \checkmark$$

4 What is the amount (volume) of metal needed to make a 10 inch length of pipe that is ½ inch thick and has inner diameter equal to 5 in?

Answer: The amount of metal necessary is the difference in volume between the outside and the inside cylinders. Notice that the *radius* of the outside cylinder is 2.5 + .5 = 3.

$$\text{Volume of outside cylinder is } \pi r^2 h = \pi(3)^2(10) = 90\pi$$
$$\text{Volume of inside cylinder is } \pi r^2 h = \pi(2.5)^2(10) = 62.5\pi$$
$$\text{Difference is } 90\pi - 62.5\pi = 27.5\pi \approx 86.35 \checkmark$$

page 221
Solid Search

Write the letter of the matching figure. Be sure to check in the back of the book to see how you did.

Cone __D__
Triangular pyramid __J__
Square pyramid __B__
Right Cylinder __A__
Oblique prism __E__
Right triangular prism __H__
Right pentagonal prism __F__
Cube __C__
Right rectangular prism __I__
Oblique cylinder __G__

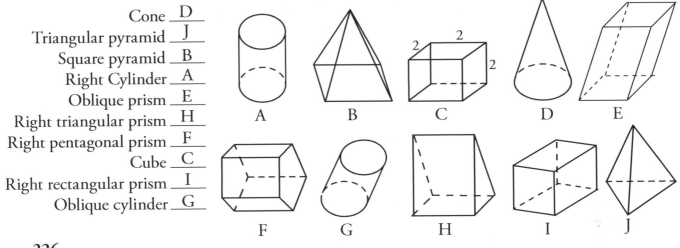

page 226
Now You Try It

1. A plane cuts a sphere with a 10 inch radius and forms a circle of intersection with a 4 inch radius. How far away from the sphere's center does the plane cut the sphere?

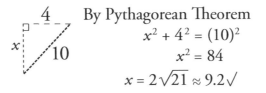

By Pythagorean Theorem
$$x^2 + 4^2 = (10)^2$$
$$x^2 = 84$$
$$x = 2\sqrt{21} \approx 9.2 \checkmark$$

2. Find the area and volume of the sphere in problem 1.

$$\text{Area} = 4\pi r^2 = 4\pi(10)^2 = 400\pi \approx 1256 \checkmark$$
$$\text{Volume} = \frac{4}{3}\pi r^3 = \frac{4}{3}\pi(10)^3 = 1{,}333\tfrac{1}{3}\pi \approx 4{,}186.7 \checkmark$$

3. A basketball with an 8 inch radius fits perfectly in a box. Find the volume of the unused space in the box.

All radiuses of a sphere are equal, so the box is a cube. The diameter of the ball is 2r = 2(8) = 16, which is also the length of each side of the box. Since each side of the box = 16, the volume of the box equals:

$$V_{cube} = s^3 = (16)^3 = 4096.$$

The volume of the ball equals:

$$V_{sphere} = \frac{4}{3}\pi r^3 = \frac{4}{3}\pi(8)^3 \approx \frac{4}{3}(512)\pi \approx 682.67\pi \approx 2144.7$$

The unused space in the box is the volume of the box minus the volume of the ball:

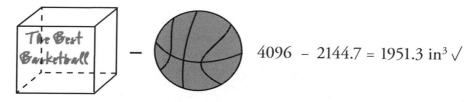

$4096 - 2144.7 = 1951.3 \text{ in}^3$ √

page 231
Now You Try It

1. The two cylinders at right are similar.

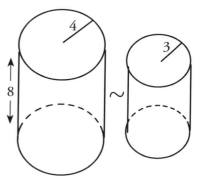

The scale factor, is the ratio of corresponding lengths from each figure. The radiuses of the cylinders are 4 and 3 (note the order of this problem is big to small).

Therefore, the scale factor (SF) for these 2 figures is: $\frac{4}{3}$

In less than 30 seconds, find the **ratios** of:

a. The heights of the cylinders.

Height is a length. The ratio of any 2 corresponding lengths is the scale factor (SF) itself, $\frac{4}{3}$ √

b. The circumferences of the bases.

Circumference is a length. The ratio of any 2 corresponding lengths is the scale factor (SF) itself, $\frac{4}{3}$ √

c. The lateral area of the cylinders.

Lateral area is an area. The ratio of any 2 corresponding areas is the scale factor squared (SF)², $\frac{(4)^2}{(3)^2} = \frac{16}{9}$ √

d. The total area of the cylinders.

Total area is an area. The ratio of any 2 corresponding areas is the scale factor squared (SF)², $\frac{(4)^2}{(3)^2} = \frac{16}{9}$ √

e. The volumes of the cylinders.

The ratio of the volumes is the scale factor cubed (SF)³,

$$\frac{(4)^3}{(3)^3} = \frac{64}{27} √$$

page 231 (continued)

Order Counts! For these problems each ratio's order is big to small:

2. Find the height of the smaller cylinder.

   The ratio of the heights must equal the scale factor. Since $SF = \dfrac{4}{3}$, let x equal the height of the smaller cylinder and create the proportion: $\dfrac{4}{3} = \dfrac{8}{x}$

   Cross multiply, $4x = (3)(8) = 24$, and solve, $x = 6\checkmark$

3. The lateral area of the smaller cylinder is 36π. Using the ratio method, find the lateral area of the larger cylinder.

   The ratio of any corresponding areas must equal the scale factor squared: $(SF)^2 = \dfrac{(4)^2}{(3)^2} = \dfrac{16}{9}$

   Now let x equal the L.A. of the large figure and create the following proportion:

   $\dfrac{16}{9} = \dfrac{x}{36\pi}$ Cross multiply, $576\pi = 9x$, and solve, $x \approx 64\pi \approx 201\checkmark$

4. The volume of the larger cylinder is 128π. Using the ratio method, find the volume of the smaller cylinder.

   The ratio of the volumes must equal the scale factor cubed. $(SF)^3 = \dfrac{(4)^3}{(3)^3} = \dfrac{64}{27}$.

   Now, let x equal the volume of the small figure and create the following proportion:

   $\dfrac{64}{27} = \dfrac{128\pi}{x}$ Cross multiply, $64x = 3456\pi$, and solve, $x = 54\pi \approx 169.6\checkmark$

INDEX

ORDER FORM

For each copy of Tutor in a Book's Geometry, please enclose 29.95
Sales Tax: California Residents only, please add 8.5%.
Shipping: In the U.S. please add $5.00 per book. Books will be sent by priority mail within 2 business days of receipt of money order/cashier's check or within 10 business days upon receipt of personal check. Please email us admin@tutorinabook.com to notify us of your order. For more information and international orders, please visit us on the internet at tutorinabook.com

Please **print** all information. Thank you.

Name: _____

Address: _____

City : _____

State: _____ Zip: _____

Mail to:
Tutor In a Book
PO Box 6178
Moraga, Ca 94570

Name: _____

Address: _____

City : _____

State: _____ Zip: _____

Mail to:
Tutor In a Book
PO Box 6178
Moraga, Ca 94570

Name: _____

Address: _____

City : _____

State: _____ Zip: _____

Mail to:
Tutor In a Book
PO Box 6178
Moraga, Ca 94570